Encyclopedia of Food Mycotoxins

Springer

Berlin
Heidelberg
New York
Barcelona
Hong Kong
London
Milan
Paris
Singapore
Tokyo

Martin Weidenbörner

Encyclopedia of Food Mycotoxins

With 96 Figures and 9 Tables

Springer

DR. MARTIN WEIDENBÖRNER
Justus-Liebig-Universität Gießen
Institut für Angewandte Mikrobiologie
Heinrich-Buff-Ring 26–32 (IFZ)
35392 Gießen
Germany

Library of Congress Cataloging-in-Publication Data

Weidenbörner, Martin
Encyclopedia of food mycotoxins / Martin Weidenbörner.
Incldes bibliographical references.
ISBN 3540675566 (alk. paper)
1. Mycotoxins-Encyclopedias. 2. Food-Toxicology-Encyclopedias. I. Title.
RA1242.M94 W454 2001
615.9'5295-dc21 00-049693

http.//www.springer.de

Springer-Verlag Berlin Heidelberg New York
a member of BertelsmannSpringer Science+Business Media GmbH

Printed in Germany

Cover design: design & production GmbH, Heidelberg
Typesetting: medio Technologies AG, Berlin
Printed on acid-free paper SPIN: 10706056 52/3020Rw – 5 4 3 2 1 0

My Children Anna and Vincent

Acknowledgement

I am gratefully indebted to Dr. Hem Chandra Jha who gave me very valuable advice on chemical problems. Furthermore I wish to thank Mrs. Rita Geißler-Plaum for the very careful preparation of the many mycotoxin molecular structures as well as the graphics. I would also like to thank Professor Dr. Hans P. van Egmond for his collegial help in the preparation of the table of mycotoxin limits in food. My thanks go to the CENA-Verlag for giving the permission to use various microscopic pictures. I appreciate the patience of my family, especially my wife, which was the prerequisite for writing this Encyclopedia. My father (a laywer) is thanked for the very careful review of the total manuscript, which cost him much time and probably also nerves.

Preface

The main emphasis of the present book is listing all foods which have been reported to be contaminated with mycotoxins (degree of contamination, concentration, country of origin/detection). To find out quickly whether a foodstuff is contaminated by a specific mycotoxin, the contaminated foods have been listed alphabetically under "Natural Occurrence" of a mycotoxin.

Products are listed by the country in which they were investigated for mycotoxin contamination. In some cases, the country of detection is **not** necessarily the country of origin, but information was lacking concerning the country of origin of such imports in the original literature. If only "imported" occurs after the country of investigation no more data were available in the original literature. Sometimes, e.g., in the case of nuts or spices, the original literature neither contained the producing country nor the addition "imported". In these cases also no indications were given in the Encyclopedia. However, in all cases where the origin of the investigated food was known, the name of the producing country was given.

The multiple listing of some countries in connection with mycotoxin contamination of food should not implicate a high rate of mycotoxin contamination of foods in these countries but primarily documents the efforts being made to detect toxic fungal metabolites in food.

The special data concerning the mycotoxin contamination of food, e.g. 6/12, means six positive (contaminated) products from a total of twelve. Means represent the mean of positively contaminated samples, except where indicated otherwise. An entry of $< x$ generally refers to the limit of detection. Values above this level are included in calculating the mean of all positive samples.

The data concerning mycotoxin contamination of food listed in the Encyclopedia based on results predominantly published in recommended journals and scientific books in this field (mainly the literature given at the end of the book). In the case of commonly isolated mycotoxins, e.g. aflatoxins, trichothecenes, it was not possible to consider all the results published.

In the literature, sometimes contradictory information about the mycotoxin spectrum of mold species can be found. Therefore, only the "safe" and food relevant mycotoxins of a species and not all known toxic metabolites were listed. This information mainly based on Frisvad J (1988) Fungal species and their specific production of mycotoxins. In: Samson RA, Reenen-Hoekstra ES (Eds) Introduction of Food-borne Fungi, pp 239–249. Centralbureau voor Schimmelcultures, Baarn (*Aspergillus* and *Penicillium* species), Marasas WFO, Nelson PE, Tousson TA (1984) Toxigenic *Fusarium* Species, Identity and Mycotoxicology. The Pennsylvania State University Press,

University Park, PA (*Fusarium* species), Samson RA, Hoekstra ES, Frisvad JC, Filtenborg O (1998) Introduction to Food-borne Fungi. Centraalbureau voor Schimmelcultures, Baarn (*Aspergillus* species and others). According to Ainsworth & Bisby's "Dictionary of the Fungi" all mycotoxigenic fungi listed in the Encyclopedia may be grouped easily to their corresponding family, order, phylum and kingdom.

The names used for all *Penicillium* species based on Pitt JI (1979) The Genus *Penicillium* and its Teleomorphic States *Eupenicillium* and *Talaromyces*, Academic Press, London.

Although in some cases more fungal species are known to produce a mycotoxin usually only the names of food relevant molds like *Aspergillus* spp., *Penicillium* spp. and/or *Fusarium* spp. are given.

Since in some cases various toxicological data of mycotoxins do exist for better comparison only the data of the per oral application in rats/mice (as far as possible) were chosen.

Gießen, Summer 2000 Martin Weidenbörner

Abbreviations

BGY	Bright greenish yellow (fluorescence)
bm	body mass
bw	body weight
conc	concentration
d	day(s)
EC	Esophageal cancer
ELISA	Enzyme linked immunosorbent assay
EU	European Union
FAO	Food and Agricultural Organization of the United Nations World Health Organization
FDA	United States Food and Drug Administration
GC	Gas chromatography
GC-MS	Gas chromatography-mass spectrometry
h	hour(s)
HPLC	High performance liquid chromatography
HTST	High temperature short time
IARC	International Agency for Research on Cancer
ip	intraperitoneal
iv	intravenous
JECFA	Joint Expert Committee on Food Additives
kGy	kilo Gray
LD_{50}	Lethal dosis of e.g. aflatoxin that will cause acute toxicity in 50 % of the target population
mc	moisture content
min	minutes
mp	melting point
mw	molecular weight
nc	no comment (not stated, unclear)
ND	Not detected
NOAEL	No observed adverse effect level
NMR	Nuclear magnetic resonance
po	per os
PTWI	Provisional tolerable weekly intake
sa	sample(s)
sc	subcutaneous
sqd	semi-quantitative determination

TLC	Thin-layer chromatography
tr	traces
UAE	United Arabic Emirates
WHO	World Health Organization of the United Nations
kg	kilogram
mg	milligram = 10^{-3} g; 1 mg/kg = 1:10^6 = ppm = parts per million
µg	microgram = 10^{-6} g; 1 µg/kg = 1:10^9 = ppb = parts per billion
l	litre
ml	millilitre = 10^{-3} l; 1 ml/l = 1:10^3
µl	microlitre = 10^{-3} ml; 1 µl/l = 1:10^6 = ppm = parts per million

A

AAL-toxins is the abbreviation for *Alternaria alternata* f. sp. *lycopersici* toxins which possess a "sphingosine-like" structure (see Figure AAL-toxins). AAL-toxins include the two fractions T_A and T_B. T_A ($C_{13}H_{53}NO_{15}$, MW = 679) consists of two esters (C_{13} or C_{14}) of 1,2,3-propane- tricarboxyclic acid and 1-amino-11,15-dimethylheptadeca-2,4,5,13,14-pentol. T_B ($C_{13}H_{53}NO_{13}$, MW = 647) consists of two esters (C_{13} or C_{14}) of 1,2,3-propane-tricarboxyclic acid and 1-amino-11,15-dimethylheptadeca-2,4,13,14-tetrol. These fractions contain four closely related compounds T_A-1, T_A-2, T_B-1 and T_B-2. Recently they were renamed alperisins A1, A2, B1, and B2. The alperisins are remarkably similar to the → fumonisins.

CHEMICAL DATA

Empirical formula: $C_{13}H_{53}NO_{15}$, molecular weight: 679 (T_A)

Empirical formula: $C_{13}H_{53}NO_{14}$, molecular weight: 663 (T_B)

FUNGAL SOURCES

Alternaria alternata f. sp. *lycopersici*

NATURAL OCCURRENCE

There are no reports on the natural occurrence of these toxins in plant products, probably because *A. alternata* f. sp. *lycopersici* is a rarely occurring pathotype of *A. alternata*. However, AAL-toxins and fumonisins (FB_1, FB_2, FB_3) occur together in spores and mycelia of *A. alternata*.

TOXICITY

Like fumonisin B_1 the AAL-toxins caused stem cancer disease in "Earlypark-7" and other susceptible tomato cultivars. In addition, AAL-toxins and the fumonisins inhibited ceramide synthase in animal cells, cell prolifeartion in rat liver and dog kidney cells.

Acacia concinna (medicinal seeds) may contain the following → mycotoxins:

→ aflatoxin B_1

incidence: nc/nc, conc. range: 80-1130 µg / kg, country: India

→ citrinin

incidence: nc/nc, conc. range: 10-760 µg / kg, country: India

Acetoxyscirpenediol 4- or → 15-acetylscirpentriol

3-Acetyldeoxynivalenol (Syn.: deoxynivalenol monoacetate) is a 3α-acetoxy-7α,15-trihydroxy-12,13-epoxytrichothec-9-en-8-one and belongs to the → trichothecenes (→ mycotoxins) (see Figure 3-Acetyldeoxynivalenol).

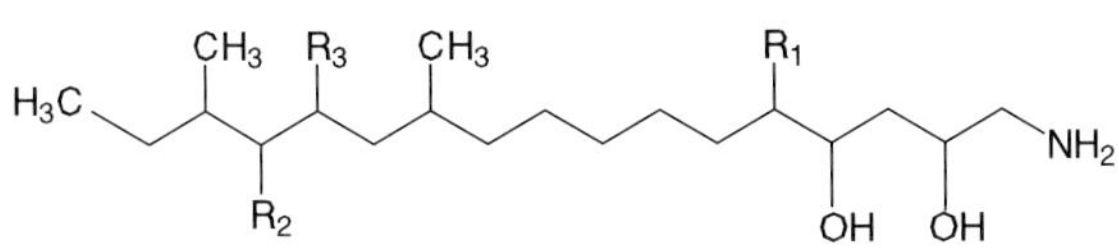

AAL-TOXIN	R_1	R_2	R_3
T_A–1	OH	OH	$–O_2C–CH_2–CH(CO_2H)–CH_2–CO_2H$
T_A–2	OH	$–O_2C–CH_2–CH(CO_2H)–CH_2–CO_2H$	OH
T_B–1	H	OH	$–O_2C–CH_2–CH(CO_2H)–CH_2–CO_2H$
T_B–2	H	$–O_2C–CH_2–CH(CO_2H)–CH_2–CO_2H$	OH

AAL-toxins. *Alternaria alternata* f. sp. *lycopersici* (AAL) toxins

Chemical Data
Empirical formula: $C_{17}H_{22}O_7$, molecular weight: 338

Fungal Sources
→ Fusarium culmorum (W.G. Smith) Sacc., → Fusarium graminearum Schwabe

Natural Occurrence
→ barley, → maize, → oats, → rye, → triticale, → wheat

Toxicity
feed refusal (rats)
LD_{50} (ip): 49.4-49.9 mg / kg bw mice (ddS strain)

Detection
ELISA, TLC, GC-MS, MS

Further Comments
Most Japanese strains of *F. graminearum* produced 3-acetyldeoxynivalenol. The same is true for Chinese strains although the 15-acetatedeoxynivalenol could be isolated from Chinese grain.
→ deoxynivalenol

15-Acetyldeoxynivalenol belongs to the → trichothecenes (→ mycotoxins) (see Figure 15-Acetyldeoxynivalenol).

Chemical Data
Empirical formula: $C_{17}H_{22}O_7$, molecular weight: 338

Fungal Sources
→ Fusarium graminearum Schwabe

Natural Occurrence
→ maize, → wheat

Toxicity
In combination with → deoxynivalenol and → zearalenone the aforementioned contaminated samples caused feed refusal in swine.

Detection
GC-MS

Further Comments
Nearly all strains of *F. graminearum* isolated in North America are able to produce this mycotoxin. This trichothecene mycotoxin occurred in naturally infected field maize samples (ca. 16,300 and 1510 µg / kg) used for feed. 15-acetyldeoxinvalenol co-occurs with → deoxynivalenol and → zearalenone.

4-Acetylnivalenol → fusarenon X

4-Acetylscirpentriol (Syn.: 15-acetylscirpentriol)

Acute aflatoxicosis → aflatoxicosis

Acute cardiac beriberi (Syn.: Shoshinkakke) A probable → mycotoxicosis which belongs to the complex of "yellow rice diseases" (→ yellow rice disease). It was first described in Japan at the end of the last century. The disease has mainly been reported from Asian countries where → rice is a staple food and has been recognized for the past three centuries. The mold damaged rice is mainly contaminated with → Penicillium citreonigrum Dierckx (synonyms *P. citreoviride*, *P. toxicarium*). → Citreoviridin the most important mycotoxin (neurotoxin) of this mold which causes a very rapid → paralysis of the respiratory muscles. In combination with → convulsion, vomition, ascending → paralysis, and lowering

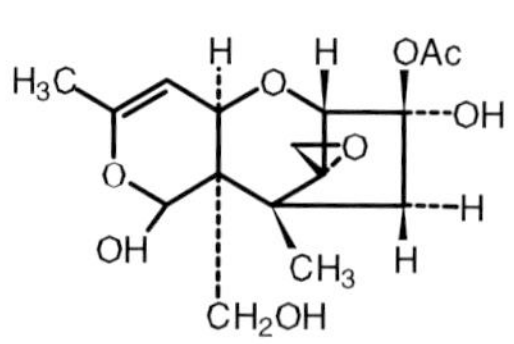

3-Acetyldeoxynivalenol

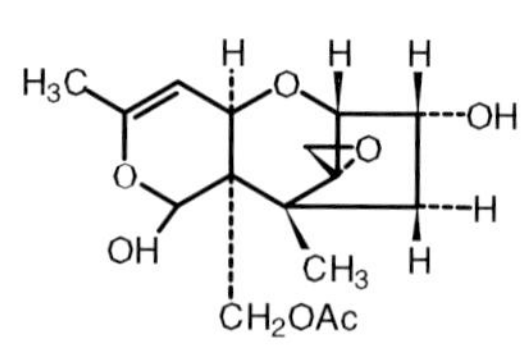

15-Acetyldeoxynivalenol

of the body temperature, the patient usually dies within a short period of 1-3 days, once the disease started. There is no method available of saving the patient from acute cardiac beriberi.
Because moldy "yellow rice" was thought to be responsible for this disease the sale of this rice was prohibited in Japan in 1910. Subsequently no more cases of acute cardiac beriberi have been reported. The disease is now of only historical interest in Japan. However, in other parts of Asia *P. citreonigrum* and its mycotoxin citreoviridin which is also produced by *P. ochrosalmoneum* may still contribute acute cardiac beriberi.
It is under discussion whether there are several types of beriberi (e.g. atropic and wet beriberi) having the same etiological origin. The difference in symptoms compared to acute cardiac beriberi may be due to dose and duration of intake of the mycotoxin. In these cases severe → paretic signs were not observed.
In contrast to acute cardiac beriberi, the cause of beriberi is a nutritional disease, an avitaminosis (vitamin B). This is proved by the following facts: the slower course of the disease, no dilation of the right ventricle, and no hypertrophy of adrenal medulla. In addition, administration of liver removed from typical shoshin-kakke patients led to the recovery of vitamin B_1-deficient animals. This indicates that adequate amounts of vitamin B_1 were present in the liver of these patients at the time of death.
However, to prove beyond doubt that citreoviridin is the cause of acute cardiac beriberi, the etiology of the chemical pathway of this neurotoxin has to be clarified.

Aflatoxicol (Abbr.: AFL, AFR_0) AFL was first reported in microorganisms (→ mycotoxins) and is the cyclopentanol derivative (2,3,6a,9a-tetrahydro-1-hydroxy-4-methoxy-cyclopenta[c]furo[3′,2′:4,5]furo[2,3-h][1]-benzopyran-11(1H)-one) of → aflatoxin B_1 (see Figure Aflatoxicol).

Chemical Data
Empirical formula: $C_{17}H_{13}O_6$, molecular weight: 313

Natural Occurrence
→ human breast milk, → pistachio nuts

Toxicity
AFL resulted from the *in vitro* and *in vivo* metabolism of AFB_1 by soluble NADPH-dependent reductases of submitochondrial liver fractions from humans and several animal species (e.g. poultry, rabbits, trouts). A microsomal AFL-dehydrogenase catalyzes the enzymatically reversible reaction. AFL therefore may represent a storage reservoir of AFB_1 that enhances the toxicity of AFB_1. Mice or rats which are relatively resistant to AFB_1 produce only very little AFL. Therefore, the minor rate of transformation might be a determinant in the susceptibility of animals to the acute toxic action of AFB_1. AFL is reported to be 18 times less toxic than AFB_1 in the duckling biliary → hyperplasia assay. In Fischer rats AFL shows nearly one half the hepatocarcinogenic potency of AFB_1. Carcinogenicity and mutagenicity (→ mutagenic) were almost the same as for AFB_1 in rainbow trout and in *Salmonella typhimurium*, respectively. Biological activity of aflatoxicol B is unknown.

Detection
see → aflatoxins

Further Comments
Two stereoisomers of AFL are known, the "A" isomer, also referred to as aflatoxin R_0, and the "B" isomer. The latter is only formed by microorganisms whereas AFR_0 also resulted from animal metabolism.

Aflatoxicol H_1 (Abbr.: $AFLH_1$) is the hydroxylated oxidative metabolite of

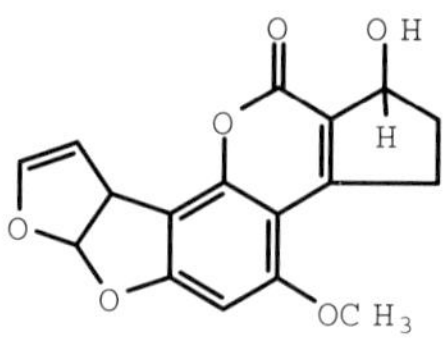
Aflatoxicol

→ aflatoxicol. It resulted from the metabolism of → aflatoxin B_1 by microsomal and soluble enzymes of primate and human liver and from → aflatoxin Q_1 incubated with cytosol enzymes.

TOXICITY
No toxicity has been reported in chick embryos and bacteria but it was → mutagenic (2% that of AFB_1) in the case of *Salmonella typhimurium*.

Aflatoxicosis is caused by → Aspergillus flavus Link and → Aspergillus parasiticus Speare due to the formation of → aflatoxins. Although these molds are of ubiquitous distribution, *A. parasiticus* predominates in tropical and subtropical countries. These → storage fungi invade seeds and → grains, particularly → peanuts, → maize (before harvest), and edible → nuts. Saprophytic growth on a wide range of foodstuffs is possible. Certain climatic conditions favour preharvest invasion and aflatoxin contamination of maize and peanuts. Countries with colder climates do not support aflatoxin production. Here, aflatoxicosis may be imported by contaminated feeds and foods. Species which are mainly affected by aflatoxins are humans, → cattle, dogs, → poultry, pigs, and trout.
The aflatoxicosis can be divided into two forms: primary aflatoxicosis with the acute and chronic forms, and secondary aflatoxicosis.
Acute aflatoxicosis results from high and moderate aflatoxin concentrations which cause the death of the animal. The main symptoms are: fatty, pale, and decolorized livers; interference of normal blood clotting mechanisms with subsequent hemorrhages (→ haemorrhage); decrease in total serum proteins and increase in certain serum enzymes of the liver; accumulation of blood in the gastrointestinal canal. In addition, lesions of the kidney (glomerular → nephritis) and congestions (→ congestion) in the lungs are possible.
The most severe case of acute aflatoxicosis has been observed in north-west India (1974). Ca. 25% of the exposed population (397 affected, 106 died) died after eating molded → maize with aflatoxin levels ranging from 6250 to 15,600 µg / kg. In contrast to females males were affected twice as often. Patients suffered from → icterus, in general vomiting and → anorexia preceded. → Ascites and → edema of the lower extremities subsequently occurred. In another case of acute aflatoxicosis (Kenya) patients showed similar clinical signs. Pathological changes in the liver were characteristic of toxic → hepatitis. In addition, three children in the Province of Taiwan, China and one child in Uganda died from acute liver necrosis. Their death was associated with the ingestion of → rice (200 µg aflatoxins / kg) and → cassava (1700 µg aflatoxins / kg), respectively, which most probably caused the disease. The reported outbreaks are only seen as the tip of the iceberg of worldwide occurring aflatoxicosis.
Chronic aflatoxicosis is caused by long term consumption of moderate to low aflatoxin concentrations. Much more serious veterinary problem may arise compared to acute aflatoxicosis. Symptoms are: liver congestions with hemorrhagic and necrotic regions; proliferation of the hepatic parenchyma and epithelial cells of the → bile duct; kidney congestion accompanied by occasional hemorrhagic → enteritis. Reduced feed efficiency and retarded growth rate are common, the

reproducive efficiency is decreased. Development of liver cancer (e.g. hatchery-reared trout) may result from long-term consumption of low levels of → aflatoxins as extremely potent hepatocarcinogenes.

Secondary aflatoxicosis (low aflatoxin concentrations) impairs the native resistance by reduction of phagocytic effectiveness of macrophages and nonspecific humoral substances (complements). The immunosuppressive effects of aflatoxins predispose animals to secondary infections by bacteria, fungi and viruses. Epidemiological studies in different parts of Africa and Asia show that aflatoxins may cause liver cancer in humans, albeit in combination with the hepatitis B virus. People e.g. living in Kenya, Mozambique, Swaziland and Thailand showed a high incidence of hepatic carcinomas. In these countries → foods and feeds are often contaminated with aflatoxins. In the Philippines AFM_1 has been detected in the 24 h urine samples of people who ingested → peanut butter containing aflatoxin. A level as high as 10-15 μg → aflatoxin B_1 in the diet seems to be sufficient for detection of → aflatoxin M_1 in urine.

Aflatoxin B_1 (Abbr.: AFB_1) is a 2,3,6a,9a-tetrahydro-4-methoxy-cyclopenta[c]furo[3′,2′:4,5]furo[2,3-h][1]-benzopyran-1,11-dione (→ mycotoxins) generally produced in the largest amount both in nature and in culture (see Figure Aflatoxin B_1).

CHEMICAL DATA

Empirical formula: $C_{17}H_{12}O_6$, molecular weight: 312

FUNGAL SOURCES

→ Aspergillus flavus Link, → Aspergillus nomius Kurtzman et al. → Aspergillus parasiticus Speare.

NATURAL OCCURRENCE

→ Acacia concinna, almonds, → ammi, → apples, → baby food, → bacon, → barley, → bean jam, → beans, → beefburger, → beer, burukutu, → beer, pito, → beer, sorghum, → Blepharis edulis, → bondakaledkai, → Brazil nuts, → bread, → buckwheat, → buckwheat flour, → cabbage, → Caesalpinea digyna, → Cassia fistula, → cardamom, → cardamom, greater, → cashew nuts, → cayenne pepper, → cereals, → cheese, → cheese, blue, → cheese, pepper, → cheese, Tilsit, → cheese rind, → cheese trimmings, → cherries, → chicken liver, → cocoa beans, → congressbele, → copra, → coriander, → corn flakes, → cumin, → curcuma, → dairy products, → duck, → emu aran, → equsi meal, → fennel, → fenugreek, figs, → galgant, → garlic, → garlic / onions, → ginger, → groundnut toffee, → ham, → hare, → hazelnuts, → hot dog, → human breast milk, → Hydnocarpus laurifolia, → Indian cassia, → ingwer, → job's-tears, → kubeba, → lemmons, → lentils, → libritos, → lineseed oil, → lineseeds, → mackarel, → maize flour, → maize grits, → mango, → meat, luncheon, → milk, → milk powder, → miso, → muesli, → nutmeg, → nuts (mixed), → oats, → oat flakes, → ogbono, → ogiliugba, → ogoro, → oil seeds, → oil seed rape, → olive oil, → olives, → oranges, → pastries, → peaches, → peanut brittle, → peanut butter, → peanut oil, → peanut products, → peas, → pecans, → persipan, → pheasants, → pig liver, → pine nuts, → Piper betle, → pipian paste, → pop corn, → rice, → rice cake, → roe deer, → rye, → sago, → salami, → sausages, → shrimp, → sorghum, → soybean, → spices, → sunflower seeds, → sunflower seed oil, → taro, → tomatoes, → tomato ketchup, → tumeric, → vegetables, walnuts, → wheat

For further information see → aflatoxins and → aflatoxin G_2.

Plant commodities which may be highly contaminated with → aflatoxins are → nuts such as → peanuts, Brazil nuts, → pistachio nuts as well as copra,

→ maize, and cottonseeds. Agricultural products with a slightly lower potential of aflatoxin contamination are → almonds, → figs, pecans, spices, and → walnuts. Animal products are less likely substrates, e.g. → milk, animal tissue.

TOXICITY
It is the strongest natural carcinogen and the main hepatocarcinogen in animals, although effects vary with species, age, sex, and general nutrition. For example trout, duckling, and pig, are highly susceptible, whereas e.g. sheep and → cattle, are more resisant. The liver is the primary organ affected (induction of liver lesions, liver carcinoma, bile duct proliferation). In Fischer rats and rainbow trout AFB_1 is the most potent hepatocarcinogen. Changes in other organs (e.g. kidneys, lung) have been observed.
From primate data the doses of AFB_1 required to cause acute → aflatoxicosis in humans were extrapolated. It was estimated that the intake of → food contaminated with 1700 μg / kg bw for a short time could be sufficient for severe liver damage while a single dose of 75,000 μg / kg bw could result in death. Apparent acute aflatoxicosis would not occur if 340 μg AFB_1 / kg bw is consumed per day. In the USA the ingestion of AFB_1 with maize and peanut products contributes to a greater risk of hepatic cancer in adults than AFM_1 in milk and → dairy products. In comparison to these agricultural products the human intake of aflatoxins by meat and meat products is negligible.
The IARC (1993) evaluated AFB_1 as a Class 1 human carcinogen.
LD_{50} (po): 5.5-7.2 mg / kg bw male rats (weight: 100 g), 17.9 mg / kg bw female rats (weight: 150 g)

DETECTION
see → aflatoxins

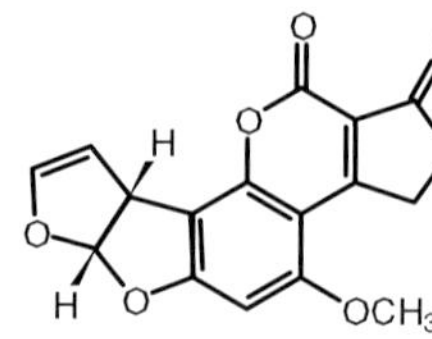

Aflatoxin B_1

FURTHER COMMENTS
Spiking commercially manufactured cigarettes with AFB_1 (100-300 μg / kg) did not result in any contamination of the gas phase or the ashes.

Aflatoxin B_2 (Abbr.: AFB_2) is the dihydro derivative of → aflatoxin B_1 (2,3,6a,8,9a-hexahydro-4-methoxy-cyclopenta[c]furo[3′,2′:4,5]furo[2,3-h][1]-benzopyran-1,11-dione) and synthesized by the reduction of the single double bond in the terminal dihydrofuran ring (→ mycotoxins) (see Figure Aflatoxin B_2).

CHEMICAL DATA
Empirical formula: $C_{17}H_{14}O_6$, molecular weight: 314

FUNGAL SOURCES
→ Aspergillus flavus Link, → Aspergillus nomius Kurtzman et al., → Aspergillus parasiticus Speare

NATURAL OCCURRENCE
AFB_2 occurs in the same commodities as AFB_1 but AFB_2 is found in smaller amounts. Via milk it is secreted as → aflatoxin M_2.

TOXICITY
This carcinogenic (?) and → genotoxic substance shows toxic properties similar to AFB_1 but has markedly reduced toxic potency in comparison to AFB_1. Instead of 3.9 μg AFB_1 50 μg AFB_2 are necessary to produce similar bile duct proliferation in ducklings. Estimated lethal dose for human beings 1-10 mg / kg.
LD_{50} (po): 84.8 μg / 50 g bw one-day old ducklings

DETECTION
see → aflatoxins

Aflatoxin B_{2a} (Abbr.: AFB_{2a}) (Syn.: AFB_1 hemiacetyl, aflatoxin W, hydroxydihydroaflatoxin B_1) represents the corresponding "water adduct" (2-hydroxy derivative) of → aflatoxin B_1 (→ mycotoxins) which resulted from the hydration of the 2,3-vinyl ether bond of this aflatoxin (2,3,6a,8,9,9a-hexahydro-8-hydroxy-4-methoxy-cyclopenta[c]furo[3′,2′:4,5]-furo[2,3-h][1]-benzopyran-1,11-dione). Conversion occurs rapidly under mildly acidic conditions. Although this blue fluorescing compound is 60-100 (200) times less toxic to ducklings it may be dehydrated to the highly toxic AFB_1. Furthermore, AFB_{2a} is a biotransformation / detoxification product of AFB_1 produced by hepatic microsomes *in vitro* of some animals (e.g. mouse, guinea-pig, avian). It is under discussion whether AFB_{2a} reacts readily with free amino groups of functional proteins (see Figure Aflatoxin B_{2a}).

CHEMICAL DATA
Empirical formula: $C_{17}H_{14}O_7$, molecular weight: 330

FUNGAL SOURCES
→ Aspergillus flavus Link, → Aspergillus parasiticus Speare

TOXICITY
In the standard duckling assay (initiation of → bile duct proliferation) both AFB_{2a} and AFG_{2a} are very much less toxic than AFB_1 (60-100 times) after oral application. In Khaki Campbell ducklings (day-old) no acute toxicity was noted at levels up to 1200 µg / duckling.

Aflatoxin B_3 (Abbr.: AFB_3) (Syn.: parasiticol) Older cultures of → Aspergillus flavus Link and → Aspergillus parasiticus Speare may contain high amounts of this 6-methoxy-7-(2′hydroxyethyl) difurocoumarin (7a,10a-dihydro-4-(2-hydroxyethyl)-5-methoxy-2H-furo[3′,2′:4,5]-furo[2,3-h]-1-benzopyran-2-one) as a possible precursor of → aflatoxins. On the other hand it seems to be the first step in the biological degradation of → aflatoxin G_1 by e.g. *Rhizopus* spp. (see Figure Aflatoxin B_3).

CHEMICAL DATA
Empirical formula: $C_{16}H_{14}O_6$, molecular weight: 302

TOXICITY
Parasiticol has the same acute toxicity to ducklings as → aflatoxin B_1. However, the tendency to cause biliary → hyperplasia is low. In chick embryo studies toxicity was only 1/100 than that of AFB_1.

Aflatoxin D_1 is a major product (10-30%) - besides the 206-molecular weight compound (3-10%) - from the reaction of aflatoxin B_1 with heated ammonium hydroxide. aflatoxins

Aflatoxin G_1 is a mycotoxin (→ mycotoxins) that has a structure very similar to that of → aflatoxin B_1 (3,4,7a,10a-tetrahydro-5-methoxy-1H,12H-furo[3′,2′:4,5]-furo[2,3-h]pyrano[3,4-c][1]-benzopyran-1,12-dione) but there are two lactone functions rather than one and the two

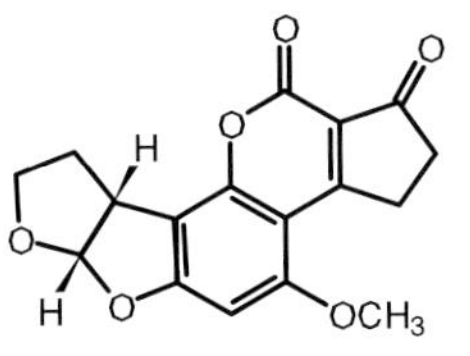

Aflatoxin B_2

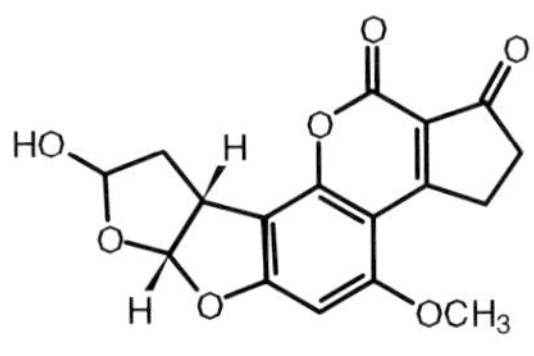

Aflatoxin B_{2a}

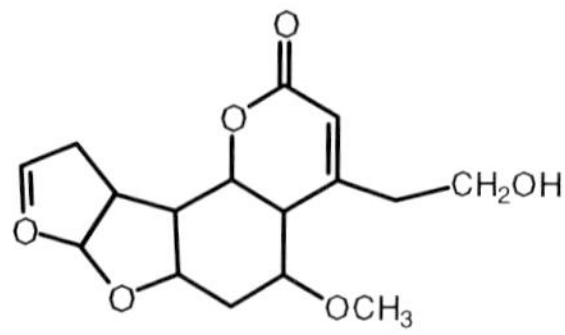

Aflatoxin B_3

dihydrofuran rings are fused in a *cis* configuration (see Figure Aflatoxin G_1).

CHEMICAL DATA
Empirical formula: $C_{17}H_{12}O_7$, molecular weight: 328

FUNGAL SOURCES
→ Aspergillus flavus Link, → Aspergillus nomius Kurtzman et al., → Aspergillus parasiticus Speare

NATURAL OCCURRENCE
Same commodities as AFB_1, in addition, → celery seeds.

TOXICITY
This carcinogenic (liver- and kidney carcinoma) and → genotoxic mycotoxin possesses a similar toxicity to that of AFB_1, although acute toxicity was less than AFB_1 but greater than AFB_2. It is a slightly less potent liver carcinogen but a slightly more potent kidney carcinogen, with a comparable carcinogenic potency to aflatoxin B_1 i.e. within a factor of 10. Ducklings treated with AFG_1 showed the same lesions as AFB_1-treated animals. The zone in affected rat liver lobule was the same as in B_1. However, a consistent pattern as seen with AFB_1 was absent. The LD_{50} in the rat was twice that of AFB_1.

DETECTION
see → aflatoxins

FURTHER COMMENTS
Optimum temperature for AFG_1 production is 30 °C.

Aflatoxin G_2 is the dihydro derivative of → aflatoxin G_1 (3,4,7a,9,10,10a-hexahydro-5-methoxy-1H,12H-furo[3´,2´:4,5]-furo[2,3-h]pyrano[3,4-c][1]-benzopyran-1,12-dione) and synthesized by the reduction of the single double bond in the terminal dihydrofuran ring (see Figure Aflatoxin G_2).

CHEMICAL DATA
Empirical formula: $C_{17}H_{14}O_7$, molecular weight: 330

FUNGAL SOURCES
→ Aspergillus flavus Link, → Aspergillus nomius Kurtzman et al., → Aspergillus parasiticus Speare

NATURAL OCCURRENCE
→ beans, → brazil nuts, → cocoa beans, → cumin, → ginger, → Indian cassia, → lemons, → maize, → mango, → olive oil, → oranges, → peanuts, → peanut brittle, → pepper, → pop corn, → rice, → sausages, → sesame seeds, → shoyu, → sunflower seeds, → tumeric, → walnuts
For further information see → aflatoxins and → aflatoxin B_1.

TOXICITY
This carcinogenic (?) and → genotoxic mycotoxins possesses the least acute toxicity of the 4 major naturally occurring → aflatoxins.
LD_{50} (po): 172.5 μg / 50 g bw one day old ducklings.

DETECTION
see → aflatoxins

Aflatoxin G_{2a} (Abbr.: AFG_{2a}) Aflatoxin G_1 is converted by strong acids to the corresponding "water adduct" (2-hydroxy derivative = AFG_{2a}) which retains its

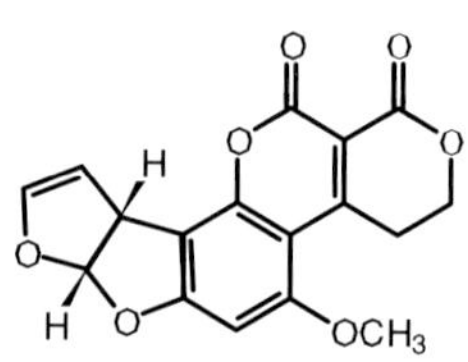

Aflatoxin G_1

toxicity (3,4,7a,9,10,10a-hexahydro-9-hydroxy-5-methoxy-1H,12H-furo[3´,2´:4,5]furo[2,3-h]pyrano[3,4-c][1]-benzopyran-1,12-dione). Livers of certain animals ingesting → aflatoxin G_1 produce AFG_{2a} which might be a detoxification mechanism (see Figure Aflatoxin G_{2a}).

Chemical Data
Empirical formula: $C_{17}H_{14}O_7$, molecular weight: 330

Fungal Sources
→ Aspergillus flavus Link, → Aspergillus parasiticus Speare

Toxicity
No significant differences in growth and characteristic liver lesions occurred in day-old Khaki Cambell ducklings (1600 µg / duckling). LD_{50} of AFB_1 in the same assay was 18.2 µg / duckling.

Aflatoxin GM_1 is a 4-hydroxylated derivative of → aflatoxin G_1 but only minor quantities have been detected in → Aspergillus flavus Link cultures.

Aflatoxin M_1 (Abbr.: AFM_1) is the 4-hydroxylated derivative of → aflatoxin B_1 (2,3,6a,9a-tetrahydro-1,9a-dihydroxy-4-methoxy-cyclopenta[c]furo[3´,2´:4,5]-furo[2,3-h][1]-benzopyran-11(1H)-one). It is found in liver, kidneys, blood, bile, feces, urine, and → milk of mammals (→ mycotoxins). Hydroxylation mainly occurs in the liver in the benzylic position at the junction of the two furan rings. It was the first → aflatoxin B_1 metabolite identified which was originally (early 1960s) found in cow´s milk. Structural elucidation was first achieved in 1966. Subsequently isolation of AFM_1 has also been reported from other kinds of milk as well as → dairy products (see Figure Aflatoxin M_1).

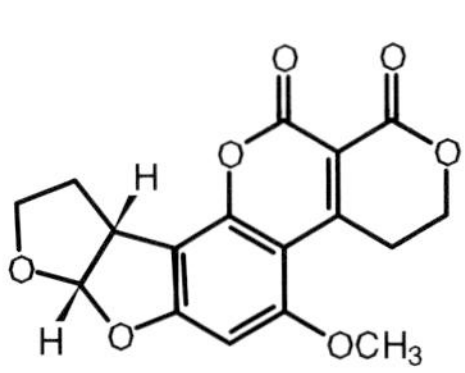

Aflatoxin G_2

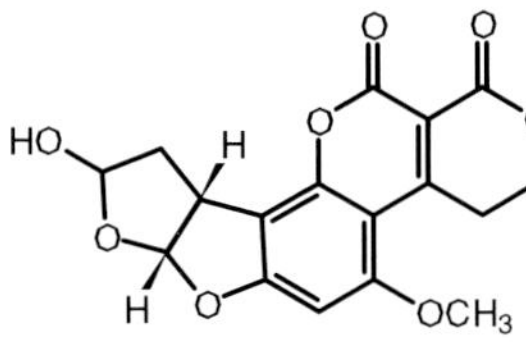

Aflatoxin G_{2a}

Chemical Data
Empirical formula: $C_{17}H_{12}O_7$, molecular weight: 328

Fungal Sources
→ Aspergillus flavus Link, → Aspergillus parasiticus Speare

Natural Occurrence
→ cheese, → cheese, blue, → cheese, Blue Haverti, → cheese, Brie, → cheese, butter, → cheese, Camembert, → cheese, Camembert & Brie, → cheese, Cheddar, → cheese, Cheshire, → cheese, Chester, → cheese, Cottage, → cheese, Comte, → cheese, Cream, → cheese, Double Gloucester, → cheese, Edam, → cheese, Emmental, → cheese, Fresh, → cheese, Gouda, → cheese, Grana Padano, → cheese, Lancashire, → cheese, Leicester, → cheese, Maribo, → cheese, Mozarella, → cheese, Parmesan, → cheese, Romadur, → cheese, Samsoe, → cheese, Stilton, → cheese, Wensleydale, → cheese, Wine, → cream, full, → human breast milk, → milk, → milk powder, → milk, pasteurized, → milk, sterilized, → milk, UHT, → milk, camel, → pistachio nuts, → soybean milk powder, → whey powder, → yogurt
Besides milk and dairy products this mycotoxin (→ mycotoxins) is also a contaminant of stored white and yellow → maize, freshly harvested yellow maize, and acid treated stored yellow maize (1-

35 μg / kg) as well as moldy → peanuts. AFM_1 has also been found in → human breast milk samples as a hydroxylated derivative of AFB_1 due to the activity of cytochrome *P*4501A2.
The ingestion of AFB_1-contaminated feed by mammals leads to the excretion of AFM_1 in milk (→ carry over ca. 0.3-3%, in dairy cows in early lactation up to 6%) and urine. 85% of dosed AFB_1 is secreted as AFM_1 via milk and urine within 48 hours. First dectection of AFM_1 within 12 hours. A milk sample taken after 96 hours was free of aflatoxin. Milk and dairy products are most probably the only toxic hazard from animal products. Concentration of AFM in body tissues is usually low with the majority of reports indicating undetectable levels in meat, blood, fat etc.

Toxicity

LD_{50} : 16.6 μg AFM_2 / day old duckling; 12 μg AFB_1 / day old duckling (simultaneous application)
A slightly less capacity in inducing → hepatic carcinoma (trout, rats) has been observed compared to AFB_1. There was inadequate evidence of the human carcinogenicity of AFM_1 (IARC 1993). AFM_1 induced hepatocarcinoma in trout and occasionally subcutaneous → sarcoma after injection.

Detection

see → aflatoxins

Further Comments

In some countries the contamination of milk with AFM_1 may follow a seasonal trend. During summer months lower contamination levels are detected because less supplementary mixed feeds are added to the diets of dairy cattle.
AFM_1 is associated with the protein fraction of the milk. This fact is responsible for the contamination of cheeses (3-5 fold enrichment).
In contrast to AFB_1, AFB_2, AFG_1, and AFG_2 (→ aflatoxins) AFM_1 also occurs in the absence of other aflatoxins.
Human exposure is primarily due to milk and milk products from animals that ingested AFB_1-contaminated feed. AFM_1 may cause problems especially in infants with a high milk consumption because of relatively low body weight, high cell activity, and partially developed immune system.
Stability: AFM_1 is stable in raw milk. Processing of contaminated milk will not result in aflatoxin-free dairy products. No reduction was established after pasteurization or processing into cheese, → yogurt, and → cream (20-40% fat). However, other reports proved a 63% reduction after pasteurization, 80% after sterilization and 85% after dry milk processing. Depending upon the time a 100% degradation of AFM_1 was achieved by UV irradation.

Aflatoxin M_2 (Abbr.: AFM_2) is the 4-hydroxylated derivative of → aflatoxin B_2 (2,3,6a,8,9,9a-hexahydro-9a-hydroxy-4-methoxy-cyclopenta[c]furo[3´,2´:4,5]-furo[2,3-h][1]-benzopyran-1,11-dione) and found in liver, kidneys, urine, and → milk of mammals (see Figure Aflatoxin M_2).

Chemical Data

Empirical formula: $C_{17}H_{14}O_7$, molecular weight: 330

Fungal Sources

→ Aspergillus flavus Link, → Aspergillus parasiticus Speare

Natural Occurrence

→ human breast milk

Toxicity

Compared to AFM_1 AFM_2 is considerably less toxic. This may be due to the lack of the double bond terminating the difuran ring system which is common in AFB_1, AFG_1, and AFM_1.

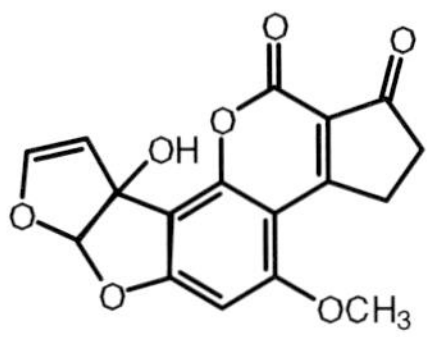

Aflatoxin M_1

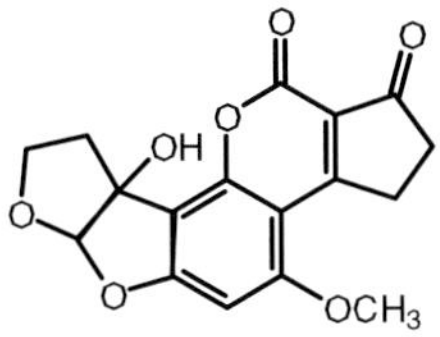

Aflatoxin M_2

LD_{50} : 62 μg AFM_2 / day old duckling; 12 μg AFB_1 / day old duckling (simultaneous application).

DETECTION
see → aflatoxins

FURTHER COMMENTS
Compared to → aflatoxin M_1 AFM_2 has a lower R_f with a violet fluorescence.

Aflatoxin M_4 (Abbr.: AFM_4) In 1986 the metabolite AFM_4 was isolated and identified in cow → milk. Contamination of commercial milk samples with this aflatoxin has been reported in France and Italy. The particular nutritional condition of the cow may be decisive for the mammary excretion of AFM_4. The name, AFM_4, derived from the fact that the hydroxyl group was located at carbon 4 of the cyclopentenone ring of AFM_1 (2,3,6a,9a-tetrahydro-2-hydroxy-4-methoxy-cyclopenta[c]furo[3´,2´:4,5]furo[2,3-h][1]-benzopyran-1,11-dione). → aflatoxins

Aflatoxin P_1 (Abbr.: AFP_1) represents the principal urinary metabolite in rhesus monkeys after intraperitoneal injection of → aflatoxin B_1. It showes considerably less toxicity than AFB_1. In mice and humans hepatic microsomes are also responsible for the metabolization of AFB_1 to AFP_1.

Aflatoxin Q_1 (Abbr.: AFQ_1) is the 3-hydroxy metabolite of → aflatoxin B_1. The major metabolic product of the metabolism in monkey, rat, and human liver microsomes preparations was approximately 18 times less toxic than AFB_1. No → mutagenic activity was detected.

Aflatoxin R_0 → aflatoxicol

Aflatoxin W (Syn.: → aflatoxin B_{2a})

Aflatoxins Aflatoxins as causing agents of the → turkey "X" disease were responsible for the death of more than 100,000 → turkey poults, aged three to six weeks, in south east England in 1960. A shipment of peanut meal ("Rosetti" meal) - imported from Brazil as a by-product from the extraction of → peanut oil, was contaminated by → Aspergillus flavus Link (but actually → Aspergillus parasiticus Speare) and contained four distinct, fluorescent highly toxic substances: → aflatoxin B_1, → aflatoxin B_2, → aflatoxin G_1, and → aflatoxin G_2 (***Aspergillus flavus* toxin** A-fla-toxin). Later it could be shown that → cyclopiazonic acid was also involved in turkey "X" disease. Besides → poultry which showed hemorrhages (→ hemorrhage) and liver necrosis frequently accompanied by lesions of the kidney e.g. ducklings, pigs, and → cattle were also affected. Toxicity of the aflatoxins comprises hepatocarcinogenicity, reduced T-cell function, diminished antibody response, and suppressed phagocyte activity. For further information see each single aflatoxin.
Aflatoxins are polycyclic, unsaturated highly substituted coumarins and one of the most important → mycotoxins.

The proposed natural synthesis of aflatoxin B_1 is as follows: acetate, norslorinic acid, averantin, averufanin, averufin versiconal hemicacetal acetate, versicolorin A, sterigmatocystin, *O*-methylsterigmatocystin, AFB_1. The first substance in the pathway to contain the essential C_2-C_3 double bond is versicolorin A.

Aflatoxins possess a fused dihydrofuran configuration that is peculiar to a limited number of compounds of natural origin. The coumarin nucleus is fused to a reactive bifuran system on one side and either a pentanone (B-aflatoxins) or a six-membered lactone (G-aflatoxins) on the other. The aflatoxin molecule probably has two reactive (toxic / carcinogenic) sites, viz. the unsatured terminal site in the bihydrofuran moiety and the lactone ring of the coumarin part.

Although approximately 20 aflatoxins have been identified only four of them, aflatoxins B_1, B_2, G_1 and G_2, occur naturally. The letters B_1, B_2, G_1, and G_2, are due to their intensive blue (B-aflatoxins) and green (G-aflatoxins) fluorescence in UV light. The subscripts indicate the relative chromatographic mobility. Two other familar aflatoxins, → aflatoxin M_1 and → aflatoxin M_2, are usually "**m**etabolites" (mammal transformation products) of AFB_1 and AFB_2. They are labeled so because of their presence in "**m**ilk" (milk toxin) previously exposed to AFB_1 and AFB_2. However, isolation of the 4-hydroxylated aflatoxins has also been reported from peanuts and → maize. The M toxins fluoresce blue to violet when exposed to long-wave UV light, but separate at a lower R_f value on TLC plates than AFB and AFG toxins. Besides the AFM-toxins further aflatoxins derived from AFB_1, AFB_2, AFG_1 and AFG_2 as metabolic products of microbial or animal systems (e.g. → aflatoxin P_1, → aflatoxin Q_1 and → aflatoxicol) or produced spontaneously in response to the chemical environment (e.g. B_{2a}, G_{2a}, and D_1).

In spite of the worldwide distribution of *A. flavus* (*A. parasiticus* predominates in warmer climates, → Aspergillus nomius Kurtzman et al. is a sporadic contaminant of → food) different factors favor aflatoxin contamination of → foods and feeds in distinct areas of the world. The → a_w of the commodity and the surrounding relative humidity as well as temperature are most decisive in storage and in the field. Preharvest invasion with *A. flavus* and subsequent aflatoxin contamination occurs in the case of peanuts and maize. Factors that promote invasion and contamination are drought stress in plants, drought-enhanced insect damage, wet weather conditions in combination with high temperatures during harvest, and use of susceptible genotypes.

Chemical Data

see: → aflatoxin B_1, → aflatoxin B_2, → aflatoxin B_{2a}, → aflatoxin G_1, → aflatoxin G_2, → aflatoxin G_{2a}, → aflatoxin B_3, → aflatoxin M_1, → aflatoxin M_2

Fungal Sources

Only 3 species, *A. flavus* Link, *A. parasiticus* and *A. nomius*, are definite producers of aflatoxins. Approximately 50% of all *A. flavus* strains synthesize aflatoxin. A higher percentage is found in warmer climates than in cooler regions, e.g. Ex-CSSR only 6 of 694 strains were aflatoxin-positive. Aflatoxins are found in the mycelium of *A. flavus* Link, in the conidia (84 mg / kg AFB_1, 566 mg / kg AFG_1) and sclerotia (135 mg / kg AFB_1, 968 mg / kg AFG_1) (see Figure Aflatoxins).

The domesticated forms of *A. flavus* and *A. parasiticus* (→ Aspergillus oryzae (Ahlburg) Cohn, *A. sojae*) have completely lost their ability to produce aflatoxins and the corresponding precursors.

Natural Occurrence

→ bakery products, → barley grits, → beer, → breakfast cereals, → cassava, → chilli, → chilli pickles, → chilli powder,

→ chocolate, → cocoa nibs, → cocoa presscake, → coconut, → coconut ice, → coconut oil, → coffee beans, → cow peas, → curry, → curry paste, → egg products, → fig paste, → fish, → foods, → garlic pickle, → garlic powder, → grains, → maize bran, → maize malt, → maize meal, → maize products, → maize starch, → maize, boiled, → maize, canned, → maize, dried, → maize, shelled, → manioc, → marzipan, → meat, → melon balls snacks, → melon seeds, → millet, → muesli ingredients, → noodles, → nuts, oil, → peach kernels, → peanut candy, → peanut mix, → peanut paste, → peanut sauce, → pigeon pea, → pistachio candy, → poppadoms, → pumpkin seeds, → small grains, → soybean flour, → tandoori, → tubers, → wine

For further information see aflatoxin B_1 and aflatoxin G_2.

Foods which show a predisposition for aflatoxin contamination include maize and → maize products, peanuts and → peanut products, → pecans, → almonds, → hazelnuts, → Brazil nuts, → pistachio nuts, and → walnuts. Small food → grains, e.g. → soybeans, → barley, → rye, → rice, and → oats, are not a major source of aflatoxin exposure if stored under suitable conditions. Other kinds of foodstuff which were found positive for aflatoxin contamination are e.g. cassava, → peas, → cowpeas, millet, → sorghum, sesame, sweetpotatoes, → spaghetti.

Of the aflatoxins present in food AFB_1, AFG_1, and AFM_1 are of primary importance and, together with aflatoxicol, represent possible health concerns. Although AFB_1, AFB_2 and AFG_1 are common in the same food sample AFB_1 predominates (60-80% of the total aflatoxin content). Generally, AFB_2, AFG_1, and AFG_2 do not occur in the absence of AFB_1. In most cases AFG_1 is found in higher concentrations than AFB_2 and AFG_2.

Maximum concentrations have been recorded from the following seeds: cotton: > 5 g aflatoxin / kg, peanuts: 1 g aflatoxin / kg, maize: 0.4 g aflatoxin / kg, pistachio nuts: 1.4 g AFB_1 / kg. Unprocessed foods of plant origin seem to be the most important potential sources of aflatoxins in the diet. In contrast, animal products are less likely substrates for aflatoxin contamination. Sugar, conventional jellies, sauerkraut, raisins and potatoes are generally free of aflatoxins.

The contamination of agricultural products with aflatoxins is not only a problem in less developed countries (tropics) but also in (warm) regions with a high developed agricultural standard (southern and sometimes mid-western USA). Plant stress, insufficient drying after harvest and storage at relatively high temperatures are the main reasons for contamination. Maize and peanuts are particularly susceptible. If these crops belong to the staple foods a higher exposure level to aflatoxins may be the consequence.

Toxicity

AFB_1, AFM_1 and aflatoxicol belong to the group of → genotoxic carcinogens with AFB_1 being the most potent. Aflatoxins with the index 1 are the most toxic ones. For this type of carcinogen, there is no threshold dose below which no tumor formation would occur. Only a zero level of exposure will result in no risk. Even very low concentrations, e.g. 1 ng aflatoxin / kg bw / day or less still contribute to the risk of liver cancer.

Besides their carcinogenic effect aflatoxins are → mutagenic, → teratogenic, and hepatogenic. In low levels they are responsible for weight gain losses, loss of reproducive capacity, and impairment of the immune systems (e.g. poultry, pigs, cattle). Conversion of AFB_1 and AFG_1 by hydroxylation to B_{2a} and G_{2a}, respectively, greatly reduces oral toxicity. The bio-

chemical effects are inhibition of adenosine triphosphatase (energy metabolism), reduction of hepatic glycogen levels (carbohydrate, lipid metabolism), binding with DNA and RNA (nucleic acid, protein metabolism).
The NOAEL for AFB_1 was estimated as 0.75 µg / kg body weight per day, using Fisher exact (statistical) test. Similarly, for aflatoxicol and AFM_1, the respective NOAELs were 1.25 and < 2.5 µg / kg bw per day.
Nutritional status of individuals exposed to aflatoxins seems to be very important for human health because malnutrition coexists in many parts of the world with aflatoxins and high incidences of liver disease, including liver cancer. Furthermore, susceptibility of monkeys to aflatoxins was significantly increased by reduced protein intake. The estimated LD_{50} for humans is about 1-10 mg AFB_1 / kg.
The carcinogenicity of aflatoxins is enhanced by e.g. gossypol, 3-methylcoumarin, cycloproprenoid fatty acids, malvalic acid and sterculic acid but possibly also by deoxynivalenol and nivalenol (synergistic effect).

Detection

ELISA, HPLC, IACA, RIA, TLC
The aflatoxin contamination (and other mycotoxins) of seeds is characterized by a negative binomial distribution function. Because very few seeds contain any significant level of aflatoxin but the majoritiy are non-contaminated, a representative sample from the lot must be taken.

Possible Mycotoxicosis

→ Aflatoxicosis (acute), → Indian childhood cirrhosis, → Kwashiorkor, → primary hepatocellular carcinoma (PHC), → Reye's syndrom

Further Comments

Production: The highest amounts of aflatoxins are synthesized in the log phase (intense sporulation), while aflatoxin production starts at the same time as the formation of conidia. Usually after six days aflatoxin production decreases. Under a given set of conditions only two or three aflatoxins are produced.
Although growth in culture yields higher AFG_1 concentration than AFB_1 in the case of natural contamination AFB_1 is frequently found in the highest concentration (e.g. "Rosetti meal" as causual agent of the "turkey X disease" contained 10,000 µg AFB_1 / kg but negligible levels of G_1).
The limiting a_w for aflatoxin production (*A. flavus*) is between 0.83 and 0.87, which is close to the minimum for growth. Synthesis increased at a_w 0.95-0.99 with optimum temperatures ranging from 25 °C to 30 °C.
Low temperatures (8-10 °C) induce production of approximately equal amounts of aflatoxins B and G. However, total production is lowered and more time required. Aflatoxin B production is stimulated by higher temperatures relative to aflatoxin G.
Optimal AFB_1 production occurred between 24-28 °C whereas 30 °C is optimal for AFG_1 formation. Only a few strains are able to synthesize aflatoxins at 7.5 °C. Fluctuating temperatures (mean 25 °C, upper limit 40-50 °C) are less favorable for aflatoxin production than a constant temperature (25 °C). Fluctuations down to 10 °C did not cause any significant effect.
CO_2- > 10% or O_2-concentration < 20% or > 90% suppresses toxin production.
The addition of cadmium, iron and molybdenium increases aflatoxin production, zinc is a prerequisite. For any given strain of fungus, the substrate influences the amount of aflatoxin produced.
Aflatoxin synthesis in the conidia of *A. flavus* is stimulated by irradation ≤ 3 kGy. A dose of 2.5 kGy enhanced synthesis of AFB_1 and AFG_1 50 times while 1 kGy was sufficient to induce afla-

toxin formation of non-producing *A. flavus* strains.

Stability: Aflatoxins are extremely heat stable compounds in the dry state up to the melting point. Moisture lowers heat stability but in general these mycotoxins are also stable during thermal processing of most food products. At elevated temperatures a partial destruction could be observed during autoclaving or roasting of nuts (40 to 60%). Cooking processes (e.g. dry heating, boiling) of cereal products, extrusion of maize meal dough (150 °C), or fermentation of dough during breadmaking cause variable losses of aflatoxins. As little as 14-26% of AFB_1 contamination of wheat was found to survive flour → milling and bread baking. However, baking temperatures are usually not sufficient to cause significant losses of aflatoxin in bread.

Decomposition occurs after exposure to sunlight, ultraviolet ligth and ionizing radiation.

Destruction or removal of aflatoxins from → food oils is achieved by alkaline treatments and refining, respectively. Aflatoxin concentration decreased in raw peanut butter and meat with increasing storage time but other studies do not report significant changes in aflatoxin levels of stored peanut meal and peanut butter. An essential stability (after one week) of AFB_1 and AFG_1 in Swiss cheese, bologna and cooked cornmeal was observed.

Cleaning and milling do not cause a general reduction of aflatoxin levels in cereal grains. A redistribution of the mycotoxins in the different fractions is most likely. E.g. dry milling of → maize usually leads to increased AFB_1 levels in the germ, hull, and degermer fines fractions. However, the ultimate distribution depends on the original amount present in the seed. Although milling of → rice and parboiled rice caused a significant decrease in AFB_1 and AFG_1 levels, aflatoxin concentrations in the → bran and polished fractions increased substantially. Increasing AFB_1 concentrations have also been found in the → flour of durum wheat from the top grade to the second. The bran contained the highest amounts. Mashing and brewing caused a partial loss of AFB_1 while distillation destroyed total aflatoxins in excess of 90%. Fermentation of AFB_1 contaminated maize under conditions used in the spirits industry led to aflatoxin-free distilled ethyl alcohol. In completely processed → beer only 18-27% of the original AFB_1 concentration was detected. Wort boiling and final fermentation steps mainly contribute to aflatoxin losses.

Detoxification: Detoxifiaction processes include degradation, destruction, or inactivation.

Physical methods: Heat - roasting temperatures (> 250 °C) are necessary for effective aflatoxin degradation; increasing the moisture content of the substrate will enhance degradation; **irradiation** - effective dose levels (X-rays, electron irradiation) cause destruction of the contaminated commodity; **adsorption** - → bentonite adsorbed aflatoxins from → milk and fluid products, hydrated sodium calcium aluminosilicate is suitable for the adsorption of AFB_1 from aqueous solutions. Similar effects have been reported for clays, charcoal, asbestos, aluminas, silicas, xeolites and aluminosilicates; **solvent extraction** - 90% aqueous acetone, 95% ethanol, hexane-ethanol, hexane-methanol, and 80% isopropyl alcohol have been used effectively.

Chemical methods:

Ammonia causes lactone ring opening of AFB_1, ultimate splitting off of the cyclopentenone part by NH_3. Several breakdown products of AFB_1 have been identified, e.g. → aflatoxin D_1 and the 206 molecular weight compound. Both substances showed a 450-fold decrease in mutagenicity (Ames test) compared to

AFB_1. The treatment prevents both acute and chronic aflatoxicosis in animals and is generally believed to be the most effective decontamination method. Ammonia treatment is used on commercial scale for the decontamination of feedstuff including corn and peanut and cottonseed meals in France, Senegal, USA (Arizona, California, Georgia, Alabama).
Acids effectively convert AFB_1 and AFG_1 to their corresponding hemiacetal forms → aflatoxin B_{2a} und → aflatoxin G_{2a}, but they have no effect on AFB_2 or AFG_2.
Oxidising agents, ozone - destruction of AFB_1 and AFG_1 but not AFB_2; **hydrogen peroxide** - destruction of aflatoxins in peanuts; in combination with riboflavin destruction of AFM_1 in milk; **bisulfite** - reaction with AFB_1 and AFG_1; **vitamin C** treatment.
Biotransformation: Microorganisms such as bacteria, actinomycetes, yeasts, molds, and algae cause degradation of aflatoxins. The most effective one, → Flavobacterium aurantiacum, removes AFB_1 (and AFM_1) from milk, maize, → maize oil, peanuts, → peanut butter, and soybeans while AFG_1 and AFM_1 are also metabolized. Other microorganisms convert or transform AFB_1 to aflatoxicol which is a very slow (3 to 4 d) and incomplete process (60% of AFB_1 is converted to aflatoxicol). However, except for ammonification (see above) the remaining methods are only of limited realistic commercial benefit.
Control: Control of aflatoxin contamination extends from growth of the crops in the field, through the storage of harvested crops, to the proper storage of prepared foods in the home.
Prevention of aflatoxin contamination of agricultural products, especially high-risk crops such as maize and peanuts, starts in the field. Growth of *A. flavus* and *A. parasiticus* is impaired / inhibited by breeding (using) resistant varieties, good agronomic practices from planting to cultivation and harvesting. Prevention of insect and mechanical damage as well as chemical plant protection favor the optimal development of the plants.
During storage low moisture content and temperature, adequate aeration, and pest control (insects, mites) inhibit aflatoxin (mycotoxin) accumulation in the harvested crops. Especially in stored maize, hot and humid storage conditions contribute to elevated aflatoxin concentrations.
In the home proper storage of prepared foods for prolonged periods at low humidity and temperature prohibits aflatoxin contamination.
At least in the EU there are uncertainties about the dietary aflatoxin intake since detailed information concerning the ingestion of typically aflatoxin-containing foods like peanuts, pistachio and Brazil nuts, figs etc. are difficult to obtain.

Agranulocytosis Absence of granules in cells in cytoplasm. → Alimentary toxic aleukia

Akakabi byo disease (Syn.: → red mold toxicosis, scabby grain intoxication)

Aleukia Absence of leukocytes (→ leukocytosis) from blood

Alimentary hemorrhagic aleukia → Alimentary toxic aleukia

Alimentary mycotoxicosis → Alimentary toxic aleukia

Alimentary toxic aleukia (Abbr.: ATA) (Syn.: septic angina, alimentary panhematopathy, alimentary toxicosis, alitoxicosis, alimentary → agranulocytosis, endemic panmyelotoxicosis, hemorrhagic (→ hemorrhage) syndrome) This → mycotoxicosis was first described in 1913 with first indications in 1891. ATA occurred sporadically during the first three decades of the century (e.g. 1924,

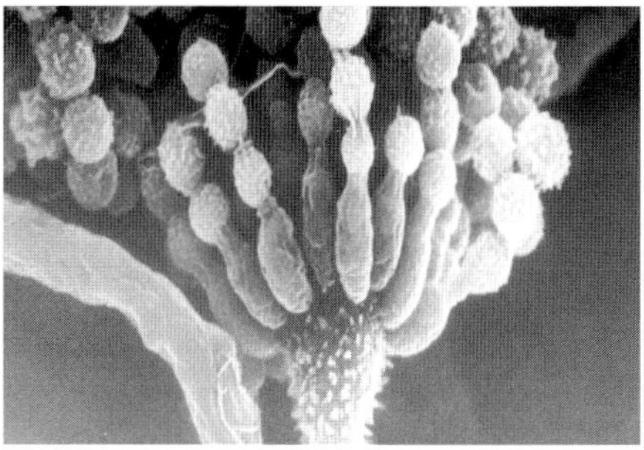

Aflatoxins. Small vesicle with phialides and conidia of *Aspergillus flavus* Link

1934) in various parts of Russia, but from 1932 on it appeared in endemic form. It claimed many victims (mortality 2-80%) especially during World War II. People of the Siberian USSR and the Orenburg district were mainly affected. During spring 1944 the morbidity in this district exceeded 10% and a high mortality was observed in 9 of 50 counties. Until the postwar years (1947) the disease caused the death of hundreds of thousands of people.
Mild winters with heavy snow in combination with frequent alternate freezing and thawing in the spring favored fungal growth in grains, especially proso → millet and → wheat (most likely to be toxic) as well as → barley, → rye, → oats, and → buckwheat. Over 3500 fungal isolates were collected from more than 1000 samples of overwintered grains. These cultures belong to more than 40 genera with 200 species. 61 isolates were → Fusarium poae (Peck) Wollenw. and 57 were → Fusarium sporotrichioides Sherb. These molds produced high amounts of → trichothecenes (e.g. → T-2 toxin, → diacetoxyscirpenol, → HT-2 toxin, → nivalenol etc.). Optimal toxin production of both *Fusarium* species occurred at 6-12 °C. Alteration of freezing and thawing temperatures caused maximal toxicity in culture.

Large numbers of people had to consume these overwintered, moldy and mycotoxin-contaminated grains (→ mycotoxins) because the shortage of manpower due to the war made harvesting impossible at the proper time. 2 to 3 weeks after ingestion of the toxic grain - at least 2 kg - the disease usually developed. A large number of victims died within 6-8 weeks after consuming at least 6 kg. However, breast-fed babies less than one year old did not show any symptoms. It seems that the toxic principle was not secreted into → human breast milk.
Ingestion of the → grains resulted in the following clinical symptoms:
First stage: burning sensation caused by inflammation of the mouth and fore-gut, emesis, diarrhoea, abdominal pain, no increase of body temperature, decrease of leukocyte counts ($\leq$ 2000 cells / mm^3); duration: 3 to 9 days.
Second stage: disorder of bone marrow functions, pronounced and progressive → aleukia; duration 3-4 weeks.
Sudden onset of the third stage: petechial hemorraghes on head (face), trunk, and limbs, necrotic changes in the mouth, throat, and esophagus, bacterial infections (septic → angina) occur, enlargement of the lymphatic glands, parenchymateous → hepatitis resulting in → jaundice (sometimes), further decrease of leukocyte counts ($\leq$ 100 cells / mm^3), significant decrease of erythrocyte and thromobocyte counts.
Constriction of the glottis (strangulation) due to edemateous swelling caused the death of one-third of the vicitims. Among survivors, intensity of toxicoses was decisive for the rate of recovery. In about 4 weeks the necrotic and hemorrhagic symptoms disappeared. However, two months or more were necessary for full recovery of the bone marrow function. Prophylaxis includes blood transfusion and administration of nucleic acid. Calcium preparations, vitamin C and K, and

sulphonamide further contribute to avoid the development of severe symptoms. Based on the closely related if not identical syndromes it was concluded that → moldy corn toxicosis and ATA have the same origin, viz. T-2 toxin and diacetoxyscirpenol, primarily produced by → Fusarium sporotrichioides Sherb.

Alkaloids → ergot alkaloids

Almond paste → marzipan

Almonds (no specification)
Contamination of maturing almonds with molds and → mycotoxins may result from kernel damage due to the navel orange worm near the time of hull split. At this time the invading molds, frequently → Aspergillus flavus Link, found moisture levels usually high enough to support growth and aflatoxin formation (→ aflatoxins).
The average probability of aflatoxin contamination in California almonds is one kernel in 26,500 unsorted in-shell nuts from the field.
Almonds may contain the following mycotoxins:
→ aflatoxin B_1
incidence: 1/110*, conc.: 93 μg / kg, country: Finland, *imported
incidence: 1/6*, conc.: 67 μg / kg, country: Finland, *imported, bitter almonds
incidence: 1/184*, conc.: ≤ 1 μg / kg, country: Finland, *imported, sliced and crushed
incidence: 7/198, conc. range: < 5 μg / kg (6 samples), 12 μg / kg (1 sa), country: Germany
incidence: 19*/23, conc. range: 39-4000 μg / kg, country: Germany, *moldy
incidence: 87/261*, conc. range: < 5 μg / kg (44 samples), 11-189 μg / kg (43 sa*), Ø conc.: 33 μg / kg, country: Germany, *ground
incidence: 77/360*, Ø conc.: 28 μg / kg, country: Germany, *ground
incidence: 2/4*, conc. range: < 5 μg / kg (1 sample), 200 μg / kg (1 sa), country: Germany, *sliced
incidence: 43/907*, Ø conc.: 23 μg / kg, country: Germany, *sliced
incidence: 15/19*, conc. range: 0.5-5 μg / kg (14 samples), 6 μg / kg (1 sa), country: UK, *ground
→ aflatoxin B_2
incidence: 1/110*, conc.: 14 μg / kg, country: Finland, *imported
incidence: 1/6*, conc.: 5 μg / kg, country: Finland, *imported, bitter almonds
→ aflatoxins (no specification)
incidence: 2*/78, conc. range: 5- > 25 μg / kg, country: Canada
incidence: 47/327, conc. range: ?, country: Germany
incidence: 77/360*, Ø conc.: 36 μg / kg, country: Germany, *ground
incidence: 43/907*, Ø conc.: 35 μg / kg, country: Germany, *sliced
incidence: 2/7*, conc. range: nc, country: UK, *ground
incidence: 15/19*, conc. range: 0.5-5 μg / kg (13 samples), 6-10 μg / kg (2 sa) (with a maximum of 10 μg / kg), country: UK, *shelled, ground
incidence: 1*/4, conc.: 0.8 μg / kg, country: UK, *aflatoxin (AFB_1, AFB_2, AFG_1, AFG_2)
incidence: 1/2*, conc.: nc, country: UK, *sugared
incidence: 1/9*, conc. range: nc, country: UK, *unblanched
incidence: 36/557, conc. range: 90 μg / kg, Ø conc.: 27 μg / kg, country: USA
incidence: 28/345, conc. range: 2-94 μg / kg, Ø conc.: 20 μg / kg, country: USA
incidence: 1/5, conc.: 10 μg / kg, country: USA
→ ochratoxin A
incidence: 1/12, conc. range: 0.2-0.49 μg / kg, country: Germany
→ nuts

Alperisins (Syn.: → AAL-toxins)

Altenuene (Abbr.: ALT) is a dibenzo-α-pyrone derivative (2,3,4,4a-tetrahydro-2,3,7-trihydroxy-9-methoxy-4a-methyl-6H-dibenzo[b,d]pyran-6-one) produced by → Alternaria spp. (→ mycotoxins) (see Figure Altenuene).

Chemical Data
Empirical formula: $C_{12}H_{14}O_6$, molecular weight: 292

Fungal Sources
→ Alternaria alternata (Fr.) Keissler, *Alternaria citrii*

Natural Occurrence
→ apples, → olives, → ragi, → sorghum, → tomatoes

Toxicity
cytotoxic
LD_{50} (ip): (50) 75-100 mg / kg bw mice. In chicks and rats no toxic effects occurred after feeding → alternariol methyl ether, → alternariol and ALT for 21 days at concentrations up to 24, 39, and 10 μg / g, respectively.

Detection
see → Alternaria mycotoxins

Alternaria (Syn.: *Macrosporium*) anamorphic → Pleosporaceae, teleomorph *Lewia* (formerly Dematiaceae)
Alternaria spp. are very common (airborne) fungi. Temperatures in the 18-22 °C range contribute to their growth. In addition, for substantial growth moisture contents of 28-34% in the substrate, i.e. water activities of ≥ → a_w 0.84, are required.
This genus may be the principal fungus in → wheat, → barley, and → sorghum, where in some years a nearly 100% infection has been recorded. Although *Alternaria* spp. cause rather limited damage (e.g. discoloration, black point of kernels) to cereal → grains mycotoxin contamination may result from infection (→ mycotoxins). Seed moisture contents of ≈ 22% due to heavy rainfall and high relative humidity at the time of harvest favor invasion.
Besides grains, → fruits and → vegetables are also subject to *Alternaria* spp. infection both pre- and postharvest. Factors that promote invasion of these fungi are: surface physical damage, fruit imperfection, overripening, cold stress. Especially commodities held in cold storage may show significant spoilage because *Alternaria* spp. grow well at low temperatures.
About 70 secondary metabolites belonging to several chemical classes, e.g. anthraquinones, cyclic peptides, dibenzopyrones, lactones, perylenequinones, tetramic acids, are produced by this genus. From feeding studies it was estimated that 68% of the *Alternaria* strains are toxic.
Important *Alternaria* toxins which contaminate food are: alternariols (→ alternariol, → alternariol methyl ether), and altenuens (→ altenuene, isoaltenuene), altertoxins (→ altertoxin I-III), → tenuazonic acid. The mycotoxin stemphyltoxin III and *Alternaria alternata* f. sp. *lycopersici* toxins (→ AAL-toxins) are known from fungal cultures and infected plant material.

Alternaria alternata f. sp. lycopersici toxins → AAL-toxins

Alternaria alternata (Fr.) Keissler (Syn.: *A. longipes, A. tenuis) A. alternata* is an extremely common saprophyte which contaminates various plants and foods.

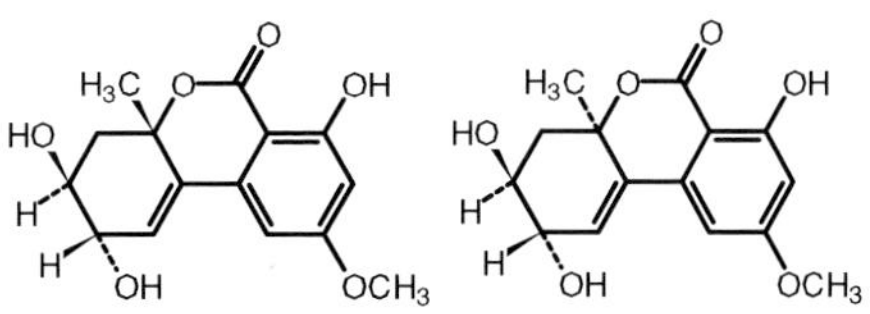

Altenuene. Altenuene and isoaltenuene

Developing → grains are infected by airborne spores of this fungus after anthesis when grain (→ wheat) moisture is as high as 70%. At high relative humidities this "black mold" causes a darkish discoloration ("black point") starting at the end of the grains (see Figure *Alternaria alternata* (Fr.) Keissler).
The relative production of alternariols and altertoxins is influenced by the water activity. Toxin production is best at water activities above → a_w 0.98. Because formation of → tenuazonic acid seems to be strain dependent a correlation between the relative toxicity of *Alternaria* infected grains and the degree of fungal invasion is doubtful.
Tobacco leaves are commonly invaded by *A. alternata* (*A. longipes*). Therefore, it has been suggested, that pulmonary exposures to → Alternaria mycotoxins might be involved in lung diseases. However, no *Alternaria* mycotoxins have been found in infected tobacco leaves.
A. alternata may produce the following → mycotoxins:
→ alternariol, → alternariol methyl ester, → altenuene, → altertoxins I-III, tenuazonic acid. In addition, *A. alternata* f. sp. *lycopersici* produces → fumonisins (FB_1, FB_2, FB_3).

Alternaria mycotoxins are produced by many *Alternaria* strains in relatively large amounts usually at the pre-harvest stage of the crop. → Tenuazonic acid, → alternariol methyl ether and → alternariol occur most frequently while the occurrence of → altenuene, isoaltenuene and → altertoxin I-III has been reported occasionally.
Alternaria toxins are divided into 3 main structural classes: dibenzo-α-pyrones (i.e. alternariol, altenuene, altenuisol, altenusin, and dehydroaltenusin), tetramic acids (i.e. tenuazonic acid), and the altertoxins (i.e. altertoxin I-III).

Natural Occurrence
→ apples, → barley, → mandarin fruits, → melon, → oats, → olives, → pecans, → pepper, → ragi, → rye, → sorghum, → sunflower seeds, → tomatoes, → triticale, → wheat
Quite high levels of alternariol, alternariol methyl ether and tenuazonic acid occurred in → apples and → tomatoes as well as → tomato paste. In → wheat flour alternariol, alternariol methyl ether, altertoxin I, and tenuazonic acid could be detected after storage at 20 °C for 28 days.

Toxicity
In comparison to the altertoxins, alternariol methyl ether is only slightly → mutagenic but the possibility of synergistic toxic effects is high. Among the *Alternaria* toxins tenuazonic acid is probably the most acutely toxic.
The involvement of *Alternaria* toxins in human and animal health disorders is under discussion. It was suggested that *A. alternata* (formerly *A. longipes*) as a common pathogen in tobacco might be involved in lung disease of man due to the exposure to its mycotoxins. However, up to now no *Alternaria* toxins have been detected in this crop. Cereal samples (→ cereals) from farms with suspected mycotoxicosis showed a higher frequency of *Alternaria* mycotoxin contamination than brands from farms with healthy ani-

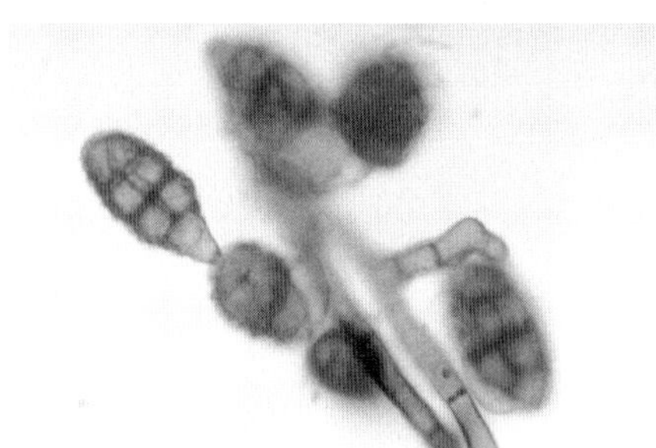

Alternaria alternata (Fr.) Keissler

mals. In addition, the contamination of fodder and feed with toxic *Alternaria* spp. could be responsible for the death of rabbits and poultry.
Although contamination of → food and feed supplies with *Alternaria* mycotoxins in the developed countries seems to be low, these levels may cause chronic rather than acute disease. Synergistic effects of cooccurring mycotoxins should be taken into account.

DETECTION
HPLC & MS, LC
The detection and analysis of *Alternaria* mycotoxins often interferes with other commonly occurring → mycotoxins (e.g. alternariol methyl ether / → zearalenone, alternariol methyl ether and alternariol / → aflatoxins). Although both *Alternaria* mycotoxins exhibit sky-blue fluorescence, it is brighter under short-wave than long-wave ultraviolet light.

POSSIBLE MYCOTOXICOSIS
Alternaria mycotoxins, especially tenuazonic acid, may responsible for the mycotoxicoses → Onyalai.

FURTHER COMMENTS
25 °C and a_w 0.98 were the optimum conditions for the production of the three *Alternaria* mycotoxins alternariol, alternariol methyl ether, and alternaric acid.

Alternaria tenuissima (Kunze ex Pers.) Wilts may produce the following → mycotoxins:
→ alternariol, → alternariol methyl ester, → tenuazonic acid.

Alternariol (Abbr.: AOH) is a dibenzo-α-pyrone derivative (3,7,9-trihydroxy-1-methyl-6H-dibenzo[b,d]pyran-6-one) produced by → Alternaria spp. (→ mycotoxins) (see Figure Alternariol).

CHEMICAL DATA
Empirical formula: $C_{14}H_{10}O_5$, molecular weight: 258

FUNGAL SOURCES
→ Alternaria alternata (Fr.) Keissler, *A. cucumerina, A. dauci, A. kikuchiana, A. solani*

NATURAL OCCURRENCE
→ apples, → barley, → mandarin fruits, → oats, → pecans, → pepper, → rye, → sorghum, → sunflower seeds, → tomatoes, → triticale, → wheat

TOXICITY
cytotoxic, fetotoxic, → teratogenic
Dosage (ip): 200 mg / kg bw mice (3 of 10 mice died)
AOH possesses a very weak acute toxicity. A synergistic effect between AOH and → alternariol methyl ether could be shown.

DETECTION
GC, HPLC, TLC

POSSIBLE MYCOTOXICOSIS
AOH may be involved in the "Fescue Foot Syndrome" (cattle).

Alternariol methyl ether (Abbr.: AME) is a dibenzo-α-pyrone derivative (→ mycotoxins) produced by → Alternaria spp. (see Figure Alternariol methyl ether).

CHEMICAL DATA
Empirical formula: $C_{15}H_{12}O_5$, molecular weight: 272

FUNGAL SOURCES
→ Alternaria alternata (Fr.) Keissler, *A. cucumerina, A. dauci, A. kikuchiana, A. solani*

NATURAL OCCURRENCE
→ apples, → barley, → mandarin fruits, → melon, → oats, → olives, → pecans, → pepper, → ragi, → rye, → sorghum, → sunflower seeds, → tomatoes, → triticale, → wheat

TOXICITY
necrotic (viscera), fetotoxic, → teratogenic, → mutagenic, and carcinogenic (?)
AME possesses a very weak acute toxicity.

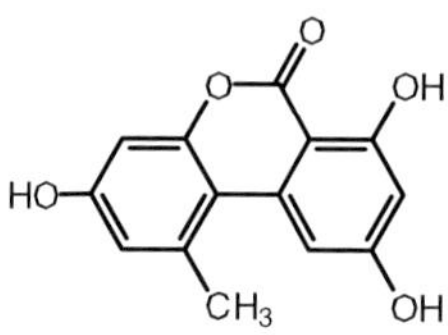

Alternariol

A synergistic effect between AME and → alternariol could be shown.
Dosage (ip): 400 mg AME / kg bw mice (1 of 10 mice died)

DETECTION
GC, HPLC, TLC
Thin-layer chromatograms and fluorescens of AME and zearalenone are similar, misidentification is possible.

Altertoxin I-III (Abbr.: ATX I, ATX II, ATX III) are 4,9-dihydroxyperylene-3,10-quinons (ATX I = 1,2,11,12,12a,12b-hexahydro-1,4,9,12a-tetrahydroxy-3,10-perylenedione; ATX II = 7a,8a,8b,8c,9,10-hexahydro-1,6,8c-trihydroxy-perylo[1,2-b]oxirene-7,11-dione; ATX III = 1a,1b,5a,6a,6b,10a-hexahydro-4,9-dihydroxy-perylo[1,2-b:7,8-b´]bisoxirene-5,10-dione) produced by → Alternaria spp. (→ mycotoxins). Although altertoxins were isolated in 1973 their correct molecular structure was not elucidated until 1986 (see Figure Altertoxin I-III). Since the altertoxins possess a high toxicity their significance in food may be comparable to that of → tenuazonic acid. The altertoxins are mainly responsible for the mutagenic activity of → Alternaria mycotoxins. Compared to the other *Alternaria* mycotoxins the altertoxins are usually produced in small quantities by → Alternaria alternata (Fr.) Keissler. This means a somewhat ameliorated risks for consumers.

CHEMICAL DATA
Empirical formula: $C_{20}H_{16}O_6$, molecular weight: 352 (ATX I)

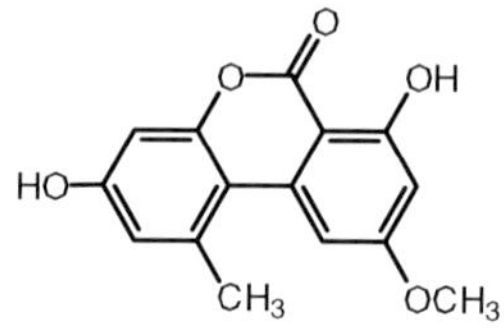

Alternariol methyl ether

Empirical formula: $C_{20}H_{14}O_6$, molecular weight: 350 (ATX II)
Empirical formula: $C_{20}H_{12}O_6$, molecular weight: 348 (ATX III)

FUNGAL SOURCES
Altertoxin I, II & III = *A. alternata*, *A. mali*, altertoxin I additionally *A. tenuissima*

NATURAL OCCURRENCE
→ altertoxin I occurs in → apples, → sorghum

TOXICITY
cytotoxic, → mutagenic
The altertoxins are very weak acute acting toxins, with an LD_{50} of 150 mg / kg bw mice. ATX-I and ATX-II were lethal to mice at the dose of 200 mg / kg bw. Treated animals showed inactivity, subendocarcial and subarachnoid hemorrhages, and blood in the cerebral ventricles. The mutagenic activity of ATX-III is approximately one tenth of that of → aflatoxin B_1. ATX-I and ATX-II possessed a lower mutagenicity.

DETECTION
see → Alternaria mycotoxins

Ammi (*Trachyspermum ammi* (Linn.) Sprague)
may contain the following → mycotoxins:
→ aflatoxin B_1
incidence: 1/7, conc.: 60 µg / kg, country: India
→ aflatoxin B_2
incidence: 1/7, conc.: 34 µg / kg, country: India
→ aflatoxin G_1

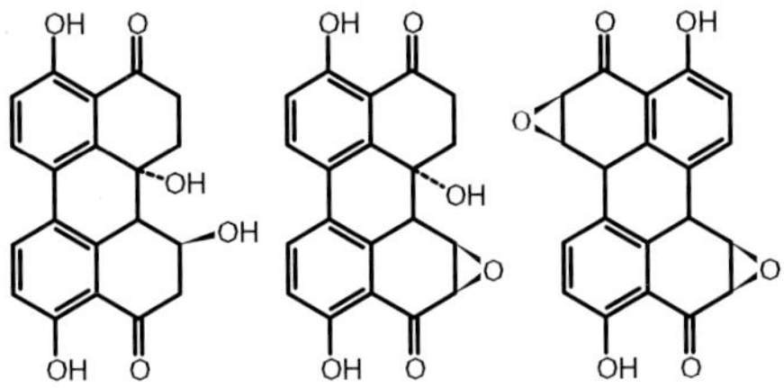

Altertoxin I-III

incidence: 1/7, conc.: 32 µg / kg, country: India

Ammoniation process Ammoniation greatly altered the biological activity of → aflatoxin B_1 (→ aflatoxins) (450-fold decrease in mutagenicity).

Anemia A below average number of erythrocytes.

Angina Any disease characterized by attacks of choking or suffocation.

Anguidine → diacetoxyscirpenol

Anorexia Loss of appetite

Antimycin → citrinin

Apiospora → Lasiosphaeriaceae

Aplastic aleukia → Alimentary toxic aleukia

Apple beverages may contain the following → mycotoxins:
→ patulin
incidence: 29/66, conc. range: 5-54 µg / l, country: Sweden

Apple butter may contain the following → mycotoxins:
→ patulin
incidence: 1/1, conc.: 1390 µg / kg, country: Finland

Apple flavor may contain the following → mycotoxins:
→ patulin
incidence: 3/14, conc. range: 6-1770 µg / kg, ∅ conc.: 607 µg / kg, country: Finland

Apple jam The ready solubility of → patulin in water and the microchannels present in jams facilitate diffusion of this mycotoxin in this substrate. In domestic consumption, often only the top moldy layer of a mold-contaminated jar is discarded which is not sufficient in the case of patulin contamination.
Apple jam (apple butter) may contain the following → mycotoxins: patulin
incidence: 1/1, conc.: 1390 µg / kg, country: Finland

Apple juice In commercial practise unsound, → Penicillium expansum Link infected → apples may partly be used for juice production. However, their portion is limited by causing off-flavors in the juice. Furthermore, the inclusion of infected apples will result in → patulin contamination. Substantial toxin reduction (90%) is achieved by simple trimming of moldy apple tissues. During juice processing patulin content is reduced by only about 20%, mainly in the concentration step (vacuum distillation at 35-40 °C) (see Table Apple juice). Mycotoxin concentration in the juice is a good quality indicator for soundness of fruits used in the process.
Although commercially processed apple juices may be contaminated by patulin, levels are usually below 100 µg / l. Juice directly made from fresh apples contained higher levels of patulin than juice prepared from concentrate.
Thermal processing is not sufficient to insure a patulin free juice. Heat treatment for 10 or 20 min at 80 °C did not cause any destruction of patulin, 10s at 90 °C

(HTST) caused an almost 20% reduction. Only little decrease in patulin content could be observed after storage for up 3-4 weeks at 22 °C (10% after 2 weeks). However, addition of → ascorbic acid, filtration or agitation with charcoal and fermentation to apple cider are highly effective in reducing patulin levels down to zero.
Apple juice may contain the following → mycotoxins: patulin
incidence: 140/241*, conc. range: 5-50 μg / kg (69 samples), 51- ≤ 1130 μg / kg, (71 sa) country: Australia, *apple and mixed → fruit juices
incidence: 1/30, conc.: 17 μg / kg, country: Brazil
incidence: 1/1, conc.: 1000 μg / l, country: Canada
incidence: 5/11, conc. range: 20-120 μg / l, country: Canada
incidence: 45/72, conc. range: ≤ 115 μg / l, ∅ conc.: 56.5 μg / kg, country: Canada
incidence: 28/61, conc. range: 20-17,700 μg / l, country: Canada
incidence: 10/51, conc. range: 5-72 μg / l, country: Finland
incidence: 8/20*, conc. range: ≤ 65 μg / l, country: Finland, *home-made
incidence: 9/13, conc. range: 100-300 μg / l, country: France
incidence: 41/66, conc. range: 2-50 μg / l, country: Germany
incidence: 4/17, conc. range: > 5.0-42.5 μg / l, country: Germany
incidence: 7/36, conc. range: 20-300 μg / l, country: Germany
incidence: 5/10*, conc. range: 60-50,000 μg / l, country: Germany, *moldy
incidence: 445/609, conc. range: ≤ 20 μg / l (286 samples), ≤ 100 μg / l (122 sa), ≤ 400 μg / l (37 sa) country: Germany
incidence: 1/33, conc.: 52 μg / kg, country: Germany
incidence: 3/20, conc. range: 106-216 μg / l, country: New Zealand
incidence: nc/140, conc. range: < 1-220 μg / l, country: Norway
incidence: 80/165, conc. range: 20-253 μg / l, ∅ conc.: 30 μg / l, country: Poland
incidence: 82/100, conc. range: 0.5-170 μg / l, ∅ conc.: 13.8 μg / kg, country: Spain
incidence: 40/49, conc. range: ≤ 70 μg / l, country: Sweden
incidence: 29/66, conc. range: 2.5-27 μg / l, country: Sweden
incidence: 19/42, conc. range: 5-50 μg / l, country: Switzerland
incidence: 20/21, conc. range: 5-20 μg / l (5 samples), > 20-50 μg / l (13 sa), > 50 μg / l (2 sa), country: Turkey
incidence: 1/2*, conc.: 5-10 μg / l, country: UK, *long life
incidence: 1/1*, conc.: 56 μg / l, country: UK, *organic
incidence: 8/13*, conc. range: 5-10 μg / l (5 samples), 16-30 μg / kg (3 sa), country: UK, *regular
incidence: 24/45, conc. range: 1-56 μg / l, country: UK
incidence: 14/20, conc. range: 1-38 μg / l, country: UK
incidence: 23/40, conc. range: ≈ 10-350 μg / l, ∅ conc.: 51 μg / l, country: USA
incidence: 8/13, conc. range: 44-309 μg / l, country: USA
incidence: 5/5, conc. range: 244-3993 μg / l, ∅ conc.: 1902 μg / l, country: USA
incidence: 50/136, conc. range: 40-440 μg / l, country: USA
incidence: 9/40, conc. range: 20,000-45,000 μg / l, country: USA
→ breakfast drinks , → cider, fruit juice, → grape juice, → soft drinks

Apple juice concentrate may contain the following → mycotoxins:
→ patulin
incidence: 8/16, conc. range: 5-50 μg / l (6 samples), > 50-646 μg / l (2 sa), country: Australia
incidence: 15/71, conc. range: ≤ 1450 μg / l, country: Finland
incidence: 27/27, conc. range: 55-610 μg / l, country: France

incidence: 79/165, Ø conc.: 30 µg / l, country: Poland
incidence: 215/215, conc. range: 7-376 µg / l, country: Turkey

Apple products (no specification) may contain the following → mycotoxins:
→ patulin
incidence: 7/105, conc. range: 11-50 µg / kg, country: Germany

Apples Patulin is the most important mycotoxin (→ mycotoxins) in apple and → apple products. It is produced by the most common → patulin-producing pathogen of apples (and pears), → Penicillium expansum Link. Fruit infection is significantly favored by surface damage. Maximum patulin levels occurred 13-14 days after inoculation with *P. expansum*. Apples and pears are usually stored at low temperatures (-1 to 4 °C) and / or modified atmosphere (1 to 5% CO_2 and 1 to 3% O_2). These precautions delay senescence and suppress postharvest decay. However, even with these common postharvest technologies, *P. expansum* can grow and produce patulin. Fungal strain as well as the fruit cultivar are decisive for the patulin rate and the amounts produced.
The mycotoxin is primarily located in areas of the spoiled apple tissue although patulin contamination in visibly healthy fruit is known. Also, penetration up to approximately 1 cm into the surrounding healthy tissue is possible. In consequence, removal of fungally decayed and surrounding tissues from apple prior to further processing significantly reduces patulin concentration in apple products. Concentration of patulin found in natural apple rots have been high as 136,000 µg / kg of fruit.
Apples may contain the following → mycotoxins:
→ aflatoxin B_1
incidence: 1/15*, conc.: 35 µg / kg, country: Germany, *moldy
→ altenuene
incidence: 5/8, conc. range: < 100-500 µg / kg, Ø conc.: 100 µg / kg, country: USA
→ alternariol
incidence: 1/20*, conc.: 160 µg / kg, country: Germany, *visibly moldy, different fruits
incidence: 7/8, conc. range.: < 100-58,800 µg / kg , Ø conc.: 7800 µg / kg, country: USA
→ alternariol methyl ether
incidence: 1/20*, conc.: 250 µg / kg, country: Germany, *visibly moldy, different fruits
incidence: 8/8, conc. range: < 100-2300 µg / kg, Ø conc.: 1000 µg / kg, country: USA
→ altertoxin I
incidence: 5/8, conc. range: nc, country: USA
patulin
incidence: 28/61*, conc. range: 20-17,700 µg / apple, country: Canada, *rotted
incidence: 5/12*, conc. range: 300-42,000 µg / kg, country: Germany, *with rotten spots
incidence: 1/16*, conc.: 2.6 µg / kg, country: Germany, *stewed
incidence: 54/104, conc. range: 1-250 µg / kg, country: Spain
→ penicillic acid
incidence: 1/6, conc.: nc, country: India
→ tenuazonic acid

Apple juice. Relative decrease in patulin contamination in the course of apple juice processing (Kubacki 1986, modified)

Processing step	losses (%)
Pasteurization I	3.4
Depectinization	1.6
Filtration	–
Pasteurization II	0.6
Concentration Vacuum distillation (35–40 °C)	18.4
Total losses	24

incidence 8/8, conc. range: 100-500 μg / kg, country: USA
→ fruits

Apricot seed paste → persipan

Arthrinium anamorphic → Lasiosphaeriaceae, teleomorph → Apiospora

Arthrinium sugarcane poisoning In China this disease is most prevalent from February to April. It caused 84 deaths in 847 cases between 1972 and 1988. A malfunction of the nervous system occurred after consumption of deteriorated sugarcane which may be contaminated by toxic fungal metabolites. The disease results in torsion spasms and may leave the victim permanently dislabled.
Besides → Fusarium moniliforme Sheldon, → Fusarium poae (Peck) Wollenw., → Penicillium aurantiogriseum Dierckx, and → Cladosporium spp. certain → Arthrinium species (*A. sacchari*, *A. saccharicola*, and *A. phaeospermum*) are discussed as the etiological fungi. From poisonous sugarcane samples 44% of the isolated fungi belonged to the latter genus.
Mice fed with *Arthrinium* culture material moved in circles and showed → paralysis of limbs. Death occurred within 3 h. The only affected organ was the brain (encephaledema). A toxic fraction of *Arthrinium* was identified as → β-nitropropionic acid. Juices of poisonous sugarcane contained this mycotoxin at levels as high as 1600 mg / kg. Such amount might be sufficient to cause human food poisoning outbreaks.
In other studies *F. moniliforme* (→ fumonisins) and *P. aurantiogriseum* (various toxic factors) predominated on mildewed sugarcane. Feeding experiments with extracts of the spoiled sugarcane or both of the fungi caused nervous disorders and death.

Ascites Accumulation of serous fluid in the abdomen.

Ascomycota → Fungi

Ascorbic acid Addition of ascorbic acid to → patulin-contaminated → apple juice removed the toxin within 3 weeks.

Asparagus The vascular and epidermal tissue of asparagus is susceptible to by → Fusarium proliferatum (Matsushima) Nirenberg, alone or together with *F. oxysporum* f. sp. *asparagi* (→ Fusarium oxysporum Schlecht. emend. Snyd. & Hansen), causing crown and root rot. Fumonisin contamination (→ fumonisins) has been reported.
Asparagus may contain the following → mycotoxins:
→ fumonisin B_1
incidence: nc/25, conc. range: $\leq$ 7400 μg* / kg, 460 μg** / kg, country: Italy, *crown, **stem
→ fumonisin B_2
incidence: nc/25, conc. range: $\leq$ 830 μg* / kg, 60 μg** / kg, country: Italy, *crown, **stem

Aspergillus anamorphic → Trichocomaceae, teleomorphs → Eurotium, → Neosartorya, → Emericella.
The genus is of ubiquitous distribution, but tends to predominate in tropical climates. Growth and metabolism of many species (e.g. → Aspergillus versicolor (Vuill.) Tiraboshi, → Aspergillus candidus Link) take place at low to very low water activities (→ Aspergillus restrictus G. Sm.). Therefore, *Aspergillus* spp. are the characteristic colonizers of stored products (see Figure *Aspergillus*). They are good indicators of previous storage conditions since each single species has its distinct minimum → a_w value. *Aspergillus* spp. is further characterized by the production of numerous toxic metabolites (→ myco-

toxins). Mycotoxin production starts at a_w levels between 0.80-0.83.
Some species are able to grow in the animal body (e.g. → Aspergillus fumigatus Fres.) and may be associated with pathogenicity.
Important mycotoxin producers are: → Aspergillus flavus Link, → Aspergillus parasticus Speare, → Aspergillus ochraceus group, *A. versicolor.* Important mycotoxins are: → aflatoxins, → citrinin, → cyclopiazonic acid, → ochratoxin A, → sterigmatocystin

Aspergillus alutaceus var. alutaceus Berkely & Curtis (formerly *A. ochraceus* K. Wilh.)
It is suggested that this is an important mycotoxin-producing fungus in cereals and the most important → ochratoxin A producer within the genus *Aspergillus.* → Peanuts and → soybeans are the main substrates. The minimum → a_w of *A. alutaceus* for OTA and → penicillic acid production is a_w 0.97-0.99 and a_w 0.85, respectively. Optimum OTA production occurs at $a_w > 0.97$.
A. alutaceus may produce → mycotoxins such as emodin, kojic acids (→ kojic acid), neoaspergillic acids, → ochratoxins, → penicillic acid, secalonic acid A (→ secalonic acids), → viomellein, → xanthomegnin.

Possible Mycotoxicosis
→ Balkan endemic nephropathy

Aspergillus candidus Link is a frequent storage fungus. It is often found on → cereals in silos where it contributes to the process of self-heating. In addition, it frequently occurs in cereals stored under a controlled atmosphere. *A. candidus* is the dominating fungus in flours (→ flour) and other → cereal products (see Figure *Aspergillus candidus* Link).
A. candidus may produce → mycotoxins such as candidulin, → kojic acid, → β-nitropropionic acid, terphenyllins, xanthoascin.

Aspergillus. *Aspergillus flavus* Link

Aspergillus clavatus Desm. prefers the humid and high temperature conditions during malting and is therefore an important fungus in malt (see Figure *Aspergillus clavatus* Desm.). It causes the "malt worker's lung disease" (an allergic respiratory disease). Carbohydrates like glucose (better than sucrose, dextrin or lactose) may be essential for → patulin production, but no formation will occur below 12 °C. The minimum water activity for patulin production of this fungus is → a_w 0.99. During malting (→ malt) of → barley and → wheat. *A. clavatus* produces not only patulin but also cytochalasin E.
A. clavatus may produce → mycotoxins such as ascladiol, cytochalasin E and "K", → kojic acid (?), patulin, tryptoquivalins and tryptoquivalons (→ tremorgenic mycotoxins)

Possible Mycotoxicosis
Ascladiol and patulin should be involved in mycotoxicosis.

Aspergillus flavus Link is a frequent mold in temperate climates. *A. flavus* has been isolated from various kinds of foodstuff but it is very common on cereal → grains and their products as well as on → spices (see Figure *Aspergillus flavus*

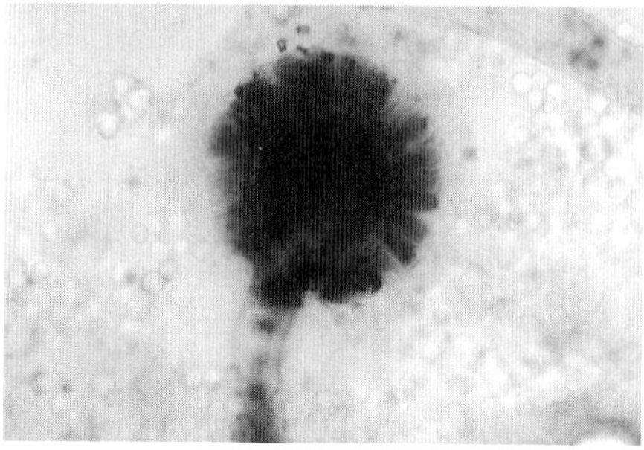

Aspergillus candidus Link

Link). Drought stress and insect damage favor the growth of this most toxic of all → Aspergillus species prior to harvesting, especially in → maize, → peanuts, and cottonseed. However, healthy plant tissue may also be invaded.

Further Comments

Although the toxicity of this fungus was described as early as 1910 by Kühl, it was not until 1960 that the → aflatoxins could be identified in peanut meal as being highly toxic metabolites of *A. flavus*. *A. flavus* may accumulate AFB_1 and AFG_1 in conidia up to 84 mg / kg and 566 mg / kg, respectively. Sclerotia may contain 135 mg AFB_1 / kg and 968 mg AFG_1 / kg. Significant variation in total aflatoxin content in conidia and sclerotia within (intrafungal) and between strains could be established.

Production: Aflatoxin production starts with the formation of conidia and is usually proportional to the weight of the mycelium produced. Using a reduced amount of inoculum resulted in a three- to 12-fold increase of aflatoxin formation. Maximum rates occur until the period of intense sporulation (≈ sixth day) when the biomass production reaches its optimal value. Subsequently and similarly to the autolysis of the mycelium a rapid decrease in aflatoxin concentration due to degradation begins.

Toxigenic potential of this fungus is influenced by the geographical origin and by the substrate from which the fungus has been isolated. 60% of the isolates (n = 1390) coming from six different countries were toxigenic. It seems that strains (n = 427) isolated in tropical regions possess a higher degree of toxicity than strains from temperate origins (46% / 15%).

Various (competing) microorganisms like *Penicillium* spp., *Aspergillus niger*, or *Trichoderma viride* inhibited aflatoxin formation but their general effects on mycotoxin production are unpredictable. Sublethal concentrations of propionic acid may stimulate aflatoxin synthesis by *A. flavus*.

Substrate: A large number of toxigenic isolates has been found on U.S. American → rice (94%) and peanuts (86%) as well as on groundnut kernels from Israel

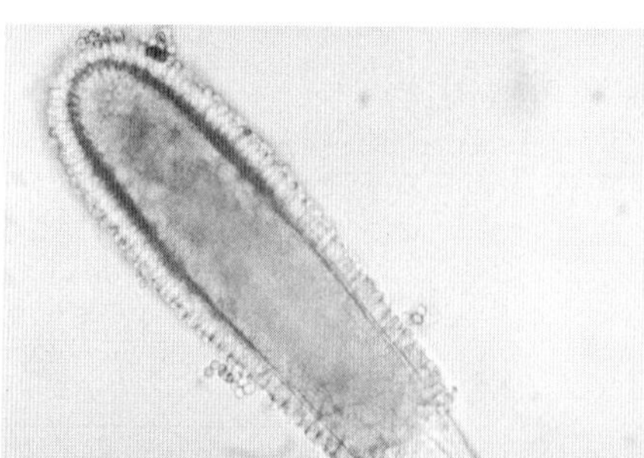

Aspergillus clavatus Desm.

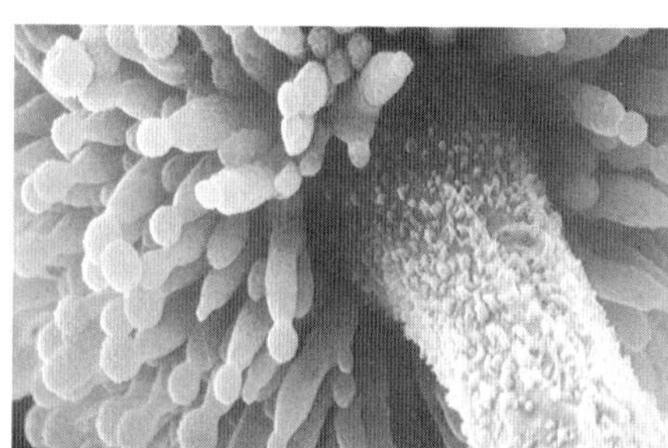

Aspergillus flavus Link

(71%). Generally *A. flavus* strains isolated from → oil seeds - especially peanuts and peanut products - showed a higher proportion of aflatoxin producers than isolates contaminating → cereals and their products. Instead of → spices - ca. 30% of the isolated *A. flavus* strains were toxic - → meat, cheeses (→ cheese), → bread or → pasta seem to be unsuitable substrates for toxigenic isolates.
Carbon sources such as glucose, mannose, sucrose, and fructose as well as glyceraldehyde all favor aflatoxin production. The same is true for nitrogen sources like ammoniacal nitrogen, glutamic acid, or uric acid. In addition, yeast extract, peptone or certain amino acids (glycine, glutamate, proline) contribute to a higher aflatoxin formation. Vitamins of the B group, e.g. thiamine, cause stimulation of aflatoxin synthesis just as cadmium, iron, magnesium and zinc do. The wheat embryo, most probably because rich in diverse nutrients, allowed the production of much higher aflatoxin levels than the testa. Yields in the range from 100,000-2,000,000 µg aflatoxin / kg substrate, depending on the used culture, are known. The largest amounts have been reported for coconut flesh (8,000,000 µg / kg).
Moisture, temperature: A moisture content of 18.3-18.5% in cereal grains and 9-10% in seeds with a high oil content like → nuts, → copra, safflower and → sunflower seeds may enable mycotoxin production. Below these values commodities are usually resistant to contamination. Temperatures between 24-28 °C are the optimum for → aflatoxin B_1 production, 30 °C favor the formation of → aflatoxin G_1. A constant temperature of 25 °C resulted in higher aflatoxin concentrations than fluctuating temperatures with a mean of 25 °C which are common in nature. 7.5 °C seems to be the lowest temperature enabling aflatoxin production whereas synthesis drops off sharply above 35 °C.
Atmosphere: Oxygen concentration as low as 1% in combination with 99% N_2 and 1% O_2, 79% N_2, and 20% CO_2, respectively, allowed aflatoxin production. However, an atmosphere consisting of 1% O_2, 19% N_2, and 80% CO_2 prevented the synthesis of aflatoxin.
A. flavus may produce → mycotoxins such as aflatoxins B_1, B_2, G_1, G_2 (although AFG_1 and AFG_2 are not generally produced), → aflatrem, aspergillic acids, aspergillomarasmins, cyclopiazonic acids (→ cyclopiazonic acid), koji acids (→ kojic acid), maltoryzin, → β-nitropropionic acid, paspalicin, paspalinine, → sterigmatocystin.

Aspergillus fumigatus Fres. is an ubiquitous species which contaminates different kinds of food like → cereals (wet stored), → peanuts, → pecans, → tomatoes (see Figure *Aspergillus fumigatus* Fres.). It frequently occurs in cereals that are in advanced state of spoilage. Low oxygen tensions are tolerated. Due to its thermophilic nature, growth is adapted to high temperatures (≤ 55 °C).
A. fumigatus may produce → mycotoxins such as fumagillin, fumigatins, fumigaclavines, → fumitremorgins A & B, gliotoxin, → kojic acid (?),→ ochratoxin A, tryptoquivalins, verruculogen.

Aspergillus glaucus group → Eurotium spp.

Aspergillus niger van Tieghem This fungus is a contaminant of various substrates of plant origin, e.g. → cereals, but it usually does not predominate in spoiled cereal grain. → Mycotoxins of the *A. niger* group (Section Nigri) have not yet been detected naturally in cereals.
A. niger may produce mycotoxins such as aspergillins, → kojic acid (?), malformins, naphthopyrones, → ochratoxin A

Aspergillus nomius Kurtzman et al. is not so common in → foods as the very important mycotoxin producers → Aspergillus flavus Link and → Aspergillus parasiticus Speare.
A. nomius may produce → mycotoxins such as → aflatoxins B_1, B_2, G_1, G_2 (consistently produced), aspergillic acids, kojic acids (→ koji acid), nominine, → tenuazonic acid.

Aspergillus ochraceus group (= Section *Circumdati*)
included are → Aspergillus alutaceus Berkley & Curtis, *A. fresenii*, *A. ostianus*, *A. petrakii*, *A. quercinus*, *A. sclerotiorum*. Fungi of this group / section produce → ochratoxin A but they are considered to be rare on grain. These fungi do not produce ochratoxin A and → penicillic acid below 12 °C.
Fungi of the *A. ochraceus* group may produce → mycotoxins such as ochratoxin A, penicillic acid, → secalonic acids, → viomellein, vioxanthin, → xanthomegnin.

Aspergillus ochraceus K. Wilh. (Syn.: → Aspergillus alutaceus Berkely & Curtis)

Aspergillus oryzae (Ahlburg) Cohn is often used for fermentating different kinds of foodstuff (e.g. koji, → miso, soya sauce, saké alcohol) in Asian countries. Although it belongs to the *A. flavus* group and shows a high similarity with

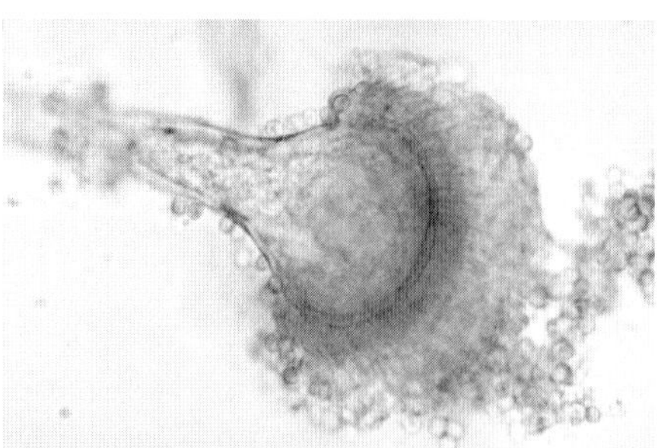

Aspergillus fumigatus Fres.

the mycotoxin producer → Aspergillus flavus Link → aflatoxins are not synthesized. *A. oryzae* may produce → mycotoxins such as aspergillomarasmin, → cyclopiazonic acid, → koji acid, maltoryzin, → β-nitropropionic acid

Aspergillus parasiticus Speare In contrast to → Aspergillus flavus Link *A. parasiticus* predominates in warmer climates (tropical and subtropical regions). Since this mold is most often associated with soil, → peanuts are mainly infected by this *Aspergillus* species. Instead of this, *A. flavus* invasion is more common in → maize. Nearly all strains of *A. parasiticus* are toxigenic while aflatoxin production is enhanced by the amino acid proline. → Aflatoxin B_1 production starts at a water activity of 0.87. Maximum aflatoxin production on sterilized → maize was observed at an → a_w of 0.90.
A. parasiticus may produce → mycotoxins such as → aflatoxins B_1, B_2, G_1, G_2 (consistently produced), aspergillic acids, koji acids (→ kojic acid), → β-nitropropionic acid, → sterigmatocystin

Aspergillus restrictus G. Sm. belongs to the important storage fungi in cereal → grains. Besides → Eurotium *halophilicum* it is the first growing fungus in → cereals stored at moisture contents that are just a little too high for safe storage (≈ 14%). This slowly growing fungus does not cause any significant rise in grain temperature. The metabolic water of this primary colonizer enables the growth of mycotoxin-producing fungi like → Aspergillus flavus Link. *A. restrictus* is often associated with storage insects such as the granary and the rice weevil which contribute to its distribution.
A. restrictus may produce → mycotoxins such as mitgilliin.

Aspergillus terreus Thom predominates in → cereals stored under airtight condi-

tions (see Figure *Aspergillus terreus* Thom). Although it produces a wide range of → mycotoxins it is not known whether they do naturally occur in cereals.
A. terreus may produce mycotoxins such as → citreoviridin, → citrinin, cytochalasin E (→ cytochalasins), flavipin?, gliotoxin, → patulin, terreic acid, teritrem A, B, C, A′, and B′.

Aspergillus terreus Thom

Aspergillus versicolor (Vuill.) Tiraboshi has frequently been isolated from moldy seeds and their products and from oil seed products (→ oil seeds). Under certain environmental conditions, e.g. airtight storage, it may predominate (see Figure *Aspergillus versicolor* (Vuill.) Tiraboshi). *A. versicolor* is the most important producer of → sterigmatocystin. Sterigmatocystin is produced in → cheese ripened at 6 °C. Toxicoses which involve *A. versicolor* are probably due to sterigmatocystin and related metabolites.
A. versicolor may produce → mycotoxins such as aspertoxin, nidulotoxin, → ochratoxin A, sterigmatocystins, versicolorins.

ATA → Alimentary toxic aleukie

Ataxia Loss of muscle coordination

Atmosphere It seems that mycotoxin production (→ mycotoxins) is more sensitive to the concentration of atmospheric gases than fungal growth. In most cases mold development and mycotoxin formation is inhibited by low O_2 concentration (< 1%) and / or elevated levels of CO_2. High CO_2 levels appeared to be more effective in controlling fungal growth and mycotoxin formation than high N_2 and low O_2 concentrations. Since fungal growth has been reported in → beer high levels of CO_2 may not be sufficient to prevent mold development and subsequent mycotoxin formation in all cases. Temperature and → a_w influence the tolerance to specific gas compositions.

a_w The water activity (a_w) of a substrate is defined as the ratio of the water vapor pressure of the food substrate (p) to the vapor pressure of pure water (p_0) at the same temperature and pressure: $a_w = p/p_0$. The a_w expresses the moisture content of a substrate (e.g. food) as an index of water available for the growth of the microorganisms. Since each fungal species / strain has minimum water requirements for growth at a given temperature and on a distinct substrate the same is true for mycotoxin production (→ mycotoxins) of the different fungi (see Table a_w). Growth and mycotoxin production in different substrates are only comparable in terms of their a_w not their water contents.

Azotemia increase of nitrogen levels in the blood

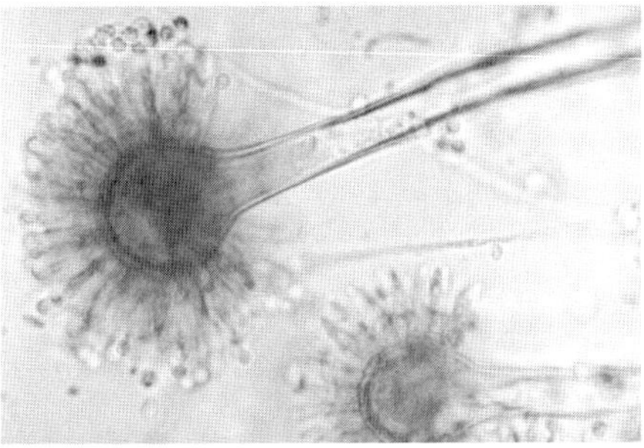

Aspergillus versicolor (Vuill.) Tiraboshi

a_w. Minimum a_w for growth and mycotoxin production by selected molds

Mold	Mycotoxin	Minimum a_w - growth -	Minimum a_w - toxin production -
Aspergillus ochraceus	penicillic acid	0.76–0.83	0.80–0.88
A. flavus	aflatoxin	0.78–0.84	0.83–0.87
A. ochraceus	ochratoxin A	0.76–0.83	0.83–0.87
Penicillium verrucosum	ochratoxin A	0.81–0.83	0.83–0.90
P. griseofulvum	patulin	0.81–0.85	0.85–0.95
A. parasiticus	aflatoxin	0.78–0.82	0.87
P. aurantiogriseum	ochratoxin A	0.79–0.85	0.87–0.90
P. patulum	patulin	0.81–0.85	0.95
P. aurantiogriseum	penicillic acid	0.79–0.85	0.97–0.99
P. expansum	patulin	0.82–0.85	0.99
A. clavatus	patulin	0.85	0.99

B

Baby cereals (no specification) may contain the following → mycotoxins:
→ deoxynivalenol
incidence: 30 products analysed, ∅ conc.: 43 μg / kg, country: Canada
ergocristine (→ ergot alkaloids)
incidence: 1/1*, conc.: 0.4 μg / kg, country: Canada, *mixed
→ cereals

Baby food may contain the following → mycotoxins:
→ aflatoxin B_1
incidence: nc*, conc. range: 4-66 μg / kg, country: France, *→ meat / vegetable preparation (→ vegetables)
incidence: nc*, conc. range: 2-26 μg / kg, country: France, *carrot preparation
→ deoxynivalenol
incidence: 14/39, conc. range: tr-90 μg / kg, country: USA
→ ochratoxin A
incidence: 2/34, conc. range: $\leq$ 0.2 μg / kg, country: Germany

Bacon may contain the following → mycotoxins:
→ aflatoxin B_1
incidence: 2*/10, conc. range: 1000-5000 μg / kg, ∅ conc.: 3000 μg / kg, country: Germany, *moldy
→ ochratoxin A
incidence: 39/206*, conc. range: 37-200 μg / kg, country: Yugoslavia, *total of smoked → meat products
→ sausages

Bakery products (no specification) may contain the following → mycotoxins:
→ aflatoxins (no specification)
incidence: 2/8, conc. range: 1-14 μg / kg, country: UK
→ citrinin
incidence: 1/2, conc.: < 150 μg / kg, country: UK
→ ochratoxin A
incidence: 3/8, conc. range: 0-80 μg / kg, country: UK
→ cereals

Baking Since most → mycotoxins are heat-stable no significant reduction in their concentration will occur during baking (see Table Influence of baking and other heat processing on stability of trichothecenes). For details see each single mycotoxin.
→ cereals

Balkan endemic nephropathy (Abbr.: BEN) A chronic kidney disease which was first described in the 1950s in the Vratza District (Bulgaria). Now the disease mainly occurs in different rural areas of Bulgaria, Romania, and Yugoslavia located within the Danube Basin. Some 20,000 people mainly of the rural population should be affected. In endemic areas up to 12% of the inhabitants suffer from this disease. Mortality rates of up to 40% have been reported. Resulting from prolonged exposure to a causal agent the affected individuals are almost exclusively between (30)-35 and (50)-55 years old. More females than males were affected.
While the onset of the disease remains unnoticed, in a progressed stage impairment of the kidney function becomes obvious. Severe → nephropathy often accompanied by urinary tract tumors are the major symptons. Tumor rate of this rarely occurring kind of cancer is nearly 100 times higher in the endemic area compared to the non-endemic ones. Histologically, this → renal disease is characterized by tubular degeneration and interstitial → fibrosis. In the more superficial part of the → cortex hyalization of the glomeruli appeared. The size of the kidneys affected is greatly reduced. A prominent and early indication of the

disease is the impairment of tubular function. The insidious onset of a normocytic, normochronic → anemia, → azotemia, and persistent → proteinuria as well as renal failure are the leading clinical symptoms. This incurable disease is treated only symptomatically which includes hemodialysis. Within 5 to 10 years BEN progresses slowly up to death. The etiology of BEN is still unknown. A positive correlation between heavy rainfall (late summer and autumn), possibly favoring growth and toxigenicity of fungi in endemic areas, and the number of people who died of nephropathy during the succeeding 2 years could be established. Therefore, involvement of fungi (e.g. → Penicillium and → Aspergillus) and certain of their → mycotoxins is discussed. The mammalian carcinogen → ochratoxin A might be the main causal agent, especially because similarities with the → mycotoxic porcine nephropathy due to this mycotoxin in Scandinavia do exist. This is corroborated by the fact that different foodstuffs, e.g. cereal → grains, produced in the endemic areas of Yugoslavia showed a significantly higher OTA contamination than products from non-endemic areas. Consequently, residues of OTA could be detected more frequently and at higher concentrations in the blood of inhabitants as well as in the → pig kidneys, → pig liver, and → pig blood of endemic areas. Besides → citrinin (→ nephrotoxin), which is also found in greater proportion and greater degree in the staple foods of affected families, a novel *Penicillium* mycotoxin, possibly a glycopeptide (molecular weight ca. 1500), might be involved in the etioloy of BEN. However, although substantial OTA contamination of food- and feedstuff has been reported no mycotoxic porcine nephropathy occurred in the endemic areas of BEN. Furthermore, data about OTA contamination of foodstuffs and the blood of BEN patients are not sufficient to prove a quantitative relationship between the degree of OTA exposure and the severity of human nephropathy. The involvement of heavy metals and / or viruses is also under discussion showing that the significance of mycotoxins in BEN still remains unresolved.

Bananas may contain the following → mycotoxins:
→ zearalenone
Incidence: 1/1, conc.: 17 µg / kg, country: India
→ fruits

Barley In years of moist weather, seeds of barley may be relatively heavily invaded by more than a dozen species of → Fusarium spp. during time of maturation. Severe invasion will result in a reddish discoloration of a portion of the kernels. In consequence, seeds fail to develop and shrivel, or the partly developed kernel deteriorates. This disease is called "scab" or "blight". Mycotoxin contamination of the kernels is possible. Barley may contain the following → mycotoxins:
→ 3-acetyldeoxynivalenol
incidence: 24/40, conc. range: $\leq$ 350 µg / kg, Ø conc.: 40 µg / kg, country: Canada
incidence: 1/6, conc.: < 200 µg / kg, country: Finland
incidence: 10/30, conc. range: 24-96 µg / kg, Ø conc.: 46 µg / kg, country: Finland
incidence: 1/30, conc.: 7 µg / kg, country: Korea
→ 15-acetyldeoxynivalenol
incidence: 24/77, conc. range: $\leq$ 400 µg / kg, Ø conc.: 40 µg / kg, country: Canada
incidence: 39/40, conc. range: 1240 µg / kg, Ø conc.: 210 µg / kg, country: Canada

→ aflatoxin B_1
incidence: 3/376, conc. range: < 10-2000 µg / kg, country: UK
aflatoxin B_1 & → aflatoxin B_2
incidence: 1*/94**, conc.: ≈ 10 µg / kg, country: Japan, *moldy, **barley and pressed barley
aflatoxin (no specification)
incidence: 13/37, conc. range: 1-5 µg / kg (9 samples), 5-20 µg / kg (3 sa), 31 µg / kg (1 sa), country: Germany
incidence: 12/137*, conc. range: 2-20 µg / kg (10 samples), > 20 µg / kg (2 sa), country: Uruguay, *and malt
→ alternariol
incidence: 1/179, conc.: 15 µg / kg, country: Germany
incidence: 1/3, conc.: 116 µg / kg, country: Poland
→ alternariol methyl ether
incidence: 12/254, conc. range: 4-25 µg / kg, ∅ conc.: 10.2 µg / kg, country: Germany
incidence: 1/3, conc.: 58 µg / kg, country: Poland
→ citrinin
incidence: 4/269, conc. range: 30-480 µg / kg, country: Sweden
incidence: 4/4*, conc. range: tr-1600 µg / kg, country: UK, *moldy
→ deoxynivalenol
incidence: 18/20, ∅ conc.: 237 µg / kg, country: Argentina
incidence: 16/90, conc. range: 7-1670 µg / kg, ∅ conc.: 270 µg / kg, country: Canada
incidence: 77/77, conc. range: 100-15,100 µg / kg, ∅ conc.: 2650 µg / kg, country: Canada
incidence: 40/40, conc. range: 30-15,790 µg / kg, ∅ conc.: 2690 µg / kg, country: Canada
incidence: 4/4, conc. range: 21-164 µg / kg, ∅ conc.: 83 µg / kg, country: China
incidence: 3/5, conc. range: 10-20 µg / kg, country: Denmark
incidence: 1/nc, conc.: 1000 µg / kg, country: Denmark
incidence: 3/6, conc. range: 1-6300 µg / kg, country: Finland
incidence: 27/30, conc. range: 10-202 µg / kg, ∅ conc.: 78 µg / kg, country: Finland
incidence: 3/3*, conc. range: 27-46 µg / kg, ∅ conc.: 38 µg / kg, country: Finland, *imported from Canada, Sweden
incidence: 1/4, conc.: 10 µg / kg, country: France
incidence: 2/10, ∅ conc.: 190 µg / kg, country: Germany
incidence: 6/25, conc. range: 150-1000 µg / kg, country: Germany
incidence: 2/13, ∅ conc.: 190 µg / kg, country: Germany
incidence: 6/123, conc. range: 10-100 µg / kg, country: Germany
incidence: 1/1, conc.: 46 µg / kg, country: Germany
incidence: 31/46*, conc. range: 20-2140 µg / kg, country: Germany, *visibly damaged
incidence: 6/7*, conc. range: 34-440 µg / kg, ∅ conc.: 137.8 µg / kg, country: Germany, *organic produce
incidence: 2/5, ∅ conc.: 195 µg / kg, country: Italy
incidence: 95/101, conc. range: 50-49,600 µg / kg, ∅ conc.: 1506 µg / kg, country: Japan
incidence: 5/5, ∅ conc.: 249 µg / kg, country: Japan
incidence: 18/20, conc. range: 0-4600 µg / kg, ∅ conc.: 900 µg / kg, country: Japan
incidence: 12/14, conc. range: 50-7840 µg / kg, ∅ conc.: 2010 µg / kg, country. Japan
incidence: 2/14, conc. range: 100-510 µg / kg ∅ conc.: 305 µg / kg, country: Japan
incidence: 12/12, conc. range: 90-1700 µg / kg, ∅ conc.: 580 µg / kg, country: Japan
incidence: 8/12*, ∅ conc: 176 µg / kg, country: Japan, *naked

incidence: 6 products* analysed, conc. range: 27-85 μg / kg, country: Japan, *parched
incidence: 1/1*, conc.: 48 μg / kg, country: Japan, *pearled
incidence: 10/14*, conc. range: 3-50 μg / kg, country: Japan, *pressed
incidence: 31/31, conc. range: ≤ 900 μg / kg, ∅ conc.: 124 μg / kg, country: Korea
incidence: 20/30, conc. range: 5-361 μg/kg, ∅ conc.: 106 μg / kg, country: Korea
incidence: 26/28*, conc. range: 4-508 μg / kg, ∅ conc.: 117 μg / kg, country: Korea, *unpolished
incidence: 24/27*, conc. range: 38-645 μg / kg, ∅ conc.: 213 μg / kg, country: Korea, *naked
incidence: 9/10*, conc. range: 29-677 μg / kg, ∅ conc.: 263 μg / kg, country: Korea, *husked
incidence: 3/11, conc. range: 168-506 μg / kg, ∅ conc.: 297 μg / kg, country: Korea
incidence: 26/44, conc. range: ≤ 1000 μg / kg, country: New Zealand
incidence: 1/6, ∅ conc.: 390 μg / kg, country: Poland
incidence: 5/8, conc. range: 10-81 μg / kg, ∅ conc.: 42 μg / kg, country: Scotland
incidence: 2/14, conc. range: 80-160 μg / kg, ∅ conc. 120 μg / kg, country: Sweden
incidence: 4/32, conc. range: 60-150 μg / kg, ∅ conc.: 90 μg / kg, country: Sweden
incidence: 1/6, conc.: 50 μg / kg, country: Sweden
incidence: 7/52, conc. range: 50-200 μg / kg, ∅ conc.: 90 μg / kg, country: Sweden
incidence: 4/4, ∅ conc.: 83 μg / kg, country: Taiwan
incidence: 2/5, conc. range: 10-30 μg / kg, ∅ conc.: 20 μg / kg, country: The Netherlands
incidence: 5/6, conc. range: 4-152 μg / kg, ∅ conc.: 58 μg / kg, country: The Netherlands
incidence: 3/3*, conc. range: 56-147 μg / kg, ∅ conc.: 110 μg / kg, country: The Netherlands, *pearled
incidence: 28/92, conc. range: 20-500 μg / kg, country: UK
incidence: 22/49, conc. range: 20-100 μg / kg, country: UK
incidence: 108/147, conc. range: 500-26,000 μg / kg, ∅ conc.: 4200 μg / kg, country: USA
incidence: nc/204, conc. range: tr-22,000 μg / kg, country: USA
incidence: 2/3, ∅ conc.: 19 μg / kg, country: Yemen

→ diacetoxyscirpenol
incidence: 16/134, conc. range: 200-17,000 μg / kg, country: Germany
incidence: 1/86, conc.: 100 μg / kg, country: USSR

3,15-diacetyldeoxynivalenol
incidence: 25/40, conc. range: ≤ 400 μg / kg, ∅ conc.: 60 μg / kg, country: Canada

→ HT-2 toxin
incidence: 37/94, conc. range: 100-10,000 μg / kg, country: Germany
incidence: 5/24*, conc. range: 210-370 μg / kg, ∅ conc.: 230 μg / kg, country: Poland, *spring barley

→ neosolaniol
incidence: 1/86, conc.: 100 μg / kg, country: USSR

→ nivalenol
incidence: 15/20, ∅ conc.: 25 μg / kg, country: Argentina
incidence: 1/6, conc.: < 100 μg / kg, country: Finland
incidence: 4/30, conc. range: 38-59 μg / kg, ∅ conc.: 46 μg / kg, country: Finland
incidence: 1/3, conc.: 44 μg / kg, country: Germany
incidence: 1/13, conc.: 40 μg / kg, country: Germany

incidence: 1/5, conc.: 23 µg / kg, country: Italy
incidence: 95/101, conc. range: 23-22,900 µg / kg, ∅ conc.: 1020 µg / kg, country: Japan
incidence: 5/7, conc. range: 90-640 µg / kg, country: Japan
incidence: 5/5, ∅ conc.: 708 µg / kg, country: Japan
incidence: 12/12, conc. range: 60-1500 µg / kg, ∅ conc.: 480 µg / kg, country: Japan
incidence: 18/20, conc. range: 0-2900 µg / kg, ∅ conc.: 700 µg / kg, country: Japan
incidence: 12/14, conc. range: 0-2320 µg / kg, ∅ conc.: 430 µg / kg, country: Japan
incidence: 2/14, conc. range: 0-270 µg / kg, ∅ conc.: 140 µg / kg, country: Japan
incidence: 12/12*, ∅ conc.: 342 µg / kg, country: Japan, *naked
incidence: 1/1*, conc. range: 220 µg / kg, country: Japan, *pearled
incidence: 13/14*, conc. range: 8-380 µg / kg, country: Japan, *pressed
incidence: 31/31, conc. range: ≤ 1100 µg / kg, ∅ conc.: 489 µg / kg, country: Korea
incidence: 28/30, conc. range: 40-2038 µg / kg, ∅ conc.: 390 µg / kg, country: Korea
incidence: 28/28*, conc. range: 17-3002 µg / kg, ∅ conc.: 546 µg / kg, country: Korea, *unpolished
incidence: 27/27*, conc. range: 85-4569 µg / kg, ∅ conc.: 1110 µg / kg, country: Korea, *naked
incidence: 10/10*, conc. range: 114-1546 µg / kg, ∅ conc.: 742 µg / kg, country: Korea, *husked
incidence: 2/11, conc. range: 189-324 µg / kg, ∅ conc.: 257 µg / kg, country: Korea
incidence: 1/4, conc.: 21 µg / kg, country: Nepal
incidence: 33/44, conc. range: ≤ 530 µg / kg, country: New Zealand
incidence: 3/6, ∅ conc.: 78 µg / kg, country: Poland
incidence: 3/8, conc. range: 7-1140 µg / kg, ∅ conc.: 391 µg / kg, country: Scotland
incidence: 4/4, conc. range: 290-976 µg / kg, ∅ conc.: 634 µg / kg, country: Taiwan
incidence: 4/6, conc. range: 30-145 µg / kg, ∅ conc.: 85 µg / kg, country: The Netherlands
incidence: 3/3*, conc. range: 17-39 µg / kg, ∅ conc.: 27 µg / kg, country: The Netherlands, *pearled
incidence: 2/3, ∅ conc.: 13 µg / kg, country: Yemen
→ ochratoxin A
incidence: 3/27, conc. range: 5-1000 µg / kg, country: Austria
incidence: 1/48, conc.: 3800 µg / kg, country: Czechoslovakia
incidence: 11/41*, conc. range: 0.05-4.9 µg / kg (8 samples), 5-14 µg / kg (3 sa), country: Denmark, *conventional
incidence: 6/20*, conc. range: 0.05-4.9 µg / kg (4 samples), 5-13 µg / kg (2 sa), country: Denmark, *ecological
incidence: 17/17, conc. range: 9-27,520 µg / kg, country: Denmark
incidence: 3/50, conc. range: 9-189 µg / kg, ∅ conc.: 80.7 µg / kg, country: Denmark
incidence: 10/68, conc. range: 0.1-206 µg / kg, ∅ conc.: 58.8 µg / kg, country: Germany
incidence: 11/165, conc. range: 100-1800 µg / kg, ∅ conc.: 634 µg / kg, country: Poland
incidence: 54/616, conc. range: 5-1200 µg / kg, country: Poland
incidence: 17*/269, conc. range: 2-20 µg / kg, country: Sweden, *14 only traces
incidence: 21/21*, conc. range: 0.1-8652 µg / kg, country: Tunesia, *and derived foods
incidence: 9/52, conc. range: ≤ 4.9-45 µg / kg, country: UK

incidence: 10/50, conc. range: ≤ 4.9-13.7 μg / kg, country: UK
incidence: 7/150, conc. range: ≤ 4.9-33.4 μg / kg, country: UK
incidence: 51/376, conc. range: < 25-5000 μg / kg, country: UK
incidence: 4/4*, conc. range: 75-11,000 μg / kg, ∅ conc.: 3038 μg / kg, country: UK, *moldy
incidence: 18/127, conc. range: tr-38 μg / kg, country: USA
incidence: 11/848, conc. range: < 15-116 μg / kg, country: USA
incidence: 23/182, conc. range: 10-29 μg / kg, country: USA
incidence: 22*/159, conc. range: ≤ 29 μg / kg, country: USA, *11 contained less than 10 μg / kg
incidence: 23/164, conc. range: < 10-29 μg / kg, country: USA
incidence: 11/103, conc. range: 0.03-17,000 μg / kg, ∅ conc.: 6.87 μg / kg, country: USA
incidence: 1/48, conc.: 3800 μg / kg, country: USSR
incidence: 8/64*, conc. range: 14-27 μg / kg, country: Yugoslavia, *area with endemic nephropathy
incidence: 1/23, conc.: 5 μg / kg, country: Yugoslavia
→ penicillic acid
incidence: 1/165, conc.: 800 μg / kg, country: Poland
→ sterigmatocystin
incidence: 2/4*, conc. range: traces, country: UK, *moldy
→ T-2 toxin
incidence: 2/6, conc. range: 50-600 μg / kg, ∅ conc.: 325 μg / kg, country: Finland
incidence: 1/18, conc.: 160 μg / kg, country: Finland
incidence: 1/?, conc.: 20 μg / kg, country: France
incidence: 1/78, conc.: 90 μg / kg, country: Germany
incidence: 2/12, conc. range: 105-165 μg / kg, ∅ conc.: 135 μg / kg, country: Germany
incidence: 6/132, conc. range: 200-14,000 μg / kg, country: Germany
incidence: 5/49, conc. range: 2-22 μg / kg, country: Norway
incidence: 12/24*, conc. range: 20-2400 μg / kg, ∅ conc.: 450 μg / kg, country: Poland, *spring barley
incidence: 1/86, conc.: 600 μg / kg, country: USSR
T-2 tetraol
incidence: 2/24*, conc. range: 10-210 μg / kg, ∅ conc.: 110 μg / kg, country: Poland, *spring barley
T-2 triol
incidence: 5/94, conc. range: 100-300 μg / kg, country: Germany
→ viomellein
incidence: 3/4*, conc. range: tr-600 μg / kg, country: UK, *moldy
vioxanthin
incidence: 3/4*, conc. range: 10-90 μg / kg, ∅ conc.: 40 μg / kg, country: UK, *moldy
→ xanthomegnin
incidence: 3/4*, conc. range: tr-450 μg / kg, country: UK, *moldy
→ zearalenone
incidence: 13/20, ∅ conc.: 5 μg / kg, country: Argentina
incidence: 2/30, conc. range: 21-30 μg / kg, ∅ conc.: 26 μg / kg, country: Finland
incidence: 2/10, ∅ conc.: 16 μg / kg, country: Germany
incidence: 3/3, ∅ conc.: 3 μg / kg, country: Germany
incidence: 21/92, conc. range: 1-1730 μg / kg, ∅ conc.: 60 μg / kg, country: Germany
incidence: 5/13, ∅ conc.: 10 μg / kg, country: Germany
incidence: 5/40, conc. range: 10-20 μg / kg, country: Germany

incidence: 24/46*, conc. range: ≤ 320 μg / kg, ∅ conc.: 24 μg / kg, country: Germany, *damaged kernels
incidence: 1/5, conc.: 56 μg / kg, country: Italy
incidence: 3/5, ∅ conc.: 9 μg / kg, country: Japan
incidence: 10/12*, ∅ conc.: 4 μg / kg, country: Japan,*naked
incidence: 1/1*, conc.: 4 μg / kg, country: Japan, *pearled
incidence: 1/13*, conc.: 6 μg / kg, country: Japan, *pressed
incidence: 1/3*, conc.: 6 μg / kg, country: Japan, *polished
incidence: 29/31, ∅ conc.: 24 μg / kg, country: Korea
incidence: 21/28*, conc. range: 3-1581 μg / kg, ∅ conc.: 110 μg / kg, country: Korea, *unpolished
incidence: 6/27*, conc. range: 40-1081, ∅ conc.: 579 μg / kg, country: Korea, *naked
incidence: 4/10*, conc. range: 183-1416 μg / kg, ∅ conc.: 552 μg / kg, country: Korea, *husked
incidence: 4/4, ∅ conc.: 18 μg / kg, country: Nepal
incidence: 15/85, conc. range: ≤ 170 μg / kg, country: New Zealand
incidence: 3/584, conc. range: 200-1200 μg / kg, ∅ conc.: 700 μg / kg, country: Poland
incidence: 8/8, conc. range: 3-33 μg / kg, ∅ conc.: 10 μg / kg, country: Scotland
incidence: 2/4, conc. range: 17-22 μg / kg, ∅ conc.: 19 μg / kg, country: Taiwan
incidence: 6/6, conc. range: 4-9 μg / kg, ∅ conc.: 7 μg / kg, country: The Netherlands
incidence: 3/3*, conc. range: 16-29 μg / kg, ∅ conc.: 22 μg / kg, country: The Netherlands, *pearled
incidence: 20/137*, conc. range: 100-200 μg / kg (12 samples), > 200 μg / kg (8 sa), country: Uruguay, *and malt
incidence: 3/3, ∅ conc.: 43 μg / kg, country: Yemen
→ cereals

Barley flour may contain the following → mycotoxins:
→ deoxynivalenol
incidence: 1/1*, conc.: 32 μg / kg, country: Germany, *whole meal
incidence: 3/6, conc. range: 8-39 μg / kg, country: Japan
→ nivalenol
incidence: 6/6, conc. range: 13-41 μg / kg, country: Japan
→ zearalenone
incidence: 6/6, conc. range: 1-4 μg / kg, country: Japan
→ flour

Barley grits may contain the following → mycotoxins:
→ deoxynivalenol
incidence: 1/1, conc.: 36 μg / kg, country: Germany
→ maize grits, → rye grits, → wheat grits

Barley malt may contain the following → mycotoxins:
aflatoxin (no specification) (→ aflatoxins)
incidence: 9/42, conc. range: 1-5 μg / kg (7 samples), 5- ≤ 14 μg / kg (2 sa), country: Germany
→ deoxynivalenol
incidence: 1/8, conc.: 70 μg / kg, country: Canada
incidence: 4/5, ∅ conc.: 40 μg / kg, country: Canada
incidence: 4/4, conc. range: 22-5840 μg / kg, ∅ conc.: 1595 μg / kg, country: Korea
incidence: 13/42, conc. range: 10-20 μg / kg (5 samples), 20-100 μg / kg (8 sa), country: UK
→ nivalenol
incidence: 4/4, conc. range: 122-436 μg / kg, ∅ conc.: 243 μg / kg, country: Korea
→ ochratoxin A

incidence: 3/50, conc. range: 9-189 μg / kg, country: Denmark
→ zearalenone
incidence: 4/4, conc. range: 2-36 μg / kg, Ø conc.: 19 μg / kg, country: Korea
→ beer, → malt

Bay leaf may contain the following → mycotoxins:
→ aflatoxin B_1
incidence: 1/1, conc.: 5.1 μg / kg, country: The Netherlands
→ spices

Bean hull poisoning Sporadically dried bean hulls (→ beans) and plants used as feed caused a high incidence of poisoning of horses in Japan (Hokkaido). The affected animals showed circular movement, motor irritation, cronic musclespasm and → tachycardia. → Jaundice, → hemorrhage of nerve cells and → renal tubular epithelium also occurred.
One of the isolated → Fusarium strains (*F. sporotrichioides* M-1-1) produced → T-2 toxin, → neosolaniol and related → trichothecences. Because the purified T-2 toxin did not cause such nervous symptoms, it was concluded that some other toxin(s) may be involved.

Bean jam may contain the following → mycotoxins:
→ aflatoxin B_1
incidence: 1/24*, conc.: 0.8 μg / kg, country: Japan, *red
incidence: 5/41*, conc. range: 0.1-0.7 μg / kg, country: Japan, *white

Beans (no specification)
Although we do not have much information about mycotoxin contamination of beans in comparison to cereal → grains several → mycotoxins such as → aflatoxins, → deoxynivalenol, → diacetoxyscirpenol, → fumonisin B_1, → ochratoxin A, → penicillic acid, → T2-toxin and → zearalenone have been detected.
Cooking of beans (*Phaseolus vulgaris* L.) naturally contaminated with OTA did not result in a total destruction of this mycotoxin. About 16-60% of the original toxin was detected after processing. Losses amounted to between 80 and 95% of the original OTA levels if soaking was included. Extending the autoclaving period did not significantly contribute to a reduction of the OTA level.
In the case of Faba beans (*Vicia faba* L.), cooking under pressure (115 °C, 2 h) caused a 20% reduction in OTA concentration. Losses in the range of 20-76% occurred in artificially contaminated beans after cooking for 20 min at 121 °C in an autoclave.
Beans may contain the following → mycotoxins:
→ aflatoxin B_1
incidence: 1/10*, conc.: 39 μg / kg, country: Brazil, *Carioquinha, dried
incidence: 1/3*, conc.: 52 μg / kg, country: Brazil, *Mulatinho, dried
incidence: 1/6*, conc.: 1.7 μg / kg, country: Egypt, **Vicia faba* L.
incidence: 4/381*, conc. range: 1.4-254 μg / kg, country: Japan, *for bean jam
incidence: 5/99*, conc. range: 1.5-12.0 μg / kg, Ø conc.: 4.5 μg / kg, country: Japan, *butter
incidence: 2/2, conc. range: 6.3-26.9 μg / kg, Ø conc.: 16.6 μg / kg, country: Japan, *red
incidence: 1/231*, conc.: 1.4 μg / kg, country: Japan, *small red
incidence: 3/37*, conc. range: 1.3-11 μg / kg, Ø conc.: 4.56 μg / kg, country: Japan, *Saltani-Saltapaya
incidence: 10*/322, Ø conc.: 213 μg / kg, country: Thailand, *total: Ø conc.: 1620 μg / kg AFB_1, AFB_2, AFG_1, AFG_2
incidence: 7*/140**, Ø conc.: 16 μg / kg, country: Thailand, *total Ø conc.: 112 μg / kg AFB_1, AFB_2, AFG_1, AFG_2, **mung

→ aflatoxin B_2
incidence: 1/6, conc.: 1.5 μg / kg, country: Egypt, *Vicia faba L.
incidence: 4/381*, conc. range: 1.2-8.5 μg / kg, country: Japan, *for bean jam
incidence: 5/99, conc. range: 0.5-2.2 μg / kg, ∅ conc.: 1.24 μg / kg, country: Japan, *butter
incidence: 2/2, conc. range: 3.5-6.9 μg / kg, ∅ conc.: 5.2 μg / kg, country: Japan, *red
incidence: 1/231, conc.: 0.4 μg / kg, country: Japan, *small red
incidence: 3/37*, conc. range: 0.4-3 μg / kg, ∅ conc.: 1.33 μg / kg, country: Japan, *Saltani-Saltapaya
→ aflatoxin G_1
incidence: 1/10*, conc.: 21 μg / kg, country: Brazil, *Carioquinha, dried
incidence: 1/3*, conc.: 31 μg / kg, country: Brazil, *Mulatinho, dried
→ aflatoxin G_2
incidence: 1/10*, conc.: 4 μg / kg, country: Brazil, *Carioquinha, dried
incidence: 1/3*, conc.: 8 μg / kg, country: Brazil, *Mulatinho, dried
aflatoxin (no specification)
incidence: 18/20*, conc. range: $\leq$ 222 μg / kg, ∅ conc.: 63 μg / kg, country: Philippines, *brown kidney
incidence: 4/7*, conc. range: $\leq$ 118 μg / kg, ∅ conc.: 58 μg / kg, country: Philippines, *lima (*Phaseolus lunatus* L.)
incidence: 29/33*, conc. range: $\leq$ 46 μg / kg, ∅ conc.: 13 μg / kg, country: Philippines, *mung (*Phaseolus aureus* Roxb.)
aflatoxins (no specification)
incidence: 9/68, conc. range: nc, country: Hong Kong
incidence: 1/12*, conc. range: nc, country: Hong Kong, *mung
incidence: 11/610, conc. range: 2-36 μg / kg, country: Japan
incidence: 2/29, conc. range: > 30 - $\leq$ 86 μg / kg, country: Philippines
incidence: 46*/64, conc. range: 1-100 μg / kg (30 samples), 100-1000 μg / kg (11 sa), > 1000 μg / kg (5 sa), country: Uganda
*15 samples contained AFB_1 (∅ conc.: 500 μg / kg), 42 AFB_2, 11 AFG_1, 1 AFG_2
→ deoxynivalenol
incidence: 2*/3, conc. range: 3100-6500 μg / kg, ∅ conc.: 4800 μg / kg, country: Taiwan, *grey and/ or pink discoloration, navy bean (*Phaseolus vulgaris* L.)
→ diacetoxyscirpenol
incidence: 2*/3, conc. range: 3300-9200 μg / kg, ∅ conc.: 6250 μg / kg, country: Taiwan, *grey and/ or pink discoloration, navy bean (*Phaseolus vulgaris* L.)
→ fumonisin B_1
incidence: 2*/3, conc. range: 500-1100 μg / kg, ∅ conc.: 800 μg / kg, country: Taiwan, *grey and/ or pink discoloration, navy bean (*Phaseolus vulgaris* L.)
→ ochratoxin A
incidence: 1/10*, conc.: 94 μg / kg, country: Brazil, *Carioquinha, dried
incidence: 1/3*, conc.: 160 μg / kg, country: Brazil, *Rosinha, dried
incidence: 4/24*, conc. range: 25-27 μg / kg, country: Bulgaria, *area with endemic nephropathy
incidence: 2/28*, conc. range: 25-50 μg / kg, country: Bulgaria, *area with endemic nephropathy
incidence: 75/157*, conc. range: 0.05-260 μg / kg, country: Bulgaria, *area with endemic nephropathy
incidence: 31/113, conc. range: 0.2-285 μg / kg, country: Bulgaria
incidence: 1/84*, conc.: 20 μg / kg, country: Canada, *beans & peas
incidence: 3/4*, conc. range: 40-2000 μg / kg, ∅ conc.: 766 μg / kg, country: Canada, **Phaseolus vulgaris* L., dried
incidence: 1/3*, conc.: 7 μg / kg, country: Egypt, *horse bean
incidence: 2/8, conc. range: 25-50 μg / kg, ∅ conc.: 37.5 μg / kg, country: USA, *red
incidence: 6/71, conc. range: 10-442 μg / kg, country: Sweden
incidence: 9/127*, conc. range: 10-442 μg / kg, country: Sweden, *brown kidney
incidence: 2/8*, conc. range: 35 μg / kg, country: USA, *black turtle soup

incidence: 6/8*, conc. range: 20-100 µg/kg, ∅ conc.: 40 µg/kg, country: USA, *great northern
incidence: 6/8*, conc. range: 25-100 µg/kg, ∅ conc.: 50 µg/kg, country: USA, *navy (*Phaseolus vulgaris* L.)
incidence: 3/8*, conc. range: 50- >1000 µg/kg, country: USA, *pinto
incidence: 2/8*, conc. range: 25 µg/kg, ∅ conc. 25 µg/kg, country: USA, *pinto
→ penicillic acid
incidence: 3/8 conc. range: 300-500 µg/kg, ∅ conc.: 550 µg/kg, country: USA, *red
incidence: 5/20, conc. range: 11-179 µg/kg, ∅ conc.: 82 µg/kg, country: USA
→ T2-toxin
incidence: 2*/3, conc. range: 5500-13,500 µg/kg , ∅ conc.: 9500 µg/kg, country: Taiwan, *grey and/ or pink discoloration, navy bean (*Phaseolus vulgaris* L.)
→ zearalenone
incidence: 1/150, conc.: 160 µg/kg, country: Yugoslavia
→ cabbage, → cowpeas, → lentils, → peas, → pigeon peas, → soybeans, → vegetables

Beauvericin (Abbr.: BEA) This cyclic lactone trimer (→ mycotoxins) with an alternating sequence at three *N*-methyl L-phenylalanyl and three D-α-hydroxyisoaleryl residues is synthesized by several → Fusarium species (see Figure Beauvericin).

CHEMICAL DATA
Empirical formula: $C_{45}H_{47}N_3O_9$, molecular weight: 725

FUNGAL SOURCES
Fusarium semitectum, F. subglutinans

NATURAL OCCURRENCE
It was first detected in Polish → maize but natural contamination of Italian and US maize has also been reported.

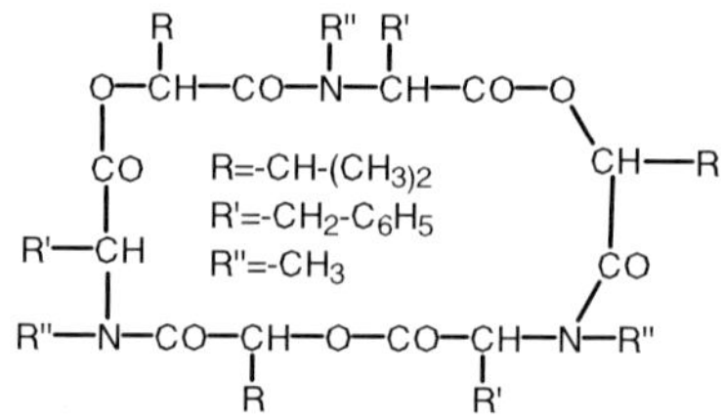

Beauvericin

TOXICITY
BEA is highly toxic for insects, as well as for murine and human cells, in which it induces apoptosis.

DETECTION
HPLC

Beefburger Detection of → aflatoxins in beefburgers results from the use of mycotoxin-contaminated → spices and/or the incorporation of aflatoxin producers.
Beefburgers may contain the following → mycotoxins:
→ aflatoxin B_1
incidence: 5/25, ∅ conc.: 8 µg/kg, country: Egypt
→ meat

Beer Beer may be contaminated with different → mycotoxins.
Some special beer types, e.g. strong beer, might be important contributors to the daily → ochratoxin A intake. The degree of contamination depends upon the quality of the initial → barley, storage conditions (< 16% mc barley malt) and the fate of OTA during malting (→ malt), brewing, and fermentation.
Although OTA does not survive malting, even if very highly contaminated barley malt is used, the addition of this mycotoxin to the mash or before fermentation (simulating use of adjuncts which are usually added at the beginning of the brewing process) revealed a possible transmission into the beer. In consequence, adjuncts such as → maize products (e.g. maize syrup and grits),

→ rice, barley and → wheat would be expected to be the source of any OTA contamination in commercial beer.
The contamination of commercial beers with → fumonisins may be attributed to the use of contaminated barley but more likely, corn-based brewing adjuncts. Often a portion of barley is replaced by adjuncts, e.g. → maize grits, which are most frequently used by the brewing factories. It could be shown that fumonisins (FB_1, FB_2) are appreciably stable towards the yeast fermentation of maize and → wort. Calculated from Canadian and imported beers the daily intake estimates for adult beer drinkers were in the range of 0.010-0.049 µg fumonisin B_1 / kg bw, assuming an average intake of 643 ml beer / day. Consumption of 950 ml beer / day (heavy beer drinkers) containing the maximum fumonisin level found will result in a daily FB_1 intake in the range of 0.24-0.60 µg / kg bw.
→ Deoxynivalenol and → nivalenol may occur in beer since the process for cleaning → grains (e.g. barley) destined for brewing is ineffecient. Contaminated kernels with near-normal size and weight cannot be selectively removed. In addition, if moldy grains such as maize, especially in developing countries, are used for local beer production, humans may be exposed to elevated levels of a number of → Fusarium mycotoxins via consumption.
Beer may contain the following mycotoxins:
→ aflatoxin B_1
incidence: 3/3, conc. range: 0.006-0.059 µg / kg, country: Mexico
→ aflatoxin B_1 & → aflatoxin B_2
incidence: 16/304*, conc. range: 1.0-2.5 µg / l, country: Kenya, *local, home brewed
→ deoxynivalenol
incidence: 28/50*, conc. range: 0.3-50.3 µg / l, ∅ conc.: 5.7 µg / kg, country: Canada, *28 Canadian and 22 imported beers
incidence: 1/49, conc.: 20 µg / l, country: France
incidence: 18/18, conc. range: $\leq$ 9 µg / l, ∅ conc.: $\approx$ 5 µg / l, country: Germany
incidence: 8/38, conc. range: 1-5.3 µg / l, ∅ conc.: 3.1 µg / l, country: Korea
incidence: 2/5*, conc. range: 3.8-10 µg / l, ∅ conc.: 6.9 µg / l, country: Korea, *imported beers
→ diacetoxyscirpenol
incidence: 5/49, conc.: ca. 10-35 µg / l, country: France
→ fumonisin B_1
incidence: 11/41, conc. range: 0.42-59 µg / l, country: Canada
incidence: 20/46, conc. range: 0.2-58.2 µg / l, country: Canada
→ fumonisin B_2
incidence: nc, conc. range: 0.45-9.2 µg / l, country: Canada
incidence: 7/46, conc. range: 0.4-11.5 µg / l, country: Canada
→ fumonisins
incidence: 14/32, conc. range: 4.8-85.5 µg / kg, ∅ conc.: 25.8 µg / l, country: Spain
→ nivalenol
incidence: 3/50*, conc. range: 0.1-0.84 µg / kg, ∅ conc.: 0.4 µg / l, country: Canada, *28 Canadian and 22 imported beers
incidence: 32/38, conc. range: < 1-20 µg / l, country: Korea
incidence. 3/5*, conc. range: 1.3-2.5 µg / l, ∅ conc.: 1.93 µg / l, country: Korea, *imported beers
→ ochratoxin A
incidence: 26/41*, conc. range: tr-0.2 µg / l, ∅ conc.: 0.061 µg / l, country: Canada, *Canadian and imported beers (11)
incidence: 21/21, conc. range: $\leq$ 0.16 µg / kg, ∅ conc.: 0.049 µg / l, country: Denmark
incidence: 5/66, conc. range: $\leq$ 0.1 µg / l, country: Germany
incidence: 80/160, conc. range: $\leq$ 0.49 µg / l, country: Germany

incidence: 6/11, conc. range: 0.03-0.08 µg / kg, country: Germany
incidence: 4/37, conc. range: 5-110 µg / l, ∅ conc.: 56.3 µg / l, country: France
incidence: 14/16, conc. range: 0.002-0.052 µg / l, ∅ conc.: 0.014 µg / l, country: UK
→ T-2 toxin
incidence: 3/49, conc. range: ca. 10-42 µg / l, country: France
→ zearalenone
incidence: 1/49, conc.: 100 µg / l, country: France
incidence: 17/140, conc. range: 300-2000 µg / l, country: Lesotho
incidence: 2/23*, conc. range: 8000-53,000 µg / l, country: Swaziland, *and other fermented products
incidence: 14-15/23, conc. range: < 90-4600 µg / l, ∅ conc.: 1410-1500 µg / l, country: Zambia
barley, → barley malt, cereals, maize, → sorghum, wheat

Beer (draft) may contain the following → mycotoxins:
→ deoxynivalenol
incidence: 2/2*, conc. range: 6.3-8.8 µg / l, ∅ conc.: 7.55 µg / l, country: Korea, *imported beers
→ nivalenol
incidence: 1/2*, conc.: 8.8 µg / l, country: Korea, *imported beers

Beer (light) may contain the following → mycotoxins:
→ nivalenol
incidence: 3/3, conc. range: 24-38 µg / l, ∅ conc.: 31.3 µg / l, country: Korea
incidence: 3/3*, conc. range: 3.3-7 µg / l, ∅ conc.: 4.6 µg / l, country: Korea, *imported beers

Beer (non-alcoholic) may contain the following → mycotoxins:
→ deoxynivalenol
incidence: 2/2*, conc. range: 18-23 µg / l, ∅ conc.: 20.5 µg / l, country: Korea, *imported beers
→ nivalenol
incidence: 1/1, conc.: 9 µg / l, country: Korea

Beer (pale) may contain the following → mycotoxins:
→ ochratoxin A
incidence: 1/28, conc. range: 0.3 µg / l, country: Germany
incidence: 7/7, conc. range: 0.01-0.033 µg / l, country: Switzerland

Beer (strong) may contain the following → mycotoxins:
→ ochratoxin A
incidence: 14/40, conc. ≤ 1.5 µg /l, ∅ conc. 0.28 µg / l, country: Germany
incidence: 9/26, conc. range: 0.35-1.53 µg / l, ∅ conc.: 1 µg / l, country: Germany
incidence: 13/32, conc. range: 0.05-0.49 µg / l, country: Germany

Beer, barley may contain the following → mycotoxins:
→ deoxynivalenol
incidence: 35/123, conc. range: ≤ 478 µg / l, ∅ conc.: 148 µg / l, country: Germany

Beer, burukutu is a Nigerian type of beverage made from guinea corn (*Sorghum* sp.) and → millet (*Penissetum* sp.) while the malt is retained. In experimental studies it could be shown that there was a → carry over of → zearalenone into the finished product from 43-62%.
Burukutu beer may contain the following → mycotoxins:
→ aflatoxin B_1 & → aflatoxin G_1
incidence: 2/2, conc. range: 253-262 µg / l, ∅ conc.: 257.5 µg / l, country: Nigeria

Beer, joala The composition of joala beer varies with the proportions of the

ingredients, which are malted → maize and / or → sorghum, → flour and hops and occasionally various → fruits such as grapes and pineapples.
Joala may contain the following → mycotoxins:
→ zearalenone
incidence: 17/40, conc. range: 300-2000 µg / l, country: Lesotho

Beer, millet may contain the following → mycotoxins:
→ aflatoxin B_1
incidence: 32/40*, conc. range: 1.7-138 µg / kg, ∅ conc.: 64 µg / kg, country: Nigeria, *native
→ aflatoxin B_1 and other → aflatoxins
incidence: 10/10, conc. range: 4- > 50 µg / l, ∅ conc.: 25 µg / kg, country: Nigeria

Beer, opaque maize is a Zambian type of beverage brewed from → maize, → millet or red → sorghum. A → carry over of → zearalenone in the range of 51.4% from starting zearalenone concentration in the finished product has been recorded in maize beer.
Maize beer may contain the following → mycotoxins:
zearalenone
incidence: nc/23, ∅ conc.: 920 µg / l, country: Zambia

Beer, pito is a Nigerian type of beverage brewed from red guinea corn (*Sorghum* sp.) and → millet (*Penissetum* sp.) or a mixture of both while the → malt is filtered off.
Pito may contain the following → mycotoxins:
→ aflatoxin B_1 & → aflatoxin G_1
incidence: 2/2, conc. range: 92-142 µg / l, ∅ conc.: 117 µg / l, country: Nigeria
→ zearalenone
incidence: 28/46, conc. range: 12.5-200 µg / l, ∅ conc.: 81.8 µg / l, country: Nigeria

Beer, sorghum may contain the following → mycotoxins:
→ aflatoxin B_1
incidence: 8/150, conc. range: 0.05-0.13 µg / l, ∅ conc.: 0.1 µg / l, country: South Africa

Beer, wheat may contain the following → mycotoxins:
→ deoxynivalenol
incidence: 50/67, conc. range: $\leq$ 569 µg / l, ∅ conc.: 245 µg / l, country: Germany
→ ochratoxin A
incidence: 1/3, conc.: 0.3 µg / l, country: Germany

Bentonite A clay (montmorillonit) with adsorptive properties used for the removal of → mycotoxins (e.g. → aflatoxins, → patulin) from → milk, → apple juice and other fluid products.
→ decontamination

Bile duct Passages for conveyance of bile in and from the liver.

Bioassays This preliminary screening system enables toxicity test of extracts made from commodities which might show a mycotoxin contamination. Bacteria, yeasts, *Tetrahymena pyriformis*, *Artemia salina* as well as larvae of trout and other fish can be used for assaying → mycotoxins while toxicity is expressed as a percentage of death in a given time. In addition, chick embryos and ducklings, as well as cell cultures (e.g. rat liver, baby hamster kidney, human epithelial), have also been used. However, lack of specificity due to other (toxic) substances coextracted with mycotoxins limits the application of bioassays.

Biscuits → Ochratoxin A is partially (ca. 60%) destroyed or immobilized during biscuit making. This reduction in OTA concentration may be explained by the high dough temperature during → baking, the low water content of these → cereal products, and / or the presence of bicarbonate in the dough.
Biscuits may contain the following → mycotoxins:
ochratoxin A
incidence: 1/9, conc. range: 0.2-0.49 µg / kg, country: Germany
incidence: 3/11*, conc. range: 0.1-1.49 µg / kg, country: Germany, *salted
→ cereals, → cookies

Black molds Molds like → Alternaria spp., → Cladosporium spp., *Epicoccum* spp. and *Helminthosporium* spp. grow on prematurely dead ears as a superficial dark mycelium and spore masses (see Figure Black molds). Spikelets with excreted honeydew are the preferred substrates. Among the black molds, especially species of the genus *Alternaria* belong to the important mycotoxin producers.

Blepharis edulis (medicinal seeds) may contain the following → mycotoxins:
aflatoxin B_1
incidence: nc/nc, conc. range: 10-1040 µg / kg, country: India

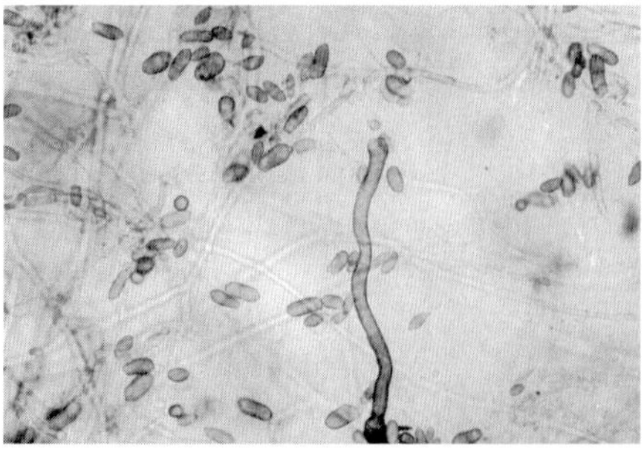

Black molds. *Cladosporium herbarum*

"Blind staggers" syndrome → Leucoencephalomalacia

Blue Castello cheese → cheese, Blue Castello

Blue cheese → cheese, Blue

Blue cheese dressing → cheese dressing, blue

Blue Haverti cheese → cheese, Blue Haverti;

Blueberries may contain the following → mycotoxins:
→ patulin
incidence: 3/16, conc. range: 75-190 µg / kg, country: Sweden
incidence: 1/12, conc.: 21 µg / kg, country: Sweden
→ fruits

Bondakaledkai is an Indian peanut (→ peanuts) based spiced snack which consists of whole seeds with an intact seed coat. Before deep-fat-frying the → nuts are covered with salt, → rice flour and a paste of red → chilli powder. Aflatoxin contamination may be due to the use of uncleaned and unpicked whole seeds along with the seed coat.
Bondakaledkai may contain the following → mycotoxins:
→ aflatoxin B_1
incidence: 22/54, conc. range: 3-1500 µg / kg, country: India
→ aflatoxin B_2
incidence: 14/54, conc. range: 3-370 µg / kg, country: India
→ congressbele, → groundnut toffee

bovinocidin → β-nitropropionic acid

Bran (no specifiaction) may contain the following → mycotoxins:
→ deoxynivalenol

incidence: 14 products analysed, Ø conc.: 170 µg / kg, country: Canada
→ ochratoxin A
incidence: 1/41, conc.: 0.1 µg / kg, country: Germany
incidence: 19/30, conc. range: 0.1-0.49 µg / kg (12 samples), 0.5-1.49 µg / kg (4 sa), 1.5-9.99 µg / kg (3 sa), country: Germany
incidence: 9/84, Ø conc.: 6.8 µg / kg, country: Germany
incidence: 5/35, conc. range: ≤ 11 µg / kg, Ø conc.: 4.5 µg / kg, country: Italy
incidence: 12/43, conc. range: ≤ 4.9 µg / kg, Ø conc.: 1.03 µg / kg, country: UK
→ cereals, → maize bran, → milling, → oat bran, → rice bran, → rye bran, → wheat bran

Brazil nuts Brown and fluorescent kernels generally contain the main part of the → aflatoxins. Since contaminated → nuts are so obviously damaged, human consumption seems unlikely. Separation of aflatoxin positive nuts is based on the assumption that the moldy nuts are lighter than the good ones. The former are removed by an air blower and by means of gravity separation.
Brazil nuts may contain the following → mycotoxins:
→ aflatoxin B_1
incidence: 33/302*, Ø conc.: 286 µg / kg, country: Germany, *in-shell
incidence: 57/135, conc. range: < 5 µg / kg (29 samples), 40-8000 µg / kg (28 sa) country: Germany
incidence: 1/2, conc.: 3200 µg / kg, country: Germany
incidence: 13/17, conc. range: < 5 µg / kg (12 samples), 5 µg / kg (1 sa), country: Germany
incidence: 10/10*, conc. range: 8-47,000 µg / kg, Ø conc.: 12,522 µg / kg, country: Germany
incidence: 16/27*, conc. range: 3-4200 µg / kg, Ø conc.: 500.2 µg / kg, country: Norway *imported
incidence: 5/23, conc. range: 0.5-5 µg / kg (4 samples), 33 µg / kg (1 sa), country: UK
→ aflatoxin B_2
incidence: 4/10*, conc. range: 0.6-883 µg / kg, Ø conc.: 517.7 µg / kg, country: Germany
incidence: 16/27*, conc. range: tr-1600 µg / kg, country: Norway, *imported
→ aflatoxin G_1
incidence: 9/10*, conc. range: 7-56,000 µg / kg, Ø conc.: 18,457 µg / kg, country: Germany
incidence: 16/27*, conc. range: 2-3250 µg / kg, Ø conc.: 478.2 µg / kg, country: Norway *imported
→ aflatoxin G_2
incidence: 3/10*, conc. range: 1.2-1000 µg / kg, Ø conc.: 533.7 µg / kg, country: Germany, * kernels visibly discolored
incidence: 16/27*, conc. range: tr-600 µg / kg, country: Norway, *imported
→ aflatoxins (AFB_1, AFB_2, AFG_1, AFG_2)
incidence: 31/69*, conc. range: 6-100 µg / kg (8 samples), 101-1000 µg / kg (7 sa), 1001-10,000 µg / kg (11 sa), > 10,000 µg / kg (5 sa), country: Sweden, *imported; edible, possibly edible and inedible nuts
aflatoxins (no specification)
incidence: 62/234, conc. range: nc, country: Germany
incidence: 33/302*, Ø conc.: 305 µg / kg, country: Germany
incidence: 4/14*, conc. range: 2-129 µg / kg, country: UK
incidence: 6/38*, conc. range: nc, country: UK
incidence: 5/23*, conc. range: 0.5-5 µg / kg (3 samples), 6-10 µg / kg (1 sa), 60 µg / kg, country: UK
*in-shell
incidence: 4/18*, conc. range: 2-129 µg / kg, country: UK

incidence: 6/12*, conc. range: ≤ 42 µg / kg, ∅ conc.: 20 µg / kg, country: USA *shelled
→ nuts

Bread (no specification)
may be contaminated by different → mycotoxins which are more or less stable during processing.
If white → flour is spiked with → ochratoxin A no decomposition of the mycotoxin occurs after baking (220 °C, 25 min).
However, levels of → aflatoxins in flour were significantly reduced during fermentation and → baking as compared to that in the finished bread. Besides oxidation during kneading, especially fermentation and hydrothermal processes during cooking caused degradation of most of the → aflatoxin B_1 (≈ 40-80%).
Infection of bread with toxigenic isolates of *Aspergillus flavus* Link resulted in aflatoxin contamination several days later, although the wrapped bread restricted fungal growth due to lack of oxygen (see Figure Bread). The pH of the bread is a decisive factor while increased concentrations of vitamin B_1, protein and salt favor aflatoxin formation.
Although → patulin contamination has been reported in spontaneously molded bread this mycotoxin reacts with sulfhydryl-containing amino acids or proteins and is therefore not stable in this substrate.
Japanese studies proved a 50% reduction in → deoxynivalenol levels compared to the original concentration by baking. However, in Canadian experiments only a 20% reduction of this mycotoxin was observed and almost none during Egyptian bread making. In Japanese bread making losses of various trichothecenes (deoxynivalenol, → nivalenol, → diacetoxyscirpenol, → neosolaniol, → T-2 toxin and → fusarenon X) amounted to ≈ 50% (artifical contamination).
Bread may contain the following mycotoxins:
→ acetyldeoxynivalenol
incidence: 4/24*, conc. range: 600-2400 µg / kg, country: India, *wheat
→ aflatoxin B_1
incidence: 4*/18**, conc. range: 5-60 µg / kg, country: Germany, *moldy, **whole meal wheat
incidence: 1*/14**, conc.: 10 µg / kg, country: Germany, *moldy, **German “Landbrot” (80% wheat and 20% rye flour)
incidence: 2*/18**, conc. range: 20-25 µg / kg, country: Germany, *moldy, **white
→ aflatoxins (AFB_1, AFB_2, AFG_1, AFG_2)
incidence: 1/4*, conc.: 3.3 µg / kg, country: UK
→ citrinin
incidence: 11*/110, conc. range: ≤ 5 µg / kg, country: Germany, *sliced packed bread, visible moldy
deoxynivalenol
incidence: 1/1*, conc.: 378 µg /kg, country: Argentina, *bran
incidence: 4/4*, conc. range: 269-384 µg / kg, ∅ conc. 327 µg / kg, country: Argentina, *homemade
incidence: 10/12*, conc. range: 198-436 µg / kg, ∅ conc.: 263 µg / kg, country: Argentina, *french
incidence: 4 products analysed*, ∅ conc.: 58 µg / kg, country: Canada, *rye bread
incidence: 11/24*, conc. range: 340-8400 µg / kg, country: India, *wheat
incidence: nc/4*, conc. range: 8-28 µg / kg, country: UK, *pitta
→ fumonisin B_1
incidence: 1/2*, conc.: 80 µg / kg, country: The Netherlands
→ fumonisins
incidence: 4/4*, conc. range: 400-3450 µg / kg, ∅ conc.: 1285 µg / kg, country: USA

incidence: 1/1*, conc.: 600 µg / kg, country: USA, *maize
nivalenol
incidence: 2/24*, conc. range: 30-100 µg / kg, ∅ conc.: 65 µg / kg, country: India, *wheat
incidence: 1/4*, conc.: 21 µg / kg, country: UK, *pitta
incidence: 20/25, conc. range: ND-240 µg / kg, country: USA
ochratoxin A
incidence: 4*/110, conc. range: $\leq$ 80 µg / kg, country: Germany, *sliced packed bread, visibly moldy
incidence: 4/57*, conc. range: 0.1-1 µg / kg, ∅ conc.: 0.07 µg / kg, country: Germany, *wheat and rye bread (German "Mischbrot")
incidence: 26/51, conc. range: $\leq$ 1.49 µg / kg, ∅ conc.: 0.17 µg / kg, country: Germany
incidence: 4/36*, conc. range: 0.2-0.3 µg / kg, country: Germany, *whole meal bread
incidence: 6/46*, conc. range: 0.2-0.9 µg / kg, country: Germany, *crisp
incidence: 33/47*, conc range: 0.1-0.49 µg / kg (16 samples), 0.5-9.99 µg / kg (17 sa), country: Germany, *crisp
incidence: 8/9*, conc. range: 0.05-0.49 µg / kg (7 samples), 0.5-1.49 µg / kg (1 sa), country: Germany, *Pumpernickel
incidence: 6/8*, conc. range: 0.1-0.49 µg / kg (3 samples), 0.5-1.49 µg / kg (3 sa), country: Germany, *toast
incidence: 1/2*, conc.: 80,000 µg / kg, country: Italy, *moldy, intended for animal feed
incidence: 63/386, ∅ conc.: 1360 µg / kg, country: Poland
incidence: 11/26*, conc.: $\leq$ 0.6 µg / kg, ∅ conc.: 0.2 µg / kg, country: Sweden, *crisp
incidence: 1/2, conc.: 0.2 µg / kg, country: Switzerland
incidence: 1/50*, conc.: 210 µg / kg, country: UK, *moldy
incidence: 3/4*, conc.: 0.2-0.8 µg / kg, country: UK, *pitta
incidence: 6/32*, conc. range: nc, country: Yugoslavia, *wheat
incidence: 1*/50**, conc.: 210 µg / kg, country: Yugoslavia, *moldy, **wheat
→ ochratoxin B
incidence: 1/2*, conc.: 9600 µg / kg, country: Italy, *moldy, intended for animal feed
incidence: 6/32*, conc. range: nc, country: Yugoslavia, *wheat
T-2 toxin
incidence: 5/24*, conc. range: 550-4000 µg / kg, country: India, *wheat
→ zearalenone
incidence: 6*/110, conc. range: $\leq$ 5 µg / kg, country: Germany, *sliced packed bread, visibly moldy
incidence: 2/2*, conc. range: 250-750 µg / kg, ∅ conc.: 500 µg / kg, country: Papua, New Guinea, *imported, wheat bread crumbs
→ cereals

Breakfast cereals may be contaminated by various → mycotoxins. This results from the fact that this kind of foodstuff is made from different kinds of → cereals and → cereal products which are often contaminated by → Fusarium, → Aspergillus and → Penicillium mycotoxins.
The detection of → deoxynivalenol in breakfast cereals proves DON contamination of the grains and its survival through processing → bread.

Bread. *Aspergillus flavus* Link on Pumpernickel

Breakfast cereals may contain the following mycotoxins:
→ aflatoxins (no specification)
incidence: 2/6*, conc. range: 1-5 μg / kg, country: UK, *→ bran-based
incidence: 1/6*, conc. range: 1-5 μg / kg, country: UK, *→ maize-based
incidence: 1/6*, conc. range: 1-5 μg / kg, country: UK, *→ oat-based
incidence: 3/5*, conc. range: 1-5 μg / kg, country: UK, *→ rice-based
incidence: 1/14*, conc. range: 1-5 μg / kg, country: UK, *→ wheat-based
deoxynivalenol
incidence: 36 products analysed, ∅ conc.: 86 μg / kg, country: Canada
incidence: 7/7*, conc. range: 30-100 μg / kg, country: UK, *bran-based
incidence: 2/3*, conc. range: traces, country: UK, *maize-based
incidence: 35/60, conc. range: ND-530 μg / kg, country: USA
incidence: 36/60, ∅ conc.: 100 μg / kg, country: USA
→ fumonisin B_1
incidence: 11/52*, conc. range: < 100-320 μg / kg, country: Canada, *maize-based
incidence: 9/17, conc. range: < 10-330 μg / kg, ∅ conc.: 130 μg / kg, country: USA
incidence: nc/3**, conc. range: 1060-3630 μg / kg, country: Zimbabwe
→ fumonisin B_2
incidence: nc/17, conc. range: < 10-70 μg / kg, ∅ conc.: 30 μg / kg, country: USA
incidence: nc/3**, conc. range: 240-910 μg / kg, country: Zimbabwe
→ fumonisin B_3
incidence: nc/3**, conc. range: 130-230 μg / kg, country: Zimbabwe
**health breakfast cereal
→ fumonisins (FB_1, FB_2, FB_3)
incidence: 12/50, conc. range: 11-194 μg / kg, ∅ con.: 29 μg / kg, country: UK
→ ochratoxin A
incidence: 13/54, conc. range: ≤ 4.9-9.8 μg / kg, ∅ conc.: 0.51 μg / kg, country: Germany
incidence: 2/26, conc. range: ≤ 0.5 μg / kg, country: Germany
incidence: 3/6*, conc. range: < 10 μg / kg, country: UK, *bran-based
incidence: 3/6*, conc. range: < 10-20 μg / kg, country: UK, *maize-based
incidence: 2/6*, conc. range: < 10 μg / kg, country: UK, *oat-based
incidence: 1/5*, conc. range: < 10 μg / kg, country: UK, *rice-based
incidence: 7/14*, conc. range: < 10-50 μg / kg, country: UK, *wheat-based
incidence: 12/243, conc. range: 5-108 μg / kg, country: UK
→ sterigmatocystin
incidence: 1/14 (wheat-based), conc.: ≤ 7 μg / kg, country: UK
→ trichothecenes* (no specification)
incidence: 5/6*, conc. range: nc, country: UK, *maize-based
incidence: 4/6*, conc. range: nc, country: UK, *oat-based
incidence: 5/13*, conc. range: nc, country: UK, *wheat-based
* max. level: ≤ 5 μg / kg
→ zearalenone
incidence: 2/6*, conc. range: < 50 μg / kg, country: UK, *maize-based
incidence: 3/14*, conc. range: < 50 μg / kg, country: UK, *wheat-based
incidence: 4/39, conc. range: 2.6-8.6 μg / kg, ∅ conc.: 4.6 μg / kg, country: USA

Breakfast drinks may contain the following → mycotoxins:
→ ochratoxin A
incidence: 2/2, conc. range: 0.1-0.3 μg / kg, ∅ conc.: 0.2 μg / kg, country: Switzerland
→ apple juice, → fruit juice, → grape juice, → soft drinks

Brick cheese → cheese, brick

Brie cheese → cheese, Brie

Buckwheat may contain the following → mycotoxins:
→ aflatoxin B_1
incidence: 23/123, conc. range: 0.1-4.2 µg / kg, country: Japan
→ aflatoxin B_2
incidence: 23/123, conc. range: 0.1-0.9 µg / kg, country: Japan
→ aflatoxin G_1
incidence: 23/123, conc. range: 0.2-0.8 µg / kg, country: Japan
→ aflatoxin G_2
incidence: 23/123, conc. range: tr-0.1 µg / kg, country: Japan
→ ochratoxin A
incidence: 1/34, conc.: 5 µg / kg, country: Germany
→ cereals

Buckwheat flour may contain the following → mycotoxins:
→ aflatoxin B_1 & → aflatoxin B_2
incidence: 1*/37, conc.: ≈ 10 µg / kg, country: Japan, *moldy
→ flour

Bulla A large blister or skin vesicle filled with fluid.

Buns may contain the following → mycotoxins:
→ deoxynivalenol
incidence: 21 products analysed, Ø conc.: 80 µg / kg, country: Canada
→ cereals

Burukutu → beer, burukutu

Butenolide is a 4-acetamido-4-hydroxy-2-butenoicacid χ-lactone (→ mycotoxins) derived from glutamic acid and associated with outbreaks of "fescue foot" of cattle in the US, Australia, and New Zealand (see Figure Butenolide).

Chemical Data
Empirical formula: $C_6H_7NO_3$, molecular weight: 141

Fungal Sources
→ Aspergillus terreus Thom, *Fusarium acuminatum* Ellis & Everh. sensu Gordon?, → Fusarium avenaceum (Fr.) Sacc.?, → Fusarium graminearum Schwabe, → Fusarium poae (Peck) Wollenw.?, *F. semitectum* ?, → Fusarium sporotrichioides Sherb., → Fusarium equiseti (Corda) Sacc. sensu Gordon

Natural Occurrence
→ wheat, → barley (11 of 34 samples were contaminated between 10-430 µg / kg)

Toxicity
LD_{50} : 43.6±1.24 mg / kg bw mice

Further Comments
Butenolide was occasionally detected in association with → neosolaniol, → T-2 toxin and → diacetoxyscirpenol. Besides → zearalenone and 12,13-epoxythrichothecene (→ trichothecenes) derivatives it belongs to the major toxic metabolites (mycotoxins) of → Fusarium spp.

Butter Manufacturing butter from naturally contaminated cream (→ cream) 18-28% of the → aflatoxin M_1 was found in the finished butter. However, the major portion occurred in the buttermilk (→ milk, butter-).

Buttermilk → milk, butter-

Byssochlamic acid belongs to the group of nonadrides characterized by the presence of anhydride groups attached to a nine membered carbocyclic ring (10-ethyl-5,9,10,11-tetrahydro-4-propyl-1H-cyclonona[1,2-c:5,6-c′]difuran-1,3,6,8(4H)-tetrone). Further members of this group are the glaucanic and glauconic acids (→ Penicillium *purpurogenum*)

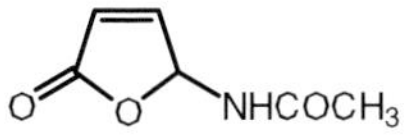

Butenolide

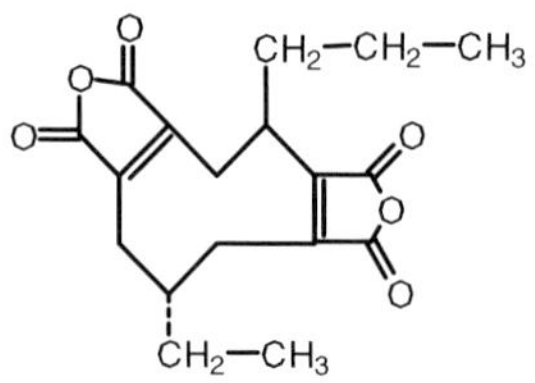

Byssochlamic acid

and the rubratoxins (*P. rubrum*) with a complex formulae (see Figure Byssochlamic acid).

CHEMICAL DATA

Empirical formula: $C_{18}H_{24}O_8$, molecular weight: 368

FUNGAL SOURCES

→ Byssochlamys spp. (*B. fulva*, *B. nivea*), → Paecilomyces variotii Bain

NATURAL OCCURRENCE

Fruit juices may be contaminated.

TOXICITY

cytotoxic, hemorrhagic (→ hemorrhage)
$LD_{50} > 2.5$ g / kg bw but < 4.9 g / kg bw mice.
Byssochlamic acid is not as toxic as → patulin.

DETECTION

TLC

FURTHER COMMENTS

Up to now, no mycotoxicoses due to the consumption of contaminated → foods have been reported. From the chemical structure it was concluded that only foods that contain fatty acids, with free glycerol present, are suitable for the production of byssochlamic acid. Therefore, byssochlamic acid is not a contaminant of margarine, → olive oil or → ham, whereas a metabolite very similar to byssochlamic acid may be formed in → butter. However, there are only few reports concerning the spoilage and contamination of foods with heat-resistant fungi like → Byssochlamys spp., → Paecilomyces variotii Bain, as well as byssochlamic acid.

Byssochlamys → Trichocomaceae, anamorph → Paecilomyces
The only two food-relevant species: *Byssochlamys fulva* Olliver & Smith* and *B. nivea* Westl.** possess very heat-resistant ascospores and may cause degradation (pectinase activity) and spoilage (mycotoxin contamination) of processed (heated) fruit products as well as canned and bottled → fruits. Reduced oxygen tensions (0.27% O_2) present in such commodities are tolerated. Adequate heat treatment for complete destruction of all inherent ascospores could impair the organoleptic quality of the product. Compared to *B. fulva*, strains of *B. nivea* possess a higher potential for → patulin production. Although *B. nivea* forms patulin in → apple juice under low O_2 levels (0.5-2%), the production of significant levels of patulin under commercial conditions is not anticipated. The minimum → a_w for patulin formation was a_w 0.950 at 37 °C after 10 days of incubation.
Byssochlamys spp. may produce → mycotoxins such as → byssochlamic acid* **, byssotoxin A*, malformins**, → patulin* **, and variotin. Up to now, no mycotoxicosis due to the consumption of foods contaminated with byssochlamic acid has been reported.

C

Cabbage (fried with pork and garlic)
may contain the following → mycotoxins:
→ aflatoxin B_1
incidence: 1/1, conc.: 748 µg / kg, country: Thailand, *total: 1299 µg AFB_1, AFB_2, AFG_1, AFG_2 / kg food
→ beans, → cowpeas, → lentils, → peas, → pigeon peas, → soybeans, → vegetables

Caesalpinea digyna (medicinal seeds)
may contain the following → mycotoxins:
→ aflatoxin B_1
incidence: nc/nc, conc. range: 80-1180 µg / kg, country: India

Carbon dioxide → atmosphere

Carcinoma A new growth or malignant tumor enclosing epithelial cells in connective tissue and tending to infiltrate and give rise to metastases.
→ Aflatoxin B_1, → aflatoxin G_1, → aflatoxin M_1, → sterigmatocystin, versicolorin A, → luteoskyrin and → rugulosin are → mycotoxins with a well-known carcinogenic potential. These toxic fungal metabolites are genotoxic and produce positive results in short-term genotoxicity assays (Ames mutagenicity test). Similar genotoxic porperties have been reported for → fusarin C and emodin which are also likely to be carcinogenic. The genotoxicity of the → trichothecenes, → ochratoxin A and → zearalenone is questionable or non-existent, but they definitely promote cancer like the → fumonisins.

Cardamom (*Elettaria cardamomum* Linn.)
may contain the following → mycotoxins:
→ aflatoxin B_1
incidence: 1/6, conc.: 20 µg / kg, country: India
→ aflatoxin B_2
incidence: 1/6, conc.: 15 µg / kg, country: India
→ aflatoxin G_1
incidence: 1/6, conc.: 12 µg / kg, country: India
→ citrinin
incidence: 1/6, conc.: 25 µg / kg, country: India
→ spices

Cardamom, greater (*Amomum subulatum* Roxb.)
may contain the following → mycotoxins:
→ aflatoxin B_1
incidence: 3/6, conc. range: 18-129 µg / kg, country: India
aflatoxin B_2
incidence: 3/6, conc. range: 14-108 µg / kg, country: India
→ aflatoxin G_1
incidence: 3/6, conc. range: 17-78 µg / kg, country: India
→ aflatoxin G_2
incidence: 3/6, conc. range: 5-76 µg / kg, country: India
→ ochratoxin A
incidence: 1/6, conc.: nc, country: India
→ spices

Cardiac beriberi → Acute cardiac beriberi

Carry over Edible tissues, → milk, and eggs of food-producing animals in general are contaminated with only low levels of → mycotoxins. This is due to the fact that only minor amounts of mycotoxins and / or their metabolites are transmitted from the feeds to this kind of foodstuff. For the extent of mycotoxin carry over and contamination the total dose level, not the duration of dose is mainly responsible.
The main important "carry over" mycotoxins are → aflatoxin B_1 / → aflatoxin M_1 in → milk and → milk products and → ochratoxin A in → meat and meat pro-

ducts (e.g. → sausages). To avoid aflatoxin contamination, many countries have a strict regulation for these mycotoxins at the feed level. OTA residues in meat are monitored by regulatory authorities especially in European countries. There are also carry over studies for → trichothecenes, → zearalenone, and → fumonisins. However, it seems that these mycotoxins are only of minor importance concerning a carry over. In addition, analytical detection may be difficult especially because the identity and hazard of the metabolites are unknown.

Cashew nuts may contain the following → mycotoxins:
→ aflatoxin B_1
incidence: 1*/6, conc.: 830 µg / kg, country: Germany, *moldy
→ aflatoxins (AFB_1, AFB_2, AFG_1, AFG_2)
incidence: 23/120, conc. range: 5-24.9 µg / kg (19 samples), > 25 µg / kg (4 sa), country: Canada
incidence: 1/3*, conc.: traces, country: Norway, *imported
→ nuts

Cassava (raw)
may contain the following → mycotoxins:
aflatoxin (no specification)
incidence: 1/1, conc.: 1700 µg / kg, country: Uganda
incidence: 23/23, conc. range: > 20 µg / kg (13 samples), country: Philippines (very high concentrations have been detected but no data were presented)
→ aflatoxins (no specification)
incidence: 4*/34, conc. range: 100-1000 µg / kg (2 samples), > 1000 µg / kg (2 sa), country: Uganda, * 2 samples contained AFB_1, 4 sa AFB_2, 2 sa AFG_1

Cassava flour may contain the following → mycotoxins:
→ ochratoxin A
incidence: 2/2, conc. range: 32-65 µg / kg, ∅ conc.: 48.5 µg / kg, country: Brazil

Cassava starch → Sago

Cassia fistula (medicinal seeds)
may contain the following → mycotoxins:
→ aflatoxin B_1
incidence: nc/nc, conc. range: 80-1110 µg / kg, country: India
→ citrinin
incidence: nc/nc, conc. range: 10-690 µg / kg, country: India

Cattle are relatively resistant against → Fusarium mycotoxins in their diet. However, elevated concentrations caused different symptons like feed refusal, gain losses, impaired → milk production, diarrhea, decreased immune response, and abnormal estrous cycles. Feeds contaminated with 200 µg or 800 µg → deoxynivalenol / kg dry matter lowered milk production with 0.45 kg and 2 kg / day, respectively. Furthermore, the higher concentration caused a delay in breeding (8 days).
→ cattle liver, → meat

Cattle liver may contain the following → mycotoxins:
→ aflatoxin B_1
incidence: 6/19, conc. range: 0.02-0.08 µg / kg (5 samples), 6.6 µg / kg, (1 sa) country: Germany
→ aflatoxin B_2
incidence: 4/19, conc. range: 0.01-0.03 µg / kg, ∅ conc.: 0.02 µg / kg, country: Germany
→ aflatoxins (no specification)
incidence: 1/19, conc.: 10.3 µg / kg, country: Germany
→ meat

Cayenne pepper may contain the following → mycotoxins:
→ aflatoxin B_1

incidence: 16/56, conc. range: < 2.5-8 µg / kg, country: Canada
incidence: 10/33, conc. range: tr-8 µg / kg, country: Canada
incidence: 3/36, conc. range: 5.0-9.8 µg / kg, ∅ conc.: 7.16 µg / kg, country: Germany
incidence: 11/22, conc. range: tr-24 µg / kg, country: Germany
→ aflatoxin B_2
incidence: 5/33, conc. range: traces, country: Canada
→ spices

Celery seeds may contain the following → mycotoxins:
→ aflatoxin G_1
incidence: 1/9, conc.: 3.7 µg / kg, country: unknown

Cereal flakes may contain the following → mycotoxins:
→ ochratoxin A
incidence: 13/51, conc. range: 0.1-0.49 µg / kg (6 samples), 0.5-1.49 µg / kg (4 sa), 1.5-9.99 µg / kg (3 sa), country: Germany
incidence: 1/5, conc.: 1 µg / kg, country: Germany
→ corn flakes, → maize flakes, → oat flakes

Cereal food (mixed)
Bsissa is a Tunesian-type of food composed of ground → barley, chick pea, and → wheat.
Bsissa may contain the following → mycotoxins:
→ ochratoxin A
incidence: 11/11, conc. range: 0.4-12,770 µg / kg, country: Tunesia

Cereal products may contain the following → mycotoxins:
→ ochratoxin A
incidence: 63/158, ∅ conc.: 0.9 µg / kg, country: Germany
incidence: 5/25, conc. range: 0.1-0.49 µg / kg (2 samples), 1.5-9.99 µg / kg (1 sa), country: Germany
incidence: 10/32, conc range: 0.1-0.49 µg / kg (6 samples), 0.5-1.49 µg / kg (4 sa), country: Germany
incidence: 2/7, conc. range: 0.1-0.49 µg / kg, country: Germany
incidence: 8/54, conc. range: 0.3-5.3 µg / kg, country: Germany
incidence: 1/40*, conc.: 2 µg / kg, country: Germany, *whole meal
incidence: 1/30, conc.: 6.2 µg / kg, country: Japan

Cereal products (whole meal) Contamination of cereal products with → mycotoxins mainly result from infection of the → grains in the field with mycotoxin producing fungi, especially → Fusarium spp. → Maize may be contaminated with aflatoxin producers (→ aflatoxins).
Under moderate conditions the → trichothecenes, mainly found in cereal grains, are relatively stable and very hard to remove from contaminated → cereals. During → milling processes they are distributed in food and feed. Food processing such as → baking and boiling in water and → oil does not cause their complete destruction. It is estimated that ca. 50% of trichothecenes remained in the final food products (e.g. → bread, → noodles).
Cereal grains may contain the following → mycotoxins:
→ deoxynivalenol
incidence: 4/4, conc. range: 255-490 µg / kg, ∅ conc.: 386 µg / kg, country: Austria

Cereals (no specification)
Cereals and → oil seeds belong to the most suitable substrates for → mycotoxins. Although mycotoxin contamination has been reported in → grains like → oats, → rice, → rye, → sorghum cer-

eals such as → barley, → maize, and → wheat seem to be more susceptible to mycotoxin formation. In general, low-grade cereals show a higher degree of mycotoxin contamination. Such grains in the developed countries normally do not enter the human food chain. However, in many developing countries the high quality cereals are often shipped abroad whereas the low-quality grains serve for human consumption.

Mycotoxin contamination (→ Fusarium mycotoxins, → trichothecenes, → zearalenone) of cereal grains usually occurs during growth and maturing of the plants while it is less common after harvest and during storage. Here, an a_w of > 0,83 (≈ > 17% moisture content) is necessary. Even aflatoxin (→ aflatoxins) contamination of cottonseed, maize, and → peanuts may occur before and during harvest although → Aspergillus flavus Link belongs to the group of → storage fungi.

Wheat, barley, and maize constitute two-thirds of the world production of cereals but similarly appear to be most affected by *Fusarium* mycotoxins. In detail, the relative vulnerability to mycotoxin formation is high for amber durum wheat, moderate for maize and six-row barley, but low for two-row barley and hard red spring wheat. Although contamination of oats, rye, and → triticale with trichothecenes and zearalenone is possible, these crops, except for some triticale varieties, seem to be more resistant or undergo no significant mycotoxin contamination.

Four trichothecenes, viz. → deoxynivalenol, → diacetoxyscirpenol, → nivalenol, and → T-2 toxin in general predominate in cereals grown in wet temperate regions like Northern Europe, parts of Northern America (e.g. Canada) and Japan. If cereals show a → zearalenone contamination there is a high probability that other *Fusarium* mycotoxins are also present. The application of fungicides may cause the accumulation of more mycotoxin in the grains by affecting the fungal metabolism.

Formation of *Aspergillus* and *Penicillium* mycotoxins is common in stored, inadequately dried agricultural products and / or rewetting of dried products, mainly by condensation, but also by flooding or when water leaks into storage bins. In cereals → aflatoxin B_1 and → aflatoxin B_2 are more often found than AFB_1, AFB_2, AFG_1, and AFG_2. However, aflatoxin contamination is a primarily a problem in maize. Rice is affected only in circumstances of poor storage in tropical and subtropical countries.

Stored ground (feed) seeds, lacking the outer protective testa, especially promote fungal growth since the rich nutrients inside are easily colonized by these storage fungi.

Cereals and → cereal products are mainly responsible for the → ochratoxin A intake at least in Europe since there is always the chance of contamination and the consumption of cereals is generally not low. In general, wheat and maize show a lower OTA contamination than rye. Mean levels of 0-2 μg OTA / kg on the EU market seem to be realistic. Temperate climatic conditions and drying with forced ambient air especially in Scandinavia favor OTA production in cereals. Other, also important, factors are mechanical injury and fungal infection, drying practice (e.g. promptness and rapidity of drying, rewetting) as well as improper storage techniques (↑ moisture contents, ↑ temperatures, ↑ oxygen, ↑ time). It is suggested that OTA contamination mainly occurs during the first period just after harvest before the a_w has decreased to a level which slows down or inhibits OTA formation. During longer periods of storage OTA production may occur if the storage conditions are unfavorable.

In cereal fractions of wheat and barley (> 2,5 mm), OTA concentrations reach

80-100% of the initial concentration in the corresponding grains. This mycotoxin is mainly present in the inner, deeper parts of the kernels but not on the surface of the grains. Chloroform extraction only removed 10-50% of the toxin from this part of the grain. → Milling results of these → grains show that the level of OTA in → flour is similar to that in → bran.

Chaff and kernels of small grain cereals (e.g. rye, wheat) may contain → Alternaria mycotoxins. The amount of such mycotoxins depends on the percentage of "black heads" due to → Alternaria alternata (Fr.) Keissler - not *A. infectoria* which is similar to *A. alternata* but a weak mycotoxin producer - at harvest time. However, cereals, e.g. wheat, without black heads or weather damage may also contain low levels of → tenuazonic acid. *Alternaria* mycotoxin formation is favored by high humidity and rainy weather before harvest. The production of *Alternaria* mycotoxins during storage is unlikely due to the low → a_w of the stored grains.

According to Frisvad (1988) the following mycotoxins may be found in cereals, maize, → peas and → beans under field conditions: → aflatoxins, → alternariol, → alternariol methyl ether, → altertoxins I-III, → butenolide, → cyclopiazonic acid, → fusarin C, → moniliformin, → tenuazonic acid, → trichothecenes, and → zearalenone. Stored cereals may be contaminated with aflatoxins, → citrinin, cyclopiazonic acid, ochratoxin A, → penicillic acid, → sterigmatocystin, → viomellein, and → xanthomegnin. The most probable mycotoxin in airtight stored cereals is → patulin.

Cereals may contain the following → mycotoxins:

→ aflatoxin B_1
incidence: 10/71*, conc. range: < 5-300 µg / kg, country: South Africa, *includes oats, wheat, barley

→ aflatoxins (no specification)
incidence: 1/52, conc.: 20 µg / kg, country: Japan

→ citrinin
incidence: 1*/52, conc.: 27 µg / kg, country: Japan, *maize flour
incidence: 4/735, conc. range: tr-6000 µg / kg, country: Poland

ergometrine (→ ergot alkaloids)
incidence: 2/2, conc. range: 0.8-6.4 µg / kg, ∅ conc.: 3.6 µg / kg, country: Canada, *wheat, rye, flax (mixture)

ergosine
incidence: 2/2, conc. range: 12-14 µg / kg, ∅ conc.. 13 µg / kg, country: Canada, *wheat, rye, flax (mixture)

ergotamine
incidence: 2/2, conc. range: 14-20 µg / kg, ∅ conc.: 17 µg / kg, country: Canada, *wheat, rye, flax (mixture)

ergocornine
incidence: 2/2, conc. range: 1.7-6.7 µg / kg, ∅ conc.: 4.2 µg / kg, country: Canada, *wheat, rye, flax (mixture)

α-ergokryptine
incidence: 2/2, conc. range: 1.1-6.6 µg / kg, ∅ conc.: 3.85 µg / kg, country: Canada, *wheat, rye, flax (mixture)

ergocristine
incidence: 2/2, conc. range: 36-40 µg / kg, ∅ conc.: 38 µg / kg, country: Canada, *wheat, rye, flax (mixture)

→ ochratoxin A
incidence: 6/315, conc. range: 3-8 µg / kg, country: Canada
incidence: 5/440, conc. range: 10-50 µg / kg, country: Canada
incidence: 19/33*, conc. range: 28-27,500 µg / kg, country: Denmark, *barley, oats
incidence: 2/151*, conc. range: 15-50 µg / kg, country: Denmark, *rye, wheat
incidence: 8/11*, conc. range: < 4.9-12.8 µg / kg, country: France, *maize, barley, oats
incidence: 1/13*, conc.: 2 µg / kg, country: France, *wheat, barley

incidence: 4/40, conc. range: < 4.9-22 µg / kg, country: France
incidence: 24/765, ∅ conc.: 11.8 µg / kg, country: Germany
incidence: 2/49, conc. range: 18-22 µg / kg, ∅ conc.: 20 µg / kg, country: Germany
incidence: 12/39*, conc. range: 0.1-2.7 µg / kg, country: Germany, *partly imported from different countries
incidence: 30/232*, conc. range: 0.1-206 µg / kg, country. Germany, *wheat, rye, oats
incidence: 18/43, conc. range: 2-304 µg / kg, country: Germany
incidence: 11/538, conc. range: 2-180 µg / kg, country: Norway
incidence: 63/784, conc. range: tr-1100 µg / kg, country: Poland
incidence: 6/100, conc. range: tr-1200 µg / kg, country: Poland
incidence: 20/296*, conc. range: 20-470 µg / kg, country: Poland, *barley, rye, wheat
incidence: 8/150, conc. range: 50-200 µg / kg, country: Poland
incidence: 158/1.353, conc. range: 5-2400 µg / kg, country: Poland
incidence: 7/84, conc. range: 16-410 µg / kg, country: Sweden
incidence: 6/47*, conc. range: 5-90 µg / kg, country: Yugoslavia, *barley, maize, wheat
→ patulin
incidence: 8/71*, conc.: nc, country: South Africa, *includes oats, wheat, barley
→ penicillic acid
incidence: 4/736, conc. range: tr-1300 µg / kg, country: Poland
→ zearalenone
incidence: 2/377, conc. range: tr-700 µg / kg, country: Poland
→ barley, → buckwheat, → grains, → maize, → millet, → oats, → rice, → rye, → sorghum, → triticale, → wheat

Chapatti may contain the following → mycotoxins:
→ deoxynivalenol
incidence: nc/4, conc. range: 6-10 µg / kg, country: UK
→ nivalenol
incidence: 1/4, conc.: 16 µg / kg, country: UK
→ ochratoxin A
incidence: 2/4, conc.: 0.5-0.9 µg / kg, ∅ conc.: 0.7 µg / kg, country: UK

Cheddar cheese → cheese, Cheddar

Cheese (no specification)
Mycotoxin producers of the genus → Penicillium, probably because of tolerance of low temperatures, are the most important contaminants on cheese during ripening and storage at low temperatures (ca. 80% of the total isolates) followed by → Aspergillus spp. (ca. 10%) and molds of other genera (ca. 10%). A hazard to human health is not necessarily if cheese exhibits some moldy spots. However, during prolonged storage periods mold growth and subsequent mycotoxin production becomes more probable. Mold-spoiled cheeses should therefore be excluded from human consumption.
To inhibit mold growth and subsequent mycotoxin contamination, the relative humidity in the curing room must be precisely and regularly checked. Some shrinkage of the cheeses is better than the development of mycotoxin producing molds. Plastic emulsions are treated with i.e. sorbate or pimaricin, which coat the cheeses so as to give them further protection against fungal infection. Hygienic measure, like cleaning and disinfecting of shelves in the curing rooms, also contribute to optimal cheese manufacture.
Although *Penicillium* spp. are well known mycotoxin producers, the most frequent → mycotoxins in cheese are the → aflatoxins, especially → aflatoxin M_1. During

production processes no destruction of AFM_1 has been reported. There are three possible contamination routes:
(i) → Carry over of → aflatoxin B_1 from cow feed into the raw → milk leads to aflatoxin M_1 accumulation.
(ii) Although cheese might be a less favorable substrate for mycotoxin production, it may still support surface growth of molds and subsequent mycotoxin formation (e.g. → sterigmatocystin); molding might occur during ripening in warehouses and after cutting and slicing during storage in shops or at home.
(iii) Contamination of → milk powder used to enrich the milk used to make cheese.
Cheese processing may result in the accumulation of aflatoxin M_1 in the curd. The affinity of AFM_1 for casein due to (possible) hydrophobic interactions with hydrophobic areas of the milk protein may be the reason. However, AFM_1 contamination of the whey in the range of 50-100% has also been reported. Decisive factors for the pattern of distribution are type and degree of milk contamination as well as milk quality, cheese processing, "contamination" of whey with AFM_1 contaminated curd, extraction technique, methodology, and expression of the results. There is an almost homogenous distribution in the concentration of AFM_1 concentration from the rind to the center (related to dry weight). The stability of AFM_1 during ripening and storage was shown in different kinds of cheese, e.g. Camembert, Cheddar, Parmesan, Swiss.
Since the mid-1980s, AFM_1 contamination in cheeses has declined, most probably as a result of strict regulations regarding aflatoxin levels in feeds in different countries.
Production of AFB_1 and AFG_1 apparently does not occur in Romadur or Camembert cheeses. However, such aflatoxins have been detected in Tilsit and Brick Cheese and in Emmentaler after they had been inoculated with aflatoxigenic molds. In addition to the aflatoxins, cheeses may be contaminated with further mycotoxins such as ochratoxin A and citrinin (see Figure Cheese). They may migrate into the cheese to a depth of at least 20 mm. In the case of → citrinin, diffusion lead to higher concentrations inside the cheese compared to its crust. Sterigmatocystin is stable in hard cheese for more than 3 months.

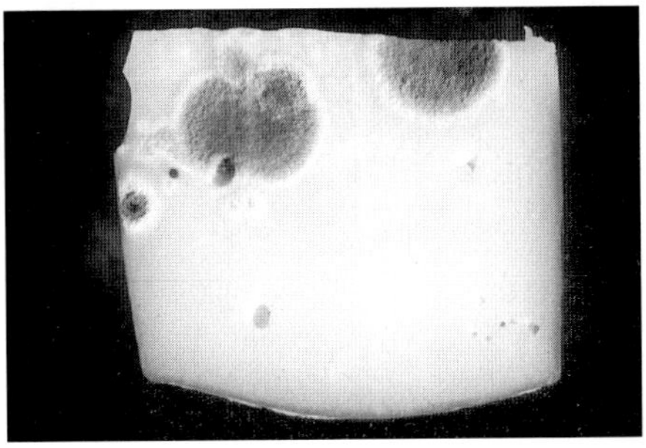

Cheese: *Penicillium* sp. on Edam cheese

Cheese may contain the following → mycotoxins:
aflatoxin B_1
incidence: 6/26, conc. range: 5-15 μg / kg, country: India
incidence: 1/248, conc.: 1 μg / kg, country: Tunesia
aflatoxin B_1 and/or → aflatoxin G_1
incidence: 79/133, conc. range: 10-50 μg / kg, country: Egypt
incidence: 16/222, conc. range: ≤ 10 μg / kg, country: Germany
aflatoxin M_1
incidence: 1/1*, conc.: 0.1 μg / kg, country: Canada
incidence: 60/60*, conc range: ≤ 0.26-0.89 μg / kg, country: Canada, *imported
incidence: 6/29, conc. range: 0.005-0.066 μg / kg, country: Czechoslovakia
incidence: 19/19*, conc. range: 0.1-0.4 μg / kg, ∅ conc.: 0.18 μg / kg, country: Denmark

incidence: 11/45, conc. range: 0.1-0.4 µg / kg, country: Denmark
incidence: 16/82, conc. range: < 0.005- > 0.25 µg / kg, country: France
incidence: 102/343, conc. range: ≤ 5.2 µg / l, country: France
incidence: 9/14*, conc. range: 0.1-0.3 µg / kg, ∅ conc.: 0.17 µg / kg, country: France
incidence: 9/34, conc. range: < 0.005- > 0.25 µg / kg, country: Germany
incidence: 2/5* conc. range: 0.15 µg / kg, ∅ conc.: 0.15 µg / kg, country: Germany
incidence: 5/22, conc. range: 0.1-0.4 µg / kg, country: Germany
incidence: 19/51*, conc. range: 0.1-0.2 µg / kg, ∅ conc.: 0.13 µg / kg, country: Ireland
incidence: 4/50, conc. range: 0.05-0.1 µg / kg, country: Italy
incidence: 130/416, conc. range: < 0.005- > 0.4 µg / kg, country: Italy
incidence: 71/83, conc. range: < 0.005- > 0.25 µg / kg, country: Italy
incidence: 5/6, conc. range: 0.05-0.4 µg / kg, country: Italy
incidence: 4/50, conc. range: 0.05-0.1 µg / kg, country: Italy
incidence: 2/4, conc. range: 0.02-0.04 µg / kg, country: Italy
incidence: 7/10, conc. range: 0.2-1.14 µg / kg, country: Italy
incidence: 12/66, conc. range: 0.28-1.3 µg / kg, country: Italy
incidence: 1/1*, conc.: 0.1 µg / kg, country: Italy
incidence: 56/126, conc. range: 0.11-0.3 µg / kg, country: Japan
incidence: 111/128, conc. range: 0.025-1.06 µg / kg, country: Japan
incidence: 13/32, conc. range: 0.012-2.52 µg / kg, country: Japan
incidence: 120/132, conc. range: 0.01-0.5 µg / kg, country: Japan
incidence: 19/80*, conc. range: 0.1-1.2 µg / kg, country: Japan, *imported
incidence: 23/43, conc. range: < 0.005- > 0.25 µg / kg, country: The Netherlands
incidence: 5/22*, conc. range: 0.15-0.5 µg / kg, country: The Netherlands
incidence: 8/40, conc. range: 0.1-0.2 µg / kg, country: The Netherlands
incidence: 30/30*, conc. range: < 0.1 µg / kg, country: New Zealand
incidence: 2/248, conc. range: 6.2-10.6 µg / kg, country: Tunisia
incidence: 86/143*, conc. range: < 0.10-0.50 µg / kg, country: UK, *imported
incidence: 1/4, conc.: 0.2 µg / kg, country: UK
incidence: 8/118*, conc. range: 0.1-1.0 µg / g, country: USA, *imported
→ aflatoxin M_4
incidence: 6/66, conc. range: 0.34-0.87 µg / kg, country: Italy
→ aflatoxins
incidence: 235/558, conc. range: < 0.25 µg / kg (143 samples), > 25 µg / kg (92 sa), country: Germany
→ citrinin
incidence: 17/44*, conc. range: < 50 µg / kg, country: UK, *retail, domestic
incidence: 3/nc, conc. range: nc, country: UK
→ mycophenolic acid
incidence: 38/100, conc. range: 20-15,000 µg / kg, country: France
→ β-nitropropionic acid
incidence: 5/18, conc. range: traces, country: USA
→ ochratoxin A
incidence: 18/44*, conc. range: ≤ 260 µg / kg, country: UK, *retail, domestic, wholesale
→ dairy products, → milk

Cheese (hard) may contain the following → mycotoxins:
→ aflatoxin M_1
incidence: 58/77, conc. range: 0.1-1.3 µg / kg, ∅ conc.: 0.43 µg / kg, country: Germany
→ mycophenolic acid
incidence: 4/48, conc. range: 10-1000 µg / kg*, country: France, *outer layer

→ patulin
incidence: 1/48, conc.: 90 μg / kg*, country: France, *outer layer
→ penicillic acid
incidence: 5/39, conc. range: ≤ 340 μg / kg*, country: France, *outer layer
→ sterigmatocystin
incidence: 3/66, conc. range: 7.5-17.5 μg / kg, country: Czechoslovakia
incidence: 9/39, conc. range: 5-600 μg / kg*, country: The Netherlands, *surface layer
incidence: 3/48, conc. range: ≤ 330 μg / kg*, country: France, *outer layer

Cheese (processed) may contain the following → mycotoxins:
→ aflatoxin B_1 and / or → aflatoxin G_1
incidence: 2/115, conc. range: nc, country: Germany
→ aflatoxin M_1
incidence: 54/134, conc. range: 0.1-0.55 μg / kg, ∅ conc.: 0.26 μg / kg, country: Germany
incidence: 14/14, conc. range: 0.1-0.3 μg / kg, ∅ conc.: 0.16 μg / kg, country: UK
→ ochratoxin A
incidence: 3/4, conc. range: 50-75 μg / kg, country: UK

Cheese (semi-hard) with a moldy crust may contain the following → mycotoxins:
→ mycophenolic acid
incidence: 7/39, conc. range: 10-5000 μg / kg, country: France
→ patulin
incidence: 4/39, conc. range: 45-355 μg / kg, country: France
→ penicillic acid
incidence: 5/39, conc. range: ≤ 710 μg / kg, country: France

Cheese (white, no further specification) may contain the following → mycotoxins:
→ cyclopiazonic acid
incidence: 2/6, conc. range: 250-370 μg / kg, ∅ conc.: 310 μg / kg, country: France

Cheese, Bhutanese may contain the following → mycotoxins:
→ ochratoxin A
incidence: 5/19, conc. range: 42-116 μg / kg, country: India

Cheese, Bleu des Causses may contain the following → mycotoxins:
→ mycophenolic acid
incidence: 3/6, conc. range: 10-1000 μg / kg, country: France

Cheese, Blue may be contaminated by different metabolic products of → Penicillium roquefortii Thom. → PR toxin is the most acutely toxic but it is produced by only a limited number of industrial strains. Formation of PR toxin depends on specific cultural conditions (↓ pH, ↓ NaCl, presence of sucrose, sufficient oxygen) which significantly differ from industrial processing methods. These are quite the opposite in Blue Cheese ripening. In addition, because of reaction with neutral and basic amino acids PR toxin is not stable in Blue Cheese. Concentrations of the formed PR-imine, a probable degradation product of PR toxin, may be rather high (≤ 42,000 μg / kg). → Roquefortine C as a frequent mycotoxin in Blue Cheese is concentrated in the moldy areas and often accompanied by roquefortine A, while roquefortine B (→ roquefortine A & B) occurs to a minor degree.
Blue cheese may contain the following → mycotoxins:
→ aflatoxin B_1 and / or → aflatoxin G_1
incidence: 2/62, conc. range: nc, country: Egypt
→ aflatoxin M_1

incidence: 5/5, conc. range: traces (4 samples), < 0.1 μg / kg (1 sa), country: Germany
→ mycophenolic acid
incidence: 4/32, conc. range: 250-500 μg / kg, country: Germany
incidence: 3/12, conc. range: 10- ≤ 1000 μg / kg, country: Germany (export to France)
→ penicillic acid
incidence: 1/110, conc.: 820 μg / kg, country: France
roquefortine A
incidence: 1/1, conc.: 785 μg / kg, country: Canada
incidence: 7/7, conc. range: 135-4700 μg / kg, ∅ conc.: 1921 μg / kg, country: Denmark
incidence: 1/1, conc.: 1833 μg / kg, country: Finland
incidence: 2/3, conc. range: 100-130 μg / kg, ∅ conc.: 115 μg / kg, country: France
incidence: 4/6, conc. range: tr-170 μg / kg, country: Germany
incidence: 5/5, conc. range: 200-360 μg / kg, country: Japan
incidence: 2/2, conc. range: tr(?)-80 μg / kg, country: UK
roquefortine B
incidence: 1/1, conc.: traces, country: Canada
incidence: 4/7, conc. range: traces, country: Denmark
incidence: 1/1, conc.: traces, country: Denmark
→ roquefortine C
incidence: 1/1, conc.: 1085 μg / kg, country: Canada
incidence: 7/7, conc. range: 60-2300 μg / kg, ∅ conc.: 982 μg / kg, country: Denmark
incidence: 1/1, conc.: 66 μg, country: Finland
incidence: 3/3, conc. range: 60-400 μg / kg, ∅ conc.: 230 μg / kg, country: France
incidence: 4/6, conc. range: 370-6800 μg / kg, ∅ conc.: 2500 μg / kg, country: Germany
incidence: 3/3, conc. range: 490-1100 μg / kg, ∅ conc.: 737 μg / kg, country: Switzerland
incidence: 12/12, conc. range: 162-651 μg / kg, ∅ conc.: 424 μg / kg, country: USA

Cheese, Blue Castello may contain the following → mycotoxins:
→ roquefortine C
incidence: 1/1, conc.: 2290 μg / kg, country: France

Cheese, Blue Haverti may contain the following → mycotoxins:
→ aflatoxin M_1
incidence: 10/10, conc. range: 0.084-0.556 μg / kg, country: Denmark

Cheese, Bresse Bleu may contain the following → mycotoxins:
→ roquefortine C
incidence: 1/1, conc.: 560 μg / kg, country: Denmark

Cheese, Brick → Aspergillus flavus Link and → Aspergillus parasiticus Speare produced → aflatoxins on Brick Cheese at 23.9 °C and 12.8 °C, respectively.

Cheese, Brie may contain the following → mycotoxins:
→ aflatoxin M_1
incidence: 6/6, conc. range: 0.058-0.414 μg / kg, ∅ conc.: 0.195 μg / kg, country: Denmark
incidence: 6/14, conc. range: 0.055-0.714 μg / kg, country: France
incidence: 2/2, conc. range: 0.024-0.029 μg / kg, ∅ conc.: 0.0265 μg / kg, country: Germany

Cheese, butter may contain the following → mycotoxins:

→ aflatoxin M_1
incidence: 5/5, conc. range: 0.025-0.041 μg / kg, ∅ conc.: 0.037 μg / kg, country: Germany
incidence: 6/7, conc. range: traces (4 samples), < 0.1 μg / kg (2 sa), country: Germany

Cheese, Camembert → Cyclopiazonic acid represents an important mycotoxin in this kind of cheese. It occurs mainly in the crust rather than in the inner part. Not yet fully ripened cheeses stored in the cold do not contain more than 500 μg cyclopiazonic acid / kg (calculation on whole cheese). A significant increase up to 5000 μg cyclopiazonic acid / kg may result from temperatures during storage that are too high. Therefore, refrigerated storage and display, together with limited shelf life are recommended to prevent the accumulation of cyclopiazonic acid. However, the actual toxicological data, in combination with consumption habits, indicate that no risk to human health in reality exists.
Camembert may contain the following → mycotoxins:
→ aflatoxin M_1
incidence: 7/7, conc. range: 0.055-0.479 μg / kg, ∅ conc.: 0.207 μg / kg, country: Denmark
incidence: 18/25, conc. range: 0.013-0.565 μg / kg, country: France
incidence: 1/100 conc.: traces, country: France
incidence: 1/1, conc.: 0.018 μg / kg, country: Germany
incidence: 19/38, conc. range: traces (14 samples), < 0.1 μg / kg (2 sa), > 0.1 μg / kg (3 sa), country: Germany
→ cyclopiazonic acid
incidence: 11/20, conc. range: 0.05-0.1 μg / kg (3 samples), 0.1-0.2 μg / kg (5 sa), 0.4-1.5 μg / kg (3 sa), country: France
incidence: 1/3, conc.: 80 μg / kg, country: Switzerland

Cheese, Camembert & Brie Camembert and Brie may contain the following → mycotoxins:
→ aflatoxin M_1
incidence: 33/65, conc. range: 0.1-0.73 μg / kg, ∅ conc.: 0.31 μg / kg, country: Germany
→ cyclopiazonic acid
incidence: 11/11, conc. range: 60-290 μg / kg, country: Germany
incidence: 11/20, conc. range: 50-1500 μg / kg, country: USA

Cheese, Cheddar → Aspergillus flavus Link and → Aspergillus parasiticus Speare produced substantial quantities of → aflatoxins on Cheddar cheese at room temperature but no natural occurrence of these → mycotoxins has been reported up to now.
Cheddar cheese may contain the following mycotoxins:
→ aflatoxin M_1
incidence: 4/4, conc. range: 0.015-0.030 μg / kg, ∅ conc.: 0.020 μg / kg, country: UK
incidence: 147/147*, conc. range: < 0.1-0.4 μg / kg, country: UK, *home made
→ citrinin
incidence: 2/2*, conc. < 100 μg / kg, country: UK, *1 mature English and 1 colored Scotch cheddar
→ ochratoxin A
incidence: 2/2*, conc. range: 260-500 μg / kg, country: UK, *1 mature English and 1 colored Scotch cheddar

Cheese, Chesire may contain the following → mycotoxins:
→ aflatoxin M_1
incidence: 36/36*, conc. range: < 0.1-0.4 μg / kg, country: UK, *home made
→ ochratoxin A
incidence: 3/5*, conc. range: ≤ 50 μg / kg, country: UK, *colored, white and red

Cheese, Chester may contain the following → mycotoxins:
→ aflatoxin M_1
incidence: 4/4, conc. range: traces (1 sample), < 0.1 µg / kg (2 sa), > 0.1 µg / kg (1 sa), country: Germany

Cheese, Comte may contain the following → mycotoxins:
→ aflatoxin M_1
incidence: 1/279*, conc. range: 1.1 µg / kg, country: Japan, *imported

Cheese, Cottage may contain the following → mycotoxins:
→ aflatoxin M_1
incidence: 1/209, conc.: 0.08 µg / l*, country: USA, *l = level reported on fluid milk basis
incidence: 15/209, conc. range: 0.05-0.4 µg / kg, country: USA

Cheese, Cream may contain the following → mycotoxins:
→ aflatoxin M_1
incidence: 9/9, conc. range: 0.037-0.134 µg / kg, ∅ conc.: 0.79 µg / kg, country: Denmark
→ penitrem A
incidence: 1/1*, conc. range: nc, country: USA, *visible moldy

Cheese, Danish Blue may contain the following → mycotoxins:
→ roquefortine C
incidence: 3/3, conc. range: 950-1700 µg / kg, ∅ conc.: 1203 µg / kg, country: Denmark

Cheese, Double Gloucester may contain the following → mycotoxins:
→ aflatoxin M_1
incidence: 10/10*, conc. range: < 0.1-0.15 µg / kg, country: UK, *home made
→ ochratoxin A
incidence: 2/2, conc. range: ≤ 50 µg / kg, country: UK

Cheese, Edam may contain the following → mycotoxins:
→ aflatoxin M_1
incidence: 31/32, conc. range: traces (11 samples), < 0.1 µg / kg (16 sa), > 0.1 µg / kg (4 sa), country: Germany
incidence: 4/4, conc. range: 0.073-0.117 µg / kg, ∅ conc.: 0.099 µg / kg, country: The Netherlands
→ ochratoxin A
incidence: 2/25, conc. range: 820-1100 µg / kg, ∅ conc.: 960 µg / kg, country: Yugoslavia

Cheese, Edam Cake may contain the following → mycotoxins:
→ sterigmatocystin
incidence: 2/66*, conc. range: 7.5-17.5 µg / kg, ∅ conc.: 12.5 µg / kg, country: Czechoslovakia, *and different other kinds of cheese

Cheese, Emmental may contain the following → mycotoxins:
→ aflatoxin M_1
incidence: 15/358*, conc. range: 0.1-1.1 µg / kg, ∅ conc.: 0.53 µg / kg, country: Japan *imported natural cheese
→ ochratoxin A
incidence: 3/3, conc. range: ≤ 50 µg / kg, country: UK

Cheese, Fresh may contain the following → mycotoxins:
→ aflatoxin M_1
incidence: 27/80, conc. range: 0.1-0.51 µg / kg, ∅ conc.: 0.23 µg / kg, country: Germany

Cheese, Goat may contain the following → mycotoxins:
→ patulin
incidence: 1/18, conc.: 30 µg / kg, country: France.
→ penicillic acid
incidence: 2/18, conc. range: ≤ 45,210 µg / kg, country: France

Cheese, Gorgonzola may contain the following → mycotoxins:
→ mycophenolic acid
incidence: 3/12, conc. range: 10-100 µg / kg, country: France
→ roquefortine C
incidence: 2/2, conc. range: 490-940 µg / kg, ∅ conc.: 715 µg / kg, country: Italy
incidence: 2/2, conc. range: 150-190 µg / kg, ∅ conc.: 170 µg / kg, country: Italy

Cheese, Gouda may contain the following → mycotoxins:
→ aflatoxin M_1
incidence: 9/9, conc. range: 0.039-0.087 µg / kg, ∅ conc.: 0.063 µg / kg, country: The Netherlands
→ sterigmatocystin
incidence: 6*/6, conc. range: nc, country: The Netherlands, *surface layer

Cheese, Gouda & Cheddar may contain the following → mycotoxins:
→ cyclopiazonic acid
incidence: nc, conc. range: 35,000-70,000 µg / kg, country: South Africa

Cheese, Grana Padano is a Parmesan-like cheese.
Grana Padano may contain the following → mycotoxins:
→ aflatoxin M_1
incidence: 219/223, conc. range: 0.005-0.1 µg / kg (203 samples), 0.101-0.25 µg / kg (15 sa), > 0.25 µg / kg (1 sa), country: Italy

Cheese, Lancashire may contain the following → mycotoxins:
→ aflatoxin M_1
incidence: 5/5*, conc. range: < 0.1-0.15 µg / kg, country: UK, *home made

Cheese, Leicester may contain the following → mycotoxins:
→ aflatoxin M_1
incidence: 6/6*, conc. range: < 0.1-0.15 µg / kg, country: UK, *home made
→ ochratoxin A
incidence: 1/2, conc.: ≤ 50 µg / kg, country: UK

Cheese, Maribo may contain the following → mycotoxins:
→ aflatoxin M_1
incidence: 3/3, conc. range: 0.087-0.412 µg / kg, ∅ conc.: 0.264 µg / kg, country: Denmark

Cheese, Mold-cured is a food product that has been consumed for centuries without causing any detrimental effects on human health. This is confirmed by long-term trials with → Penicillium roquefortii Thom and → Penicillium camembertii Thom as well as Camembert (→ cheese, Camembert) and Blue cheese (→ cheese, Blue). No harmful effects could be demonstrated in experimental animals. Because of the mycotoxicological potential of the starter cultures the following points should be considered: i) the use of non-toxic starter cultures, ii) provision of optimal conditions during manufacture and ripening, iii) sanitary precautions to prevent unwarranted mold growth.

Cheese, Moravian Block may contain the following → mycotoxins:
→ sterigmatocystin
incidence: 1/66, conc.: 7.5 µg / kg, country: Czechoslovakia

Cheese, Mozarrella may contain the following → mycotoxins:
→ aflatoxin M_1
incidence: 4/4, conc. range: 0.181-0.433 µg / kg, ∅ conc.: 0.334 µg / kg, country: Denmark
incidence: 5/5, conc. range: 0.028-0.252 µg / kg, ∅ conc.: 0.091 µg / kg, country: Germany

Cheese, Parmesan may contain the following → mycotoxins:
→ aflatoxin M_1
incidence: 18/200, conc. range: 0.035-0.190 μg / kg, country: Italy

Cheese, pepper may contain the following → mycotoxins:
→ aflatoxin B_1 and / or → aflatoxin G_1
incidence: 1/1, conc.: "high", country: France

Cheese, Romadur may contain the following → mycotoxins:
→ aflatoxin M_1
incidence: 35/50, conc. range: traces (19 samples), < 0.1 μg / kg (8 sa), > 0.1 μg / kg (8 sa), country: Germany

Cheese, Roquefort may contain the following → mycotoxins (see Figure Roquefort):
→ mycophenolic acid
incidence: 4/5, conc. range: 250-5000 μg / kg, ∅ conc.: 3375 μg / kg, country: France
→ roquefortine C
incidence: 3/3, conc. range: 200-1330 μg / kg, ∅ conc.: 670 μg / kg, country: France
incidence: 21/25, conc. range: 10-≤ 15,000 μg / kg, country: France

Cheese, Samsoe may contain the following → mycotoxins:

Roquefort. *Penicillium roquefortii* in Roquefort cheese

→ aflatoxin M_1
incidence: 5/5, conc. range: 0.07-0.504 μg / kg, ∅ conc.: 0.214 μg / kg, country: Denmark

Cheese, Stilton may contain the following → mycotoxins:
→ aflatoxin M_1
incidence: 8/8*, conc. range: 0.1-0.3 μg / kg, country: UK, *home made
→ roquefortine
incidence: 2/2, conc. range: 970-3400 μg / kg, ∅ conc.: 2185 μg / kg, country: UK

Cheese, Swiss may contain the following → mycotoxins:
→ penicillic acid
incidence: 4/33, conc. range: ≤ 500 μg / kg, country: USA

Cheese, Tilsit Washing of a Tilsit cheese previously inoculated with → Aspergillus flavus Link and → Aspergillus parasiticus Speare, seems to cause → aflatoxins to diffuse from the surface layer into the body of the cheese.
Tilsit cheese may contain the following → mycotoxins:
→ aflatoxin B_1
incidence: 1*/1, conc.: 7 μg / kg, country: Germany, *total: 9 μg aflatoxin / kg,
incidence: 18/24, conc. range: traces (7 samples), < 0.1 μg / kg (10 sa), > 0.1 μg / kg (1 sa), country: Germany

Cheese, Wensleydale may contain the following → mycotoxins:
→ aflatoxin M_1
incidence: 5/5*, conc. range: < 0.1-0.2 μg / kg, country: UK, *home made
→ ochratoxin A
incidence: 1/1, conc.: ≤ 50 μg / kg, country: UK

Cheese, Wine may contain the following → mycotoxins:

→ aflatoxin M_1
incidence: 3/7, conc. range: traces (2 samples), > 0.1 µg / kg (1 sa), country: Germany

Cheese cake may contain the following → mycotoxins:
→ ochratoxin A
incidence: 1*/3, conc.: 1075 µg / kg, country: Poland, *moldy

Cheese dressing, blue may contain the following → mycotoxins:
→ roquefortine C
incidence: 2/2, conc. range: 18-72 µg / kg, ∅ conc.: 45 µg / kg, country: USA

Cheese rind may contain the following → mycotoxins:
→ aflatoxin B_1 and / or → aflatoxin G_1
incidence: 6/34, conc. range: nc, country: Romania

Cheese trimmings (no specification) may contain the following → mycotoxins:
→ aflatoxin B_1 and / or → aflatoxin G_1
incidence: 1/1, conc.: nc, country: USA
→ ochratoxin A
incidence: 1/1, conc.: nc, country: USA

Cherries (sweet)
may contain the following → mycotoxins:
→ aflatoxin B_1
incidence: 1/8*, conc.: 5 µg / kg, country: Germany, *moldy
→ fruits

Chicken No natural contamination of Broiler-type chickens with → aflatoxins has yet been reported. Feeding results indicate a rapid tissue clearance (4 days) after the removal of the aflatoxins from the diet although the → mycotoxins were deposited in all tissues, especially gizzards, liver, and kidneys. However, contamination with → ochratoxin A is evident.

Chicken may contain the following mycotoxins:
ochratoxin A
incidence: 36/65, conc. range: ≤ 0.18 µg / kg, ∅ conc.: 0.03 µg / kg, country: Denmark
→ meat

Chicken, yolk may contain the following → mycotoxins:
→ ochratoxin A
incidence: nc/nc, conc. range: 1.6-4 µg / kg, country: Germany

Chicken liver Feed tissue ratios of → aflatoxin B_1 to AFB_1 and → aflatoxin M_1 are much higher for kidney and liver than for muscle.
The liver may contain the following → mycotoxins:
aflatoxin B_1
incidence: 1/5, conc.: < 5 µg / kg, country: Germany
→ meat

Chilli → Pepper (red) , → spices

Chilli pickles may contain the following → mycotoxins:
→ aflatoxins (AFB_1, AFB_2, AFG_1, AFG_2)
incidence: nc/4, conc. range: 1-58.5 µg / kg, country: UK
→ fumonisins (FB_1, FB_2)
incidence: 1/4, conc.: 121 µg / kg, country: UK
→ ochratoxin A
incidence: 3/4 conc. range: 0.5-1.2 µg / kg, country: UK
→ spices

Chilli powder may contain the following → mycotoxins:
→ aflatoxins (AFB_1, AFB_2, AFG_1, AFG_2)
incidence: nc/4, conc. range: 1.1-5.4 µg / kg, country: UK
→ diacetoxyscirpenol

incidence: nc/4, conc. range: 47-81 µg / kg, country: UK
→ deoxynivalenol
incidence: 1/4, conc.: 8 µg / kg, country: UK
→ HT-2 toxin
incidence: 1/4, conc.: 24 µg / kg, country: UK
→ ochratoxin A
incidence: nc/4, conc. range: 1.6-50.4 µg / kg, country: UK
→ zearalenone
incidence: nc/4, conc. range: 4.5-15.4 µg / kg, country: UK
→ spices

Chilli sauce may contain the following → mycotoxins:
→ nivalenol
incidence: 1/4, conc.: 15 µg / kg, country: UK
→ ochratoxin A
incidence: 1/4, conc.: 3.3 µg / kg, country: UK
→ zearalenone
incidence: 1/4, conc.: 7.1 µg / kg, country: UK
→ spices

Chips → maize chips

Chocolate may contain the following → mycotoxins:
→ aflatoxins
incidence: 1*/36, conc.: 5 µg AFB_1 resp. 10 µg aflatoxins / kg, country: Germany, *containing → Brazil nuts

Cider Due to alcoholic fermentation (*Saccharomyces cerevisiae*) cider is usually free of → patulin. In Canada and the USA this term is also used for not fermented → apple juice which can be misleading.
Cider may contain the following → mycotoxins:
patulin
incidence: 9/13, conc. range: 100-300 µg / l, country: France
→ apple juice

Cirrhosis Disease of the liver characterized by excessive → fibrosis.

Citreoviridin is an unsaturated lactone (2,5-anhydro-1,6-dideoxy-2-c-[(1E,3E,5E,7E)-8-(4-methoxy-5-methyl-2oxo-2H-pyran-6-yl)-2-methyl-1,3,5,7-octatetraenyl]-4-c-methyl, → mycotoxins) which was isolated in 1947 from → Penicillium citreonigrum Dierckx (formerly *P. toxicarium*), a contaminant of yellow rice (see Figure Citreoviridin).

Chemical Data
Empirical formula: $C_{23}H_{30}O_6$, molecular weight: 402

Fungal Sources
→ Penicillium spp. (e.g. *P. citreonigrum, P. miczynskii, P. manginii, P. smithii* (syn. *P. corynephorum*), *Eupenicillium ochrosalmoneum*, → Aspergillus terreus Thom

Natural Occurrence
→ pecans, probably in "yellow rice"
Toxin formation on → rice is favored by low temperatures and high humidity. These climatic conditions predominate in the northern part of Japan. Rice ("soft → grains") grown in this area often shows a contamination with *P. citreonigrum*, a fungus that is a major source of this mycotoxin. Citreoviridin has also been isolated from naturally contaminated moldy pecan fragments (→ pecans) and from standing → maize in the field (USA). The natural occurrence of this toxin in → food has rarely been reported because adequate analytical methods and sources of standard for this mycotoxin are not available, generally.

Toxicity
acute toxic, neurotoxic, paralytic, potent inhibitor of ATPase

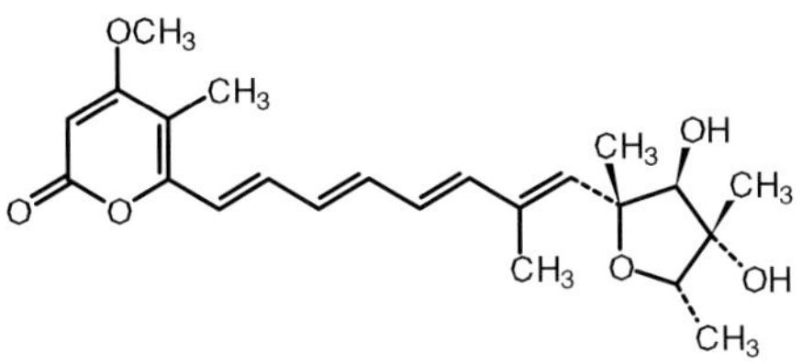

Citreoviridin

The symptoms include early onset of a progressive → paralysis in the extremities of laboratory animals. Similarly → convulsions, vomiting and impairment of the respiratory center occurred. In a later stage, the disease is characterized by → hypothermia, flaccid paralysis and cardiovascular disturbances. Along with → dyspnea, gasping and coma respiratory arrest leads to death. These symptoms are very similar to those in human patients who consume rice as a staple food and suffer from → acute cardiac beriberi.

LD_{50} (po): 3.6 mg / kg bw rat

DETECTION
TLC

POSSIBLE MYCOTOXICOSIS
This highly toxic fungal metabolite is associated in the complex of → yellow rice disease in Japan and represents a (possible) causative agent in acute cardiac beriberi in humans.

Citrinin (Syn.: antimycin, monascidin A) is a (3**R**-*trans*)-4,6-dihydro-8-hydroxy-3,4,5-trimethyl-6-oxo-3H-2-benzopyran-7-carboxylic acid which is derived from the condensation of five acetate and the introduction of three one-carbon units (see Figure Citrinin). This major "yellow rice" toxin (→ yellow rice disease) was first isolated from → Penicillium citrinum Thom in 1931. The mold occurs most frequently in "yellow rice" and produces copious quantities of this yellow toxic metabolite. Citrinin, therefore, was first implicated in the "yellow rice" syndrome in Japan.

CHEMICAL DATA
Empirical formula: $C_{13}H_{14}O_5$, molecular weight: 250

FUNGAL SOURCES
e.g. → Aspergillus spp. (e.g. → Aspergillus candidus Link, *A. carneus*, → Aspergillus terreus Thom), *Monascus purpureus*, *M. ruber*, → Penicillium spp. (e.g. → Penicillium citreonigrum Dierckx, → Penicillium citrinum Thom, → Penicillium expansum Link, → Penicillium verrucosum Dierckx chemotype II).

NATURAL OCCURRENCE
→ Acacia concinna, → bakery products, → barley, → bread, → cardamom, → Cassia fistula, → cereals, → cheese, → cheese, Cheddar, → confectionery, → coriander, → cumin, → fennel, → flour, → Hydnocarpus laurifolia, → maize, → maize flour, → meat, → oil seed rape, → pastries, → peanuts, → pepper, → pig kidneys, → Piper betle, → rice, → rye, → shoyu, → triticale, → tumeric, → wheat, → wheat grits

Citrinin mainly occurs in rice and other cereals. Different kinds of foodstuff, especially → grains, often are contaminated with both citrinin and → ochratoxin A. Since citrinin is more readily lost in analytical procedures, it seems to occur much less frequently than ochratoxin A. In general, significantly higher citrinin concentrations, compared to OTA levels, occur. Although citrinin represents a contaminant of different kinds of food products, it seems unlikely that it does constitute a human health problem.

TOXICITY
Fetotoxic, embryocidal, → mutagenic (?) and mildly → teratogenic, nephrotoxic, hepatotoxic
antibacterial, antifungal, antiprotozal, phytotoxic

In the view of kidney damage and the development of → renal tumors, a probable synergistic effect with ochratoxin A is important.
LD_{50} (po): 50 mg / kg bw rats

Detection
HPLC, NMR, spectrofluorometric determination, TLC

Possible Mycotoxicosis
→ Mycotoxic porcine nephropathy, → Balkan endemic nephropathy (citrinin and ochratoxin A); → Yellow rice disease (citrinin, → citreoviridin, other → Penicillium toxins)

Further Comments
Citrinin was mainly located in the spore wall and may be a major component of the spores of *P. verrucosum*. It was suggested that this mycotoxin, which is released in an aqueous environment, may have important function(s) in spore survival.
Stability: Citrinin was fairly stable in air or oven dried whole → maize kernels inoculated with → Penicillium spp. over a period of a year. However, during mashing this mycotoxin is degraded and therefore, → beer is citrinin-free. Because citrinin is more heat sensitive than OTA, heat treament of contaminated food will significantly reduce the citrinin level. The instability of citrinin may explain its absence from → apple juice and other → apple products. However, heating with water (ca. 140 °C) yielded a decomposition product as toxic as or even more toxic than citrinin.
Citrinin ist unstable during prolonged exposure to light or heat.
Milling: Compared to the milled product, citrinin is accumulated in the bran and polish fraction of rice. A positive correlation between highly contaminated samples and levels found in the aforementioned fractions could be established. Citrinin probably survives milling at least to some extent because maize flour (e.g.

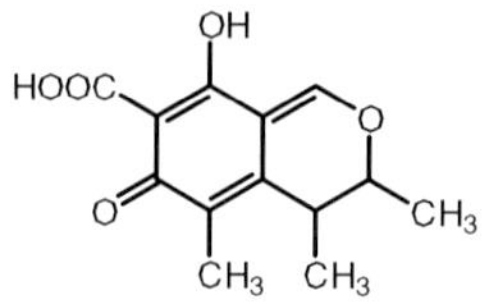

Citrinin

Thailand) was contaminated in the range of 10-98 µg / kg.
Production: Citrinin production (→ Pencillium viridicatum Westling) occurred on bread at a minimum → a_w of 0.80, optimum a_w 0.92. Durum wheat (15% and 19% moisture) enabled citrinin production after 24 weeks with maximal production after 48 weeks. Citrinin formation may occur on every kind of meat and therefore should be regarded as a serious toxin.

Cladosporium anamorphic → Mycosphaerellaceae, teleomorphs *Mycosphaerella, Venturia*
Cladosporium spp. may grow on chilled and overwintered grain. *C. herbarum* together with *C. fagi* may be associated with some forms of → alimentary toxic aleukia. The → mycotoxins epicladosporic and fagicladosporic acid (see Figure Cladosporium) may be responsible for the toxicity of → grains which have been exposed to cold winter climatic conditions since they are frequently infected by these two fungi.

Clavacin (Syn.: → Patulin)

Clavatin (Syn.: → Patulin)

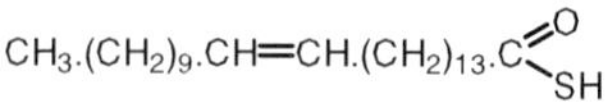

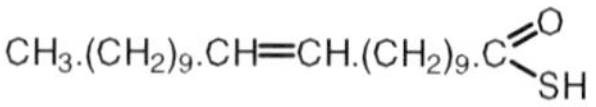

Cladosporium. Epicladosporic acid and fagicladosporic acid

Claviceps → Clavicipitaceae
Fungi of this genus grow parasitically in the spikes of → cereals, especially → rye, and grasses. During overwintering 2 - 4 cm long granules are formed, called → ergots. The most important species is *C. purpurea* which mainly infects → rye. Minor infections also occurred on → barley, → maize, → oats, and → wheat.
→ Clavine alkaloids, → ergot alkaloids, → ergotism, ergots

Clavicipitaceae → Hypocreales

Claviformin (Syn.: → patulin)

Clavine alkaloids In contrast to the well-known lysergic acid derivatives (→ ergot alkaloids), the carboxyl group has been reduced to a hydroxymethyl or a methyl group. → Sclerotia of → Claviceps species which occur on wild grasses in Africa and in the Far East contain substantial amounts of these alkaloids. Only trace amounts are found in the sclerotia and saprophytic cultures of *C. purpurea* and *C. paspali*. Important clavine alkaloids are e.g. fumigaclavine A & B.

Cocoa beans In the Central American countries like Costa Rica, outdoor drying of cocoa beans on movable rail and wheel beds (ca. 10 m^2) is the usual practice. Pushing the cocoa beans under a crude roof and storage in a layer-type fashion is a good protection against the rain. Too-wet stored or rewetted cocoa beans are prone to mold growth and subsequent mycotoxin contamination.
A significant destruction of ochratoxin A occurred during the processing of cocoa beans to dark → chocolate.
Cocoa beans may contain the following → mycotoxins:
→ aflatoxin B_1
incidence: 1/40*, conc.: 5 µg / kg, country: Norway, *imported
→ aflatoxin B_2
incidence: 1/40*, conc.: traces, country: Norway, *imported
→ aflatoxin G_1
incidence: 1/40*, conc.: 4 µg / kg, country: Norway, *imported
→ aflatoxin G_2
incidence: 1/40*, conc.: traces, country: Norway, *imported
→ aflatoxin
incidence: 3/91, conc. range: 2-20 µg / kg (1 sample), > 20 µg / kg (2 sa), country: Uruguay
→ aflatoxins (no specification)
incidence: 2*/47, conc. range: 5-9.9 µg / kg, country: Canada, *AFB_1, AFB_2, AFG_1, AFG_2
incidence: 1/14, conc.: > 4 µg / kg, country: Ghana
incidence: 1/6, conc.: > 4 µg / kg, country: Malaysia
incidence: 5/14, conc. range: > 4 µg / kg, country: Nigeria
incidence: 4/6, conc. range: > 4 µg / kg, country: Papua New Guinea
incidence: 2/9, conc. range: ≤ 17 µg / kg, country: Philippines
incidence: 1/4, conc.: > 4 µg / kg, country: Trinidad
→ ochratoxin A
incidence: 2/3, conc. range: > 60 µg / kg, country: Ecuador
incidence: 4/14, conc. range: > 60 µg / kg, country: Ghana
incidence: 2/5, conc. range: > 60 µg / kg, country: Grenada
incidence: 1/2, conc.: > 60 µg / kg, country: Ivory Coast
incidence: 1/6, conc.: > 60 µg / kg, country: Malaysia
incidence: 1/14, conc.: > 60 µg / kg, country: Nigeria
incidence: 1/1, conc.: > 60 µg / kg, country: Venezuela
→ coffee beans

Cocoa beans (raw) may contain the following → mycotoxins:
→ aflatoxins (no specification)
incidence: 7/56, conc.: < 5 μg / kg (6 samples), 5-10 μg / kg (1 sa), country: UK
→ ochratoxin A
incidence: 10/56, conc. range: < 100 μg / kg (5 samples), 101-200 μg / kg (4 sa), 201-500 μg / kg (1 sa), country: UK

Cocoa beans (roasted) may contain the following → mycotoxins:
aflatoxin (no specification)
incidence: 6/8, conc. range: ≤ 28 μg / kg, ∅ conc.: 18 μg / kg, country: Philippines
→ aflatoxins (no specification)
incidence: 7/19, conc.: < 5 μg / kg (6 samples), 5-10 μg / kg (1 sa), country: UK
→ ochratoxin A
incidence: 3/19, conc. range: 100 μg / kg, country: UK

Cocoa nibs may contain the following → mycotoxins:
→ aflatoxins (no specification)
incidence: 1/2, conc.: 11-20 μg / kg, country: UK
→ ochratoxin A
incidence: 1/2, conc.: 101-200 μg / kg, country: UK

Cocoa presscake may contain the following → mycotoxins:
→ aflatoxins (no specification)
incidence: 1/4, conc.: < 5 μg / kg, country: UK
→ ochratoxin A
incidence: 1/4, conc.: 101-200 μg / kg, country: UK

Cocoa products may contain the following → mycotoxins:
→ ochratoxin A
incidence: 1/20, conc.: ≤ 0.6 μg / kg, country: Germany

Coconut (processed)
Coconut is an excellent medium for the growth of → Aspergillus spp. and subsequent aflatoxin accumulation.
Coconut may contain the following → mycotoxins:
aflatoxin (no specification) (→ aflatoxins)
incidence: 20/29, conc. range: ≤ 26 μg / kg, ∅ conc.: 11 μg / kg, country: Philippines
→ nuts

Coconut ice may contain the following → mycotoxins:
→ aflatoxins (no specification)
incidence: 1/4, conc. range: nc, country: UK

Coconut oil (crude)
derived from moldy coconut may contain high levels of → aflatoxins and even commercially avaiable coconut oil from processed → copra may be contaminated by low to medium aflatoxin levels. Only by refining can the aflatoxin and the pigments be removed from the → oil but the expense of this method limits its use in poorer countries. Contaminated oil may effectively be decontaminated by exposure to sunlight.
Coconut oil may contain the following → mycotoxins:
aflatoxin (no specification)
incidence: 3/3, conc. range: ≤ 9 μg / kg, ∅ conc.: 3 μg / kg, country: Philippines
→ oil, → olive oil, → peanut oil, → sunflower seed oil

Coffee The manufacture of coffee cherry includes several steps: harvesting, direct drying or pulping/fermentation and drying, hulling, cleaning, sorting = producer country; decaffeination (alternatively), blending, roasting, industrial extraction (alternatively), packaging = producer or consumer country
Coffee may be an important contributor to → ochratoxin A (→ Aspergillus ochra-

ceus group) intake ($\approx$ 20%) in humans. A mean level of 0.5-1.5 µg OTA / kg has been detected in the roasted coffee sold on the EU market. Transmission of OTA to the final brew is possible.
Mycotoxin contamination of the beans mainly occurs during green coffee processing, and / or transportation. It seems that superficial OTA contamination is higher than deep bean contamination. Together with the chaff this portion is eliminated during roasting.
Industrial decaffeination may cause a 60 % reduction of ochratoxin A in a naturally-contaminated sample. During roasting, as well as during brewing, partial to almost complete OTA destruction has been observed.
Although OTA levels as low as 0.1 µg / kg coffee can now be easily detected, detection of single contaminated beans is difficult because of the extremely inhomogenous distribution of the mycotoxin in the batch. A suitable sampling procedure for OTA detection in green coffee is lacking. Highly contaminated batches of green coffee possess musty / moldy off-flavors which are carried through to the finished product and beverage. Because such batches are rejected by the coffe trade,the amount of OTA contamination in commercial roast, ground and instant coffee products is usually low.
The daily intake of four cups of coffee (24 g roasted & ground) contributes on average 19 ng OTA / day, 8 g instant coffee = 10 ng OTA / day. The resulting weekly OTA consumption constitutes not more than 2% of the PTWI of 100 ng / kg set by the Joint FAO / WHO Expert Committee on Food Additives.
Coffee may contain the following
→ mycotoxins:
→ ochratoxin A
incidence: 7/22*, conc. range: 0.2-4 µg / kg, country: Australia, *pure soluble
incidence: 2/2*, conc. range: 1.9-4.8 µg / kg, country: Czech Republic, *pure soluble
incidence: 2/2*, ∅ conc.: 1.6 µg / kg, country: Czech Republic, *adulterated soluble
incidence: 11/11*, conc. range: ≤ 3.2 µg / kg, ∅ conc.: 0.51 µg / kg, country: Denmark, *roasted
incidence: 20*, conc. range: 0-5.5 µg / kg, ∅ conc.: 1.1 µg / kg, country: Europe**, *instant, decaffeinated
incidence: 10*, conc. range: 0-1 µg / kg, ∅ conc.: 0.5 µg / kg, country: Europe**, *instant, mixed
incidence: 119*, conc. range: 0-27.2 µg / kg, ∅ conc.: 1.4 µg / kg, country: Europe**, *instant, regular
incidence: 39*, conc. range: 0-2.8 µg / kg, ∅ conc.: 0.7 µg / kg, country: Europe**, *roasted and ground, decaffeinated
incidence: 445*, conc. range: 0-8.2 µg / kg, ∅ conc.: 0.8 µg / kg, country: Europe**, *roasted and ground, regular
**collaborative study of different European countries
incidence: 2/4* **, conc. range: 10-90 µg / kg, ∅ conc.: 50 µg / kg, country: Germany, *moldy, **raw
incidence: 4/14*, conc.: ≤ 4.9 µg / kg, country: Germany, *roasted
incidence: 25/30*, conc. range: ≤ 4.9-7.54 µg / kg, ∅ conc.: 1.43 µg / kg, country: Germany, *roasted
incidence: 1/29*, conc.: 3 µg / kg, country: Germany, *roasted and raw
incidence: 5/9*, conc. range: 0.3-2.2 µg / kg, country: Germany, *pure soluble
incidence: 6/6*, conc. range: 0.5-1.6 µg / kg, country: Greece, *pure soluble
incidence: 14/14, conc. range: 0.5-6.5 µg / kg, country: Hungary, *pure soluble
incidence: 1/1, ∅ conc.: 1.2 µg / kg, country: Hungary, *adultered soluble
incidence: 2/7* **, conc. range: 3.2-4.4 µg / kg, ∅ conc.: 3.8 µg / kg, country: Indonesia, *commercial, **roasted (a total of 68 samples has been investigated

in Japan, 5 samples (2 from Indonesia, 3 from Yemen) contained OTA
incidence: 5/68, conc. range: 3.2-17 µg / kg, country: Japan
incidence: 21/22, conc. range: 0.2-3.5 µg / kg, country: Russia, *pure soluble
incidence: 12/12*, ∅ conc.: 6.93 µg / kg, country: Russia, *adulterated soluble
incidence: 6/6, conc. range: 0.3-3.6 µg / kg, country: Salvador, *pure soluble
incidence: 4/4, conc. range: 1.5-5.3 µg / kg, country: Slovakia, *pure soluble
incidence: 16/40*, conc. range: 1-7.8 µg / kg, country: Switzerland, *brew
incidence: 2/3, conc. range: 0.2-0.3 µg / kg, country: Switzerland, *pure soluble
incidence: 3/3, conc. range: 1.3-1.9 µg / kg, country: Thailand, *pure soluble
incidence: 64/80*, conc.: 0.1-8.0 µg / kg, country: UK, *soluble
incidence: 17/20*, conc. range: 0.2-2.1 µg / kg, country: UK, *roasted and ground, regular
incidence: 2/4, conc. range: 0.3-0.4 µg / kg, country: unknown
incidence: 9/13*, conc. range: 0.1-1.2 µg / kg, ∅ conc.: 0.41 µg / kg, country: USA, *import from South America
incidence: 3/6, conc. range: 1.5-2.1 µg / kg, country: USA, *pure soluble
incidence: 3/10* **, conc. range: 6.5-17 µg / kg, ∅ conc.: 10.1 µg / kg, country: Yemen, *commercial, **roasted (a total of 68 samples has been investigated in Japan, 5 samples (2 from Indonesia, 3 from Yemen) contained OTA

Coffee beans (green)
may contain the following → mycotoxins:
→ aflatoxin (no specification)
incidence: 2/201, conc. range: 3-12 µg / kg ∅ conc.: 7.5 µg / kg, country: USA
→ ochratoxin A
incidence: 1/1, conc.: 8 µg / kg, country: Austria
incidence: 3/7, conc. range: ca. 20-360 µg / kg, country: Brazil
incidence: 17/139, conc. range: ca. 20 µg / kg (13 samples), 35 µg / kg (2 sa), 50 µg / kg (2 sa), country: Colombia
incidence: 1/1*, conc.: 0.5 µg / kg, country: India, *commercial
incidence: 2/2*, conc. range: 0.5-1 µg / kg, ∅ conc.: 0.75 µg / kg, country: Indonesia, *commercial
incidence: 2/2, conc. range: ≤ 2.2 µg / kg, country: Ireland
incidence: 19/29*, conc. range: 0.2-15 µg / kg, country: Italy, *commercial
incidence: 3/68, conc. range: 20-80 µg / kg, ∅ conc.: 40 µg / kg, country: Italy
incidence: 4/5*, conc. range. < 20-400 µg / kg, country: Italy, *molded
incidence: 9/40*, conc. range: 0.5-23 µg / kg, country: Italy, *commercial
incidence: 1/12, conc.: ca. 20 µg / kg, country: Ivory Coast
incidence: 1/3*, conc.: 3.8 µg / kg, country: Ivory Coast, *commercial
incidence: 4/22, conc. range: 9.8-46 µg / kg, country: Japan
incidence: 1/1*, conc.: 1.8 µg / kg, country: Kenia, *commercial
incidence: 13/25*, conc. range: 1.2-56 µg / kg, country: Switzerland, *commercial
incidence: 7/7, ∅ conc.: 3.9 µg / kg, country: Thailand
incidence; 2/14, conc. range: ≤ 7 µg / kg, country: The Netherlands
incidence: 1/1*, conc.: 5.2 µg / kg, country: Togo, *commercial
incidence: 1/2, conc.: ca. 20 µg / kg, country: Uganda
incidence: 2/2*, conc. range: 1.5-23 µg / kg, country: Uganda, *commercial
incidence: 9/31*, conc. range: < 10-200 µg / kg, country: UK, *commercial
incidence: 2/201, conc. range: 24-96 µg / kg, ∅ conc.: 60 µg / kg, country: USA
incidence: 19/267*, conc. range: 20-360 µg / kg, country: USA, *imported, hand-cleaned coffee beans

incidence: 3/68*, conc. range: tr-80 µg / kg, country: USA, *imported, commercial
incidence: 9/19*, conc. range: 0.1-4.6 µg / kg, Ø conc.: 1.41 µg / kg, country: USA, *import form South America
incidence: 1/2, conc.: 1.3 µg / kg, country: Zaire
→ sterigmatocystin
incidence: 1*/502, conc.: 1200 µg / kg, country: Italy, *very moldy
incidende: 1*/2, conc.: 1143 µg / kg, country: South Africa, *condemned as unfit for human consumption
→ cocoa beans

Comte cheese → cheese, Comte

Confectionery may contain the following → mycotoxins:
→ citrinin
incidence: 1/1, conc.: < 100 µg / kg, country: UK
→ ochratoxin A
incidence: 1/1, conc.: traces, country: UK
→ marzipan, → nuts, → persipan

Congestion having an abnormal accumulation of blood.

Congressbele is an Indian peanut (→ peanuts) based spiced snack which consists of the kotyledons of the groundnuts. After light frying in small quantity of oil the kotyledons are spiced with → turmeric powder, → pepper and salted. A lower aflatoxin contamination, compared to → bondakaledkai, may result from a certain degree of cleaning from infested seeds.
Congressbele may contain the following → mycotoxins:
→ aflatoxin B_1
incidence: 9/41, conc. range: 6-1100 µg / kg, country: India
→ aflatoxin B_2
incidence: 5/41, conc. range: 4-700 µg / kg, country: India
→ groundnut toffee, → bondakaledkai

Convulsions Violent irregular movement of a limb or limbs, or of the body, caused by contraction of muscles.

Cookies may contain the following → mycotoxins:
→ deoxynivalenol
incidence: 35 products analysed, Ø conc.: 120 µg / kg, country: Canada
→ biscuits, → cereals

Copra (and copra meal)
contained the following → mycotoxins:
→ aflatoxin B_1
incidence: 10/16, conc. range: 10-100 µg / kg, Ø conc.: 39 µg / kg, country: Germany
→ aflatoxin B_2
incidence: 3/16, conc. range: 5-10 µg / kg, Ø conc.: 8.3 µg / kg, country: Germany
aflatoxin (no specification)
incidence: 129/182, conc. range: $\leq$ 513 µg / kg, Ø conc.: 39 µg / kg, country: Philippines
→ aflatoxins (no specification)
incidence: 7/105, conc. range: 30-120 µg / kg, Ø conc.: 42.8 µg / kg, country: India
incidence: 63/72*, conc. range: tr-200 µg / kg, Ø conc.: 46 µg / kg, country: USA,
incidence: 10/16*, conc. range: 10-100 µg / kg, Ø conc.: 37 µg / kg, country: USA, *imported
ochratoxin A
incidence: 1/384, conc.: 50 µg / kg, country: India
→ coconut, → nuts

Coriander may contain the following → mycotoxins:
→ aflatoxin B_1
incidence: 1/15, Ø conc.: 8 µg / kg, country: Egypt

incidence: 2/10, conc. range: tr-5.2 μg / kg, country: Germany
incidence: 6/9, conc. range: 25-230 μg / kg, country: India
incidence: 3/10, conc. range: 19-37 μg / kg, ∅ conc.: 25.7 μg / kg, country: India
incidence: 1/9, conc.: 45.5 μg / kg, country: Morocco
→ aflatoxin B_2
incidence: 6/9, conc. range: 20-72 μg / kg, country: India
incidence: 1/10, conc.: 5 μg / kg, country: India
→ aflatoxin G_1
incidence: 1/15, ∅ conc.: 2 μg / kg, country: Egypt
incidence: 6/9, conc. range: 13-40 μg / kg, country: India
incidence: 3/10, conc. range: 3-4 μg / kg, ∅ conc.: 3.7 μg / kg, country: India
→ aflatoxin G_2
incidence: 6/9, conc. range: 14-35 μg / kg, country: India
→ aflatoxins (no specification)
incidence: 4/10, conc. range: 10-75 μg / kg, country: India
incidence: 1/3*, conc.: 0.7 μg / kg, country: UK, *AFB_1, AFB_2, AFG_1, AFG_2
→ citrinin
incidence: 1/9, conc.: 34 μg / kg, country: India
→ deoxynivalenol
incidence: 1/4, conc.: 21 μg / kg, country: UK
→ ochratoxin A
incidence: 1/9, conc.: nc, country: India
incidence: 1/3, conc.: 4 μg / kg, country: UK
→ zearalenone
incidence: 1/9, conc.: nc, country: India
incidence: nc/4, conc. range: 3.6-6.7 μg / kg, country: UK
→ spices

Corn → Maize

Corn flakes may contain the following → mycotoxins:
→ aflatoxin B_1
incidence: 1/2, conc.: < 5 μg / kg, country: UK
→ fumonisin B_1
incidence: 1/2, conc.: 10 μg / kg, country: Italy
incidence: 2/12, conc. range: 50-100 μg / kg, ∅ conc.: 60 μg / kg, country: Spain
incidence: 1/12, con.: 55 μg / kg, country: Switzerland
incidence: 4/17, conc. range: 140-1281 μg / kg, ∅ conc.: 497 μg / kg, country: Thailand
incidence: 1/5, conc.: 1430 μg / kg, country: The Netherlands
→ fumonisin B_2
incidence: 3/17, conc. range: 120-466 μg / kg, ∅ conc.: 166 μg / kg, country: Thailand
→ fumonisins
incidence: 8/8, conc. range: < 20-760 μg / kg, country: Germany and unknown origin
incidence: 4/6, conc. range: ≤ 400 μg / kg, country: USA
→ ochratoxin A
incidence: 2/13, conc. range: 0.1-0.19 μg / kg, country: Germany
incidence: 1/34, conc.: 0.4 μg / kg, country: Germany
→ sterigmatocystin
incidence: 1/2, conc.: nc, country: UK
→ zearalenone
incidence: 1/1, conc. range: 13-20 μg / kg, country: Canada
→ cereal flakes, → maize flakes, → oat flakes

Cortex Outer layer of an organ.

Cow After oral dosing, the residues of → aflatoxin B_1 and → aflatoxin M_1 can be found in the liver and kidneys for up to 7 days. After withdrawal from the contami-

nated diet, the cattle tissue was completely free of → aflatoxins within 18 days. Probably, a longer withdrawal period is necessary for the cow´s meat than for → pork.
In vitro and *in vivo* studies show a rapid detoxification of → ochratoxin A in ruminants by the action of rumen inherent proteolytic enzymes that cleave phenylalanine from the isocoumarin of the OTA molecule.
→ meat

Cow kidney Feeding experiments with two milking cows (317-1125 µg → ochratoxin A / kg feed for 11 weeks) resulted in the contamination of the kidneys of one of the cows (5 µg OTA / kg). Neither ochratoxin α nor OTA was found in any tissue or in → milk.

Cowpeas may contain the following → mycotoxins:
aflatoxin (no specification) (→ aflatoxins)
incidence: 10/16, conc. range: ≤ 86 µg / kg, ∅ conc.: 16 µg / kg, country: Philippines
→ ochratoxin A
incidence: 5/31, ∅ conc.: 34 µg / kg, country: Senegal
→ beans, → cabbage, → lentils, → peas, → pigeon peas, → soybeans, → vegetables

Crackers may contain the following → mycotoxins:
→ deoxynivalenol
incidence: 20 products analysed, ∅ conc.: 270 µg / kg, country: Canada
→ cereals

Cranberries may contain the following → mycotoxins:
→ patulin
incidence: nc, conc. range: ≤ 265 µg / kg, country: Sweden
→ fruits

Cream (full)
may contain the following → mycotoxins:
→ aflatoxin M_1
incidence: 19/28, conc. range: tr- > 2 µg / kg, country: Germany
→ milk

Cream cheese → cheese, cream

Croissant butter may contain the following → mycotoxins:
→ deoxynivalenol
incidence: 8/8, conc. range: 326-648 µg / kg, ∅ conc.: 453 µg / kg, country: Argentina
→ cereals, → milk

Croissant fat may contain the following → mycotoxins:
→ deoxynivalenol
incidence: 4/5, conc. range: 336-563 µg / kg, ∅ conc.: 377 µg / kg, country: Argentina
→ cereals, → milk

Cumin may contain the following → mycotoxins:
→ aflatoxin B_1
incidence: 2/20*, conc. range: 0.29-0.96 µg / kg, ∅ conc.: 0.625 µg / kg, country: Egypt, *different → spices
incidence: 2/8, conc. range: 24-104 µg / kg, ∅ conc.: 64 µg / kg, country: India
→ aflatoxin B_2
incidence: 2/8, conc. range: 12-78 µg / kg, ∅ conc.: 45 µg / kg, country: India
→ aflatoxin G_1
incidence: 2/8, conc. range: 8-45 µg / kg, ∅ conc.: 26.5 µg / kg, country: India
→ aflatoxin G_2
incidence: 1/8, conc.: 30 µg / kg, country: India
→ citrinin
incidence: 1/8, conc.: 22 µg / kg, country: India
→ spices

Curcuma may contain the following → mycotoxins:
→ aflatoxin B_1
incidence: 6/7, conc. range: < 2.5-3.8 µg / kg, country: Canada
→ spices

Curry may contain the following → mycotoxins:
→ aflatoxins (AFB_1, AFB_2, AFG_1, AFG_2)
incidence: 10/29*, conc. range: 1-3.9 µg / kg (8 samples), 4-10 µg / kg (2 sa), country: UK, *imported
incidence: nc/3**, conc. range: 0.8-61.2 µg / kg, country: UK
incidence: 1/3***, conc.: 0.4 µg / kg, country: UK
→ diacetoxyscirpenol
incidence: 1/3****, conc.: 25 µg / kg, country: UK
→ fumonisins (FB_1, FB_2)
incidence: nc/3**, conc. range: 15-16 µg / kg, country: UK
incidence: 1/3****, conc.: 230 µg / kg, country: UK
→ fusarenon X
incidence: 1/3****, conc.: 7 µg / kg, country: UK
→ HT-2 toxin
incidence: 1/3****, conc.: 17 µg / kg, country: UK
→ neosolaniol
incidence: 1/3****, conc.: 9 µg / kg, country: UK
→ nivalenol
incidence: nc/3**, conc. range: 15-50 µg / kg, country: UK
incidence: nc/3***, conc. range: 9-67 µg / kg, country: UK
incidence: 1/3***, conc.: 14 µg / kg, country: UK
→ ochratoxin A
incidence: 3/3, conc. range: 5-33 µg / kg, country: Austria
incidence: 2/3**, conc. range: 2.3-21.3 µg / kg, ∅ conc.: 11.8 µg / kg, country: UK
incidence: nc/3***, conc. range: 1.8-9.4 µg / kg, country: UK
incidence: nc/3****, conc. range: 1.2-5.4 µg / kg, country: UK
incidence: 4/4, conc. range: ≤ 4.9-5.4 µg / kg, country: UK
→ T-2 toxin
incidence: 1/3****, conc.: 13 µg / kg, country: UK
→ zearalenone
incidence: nc/3**, conc. range: 1.2-10.8 µg / kg, country: UK
incidence: 1/3****, conc.: 5.2 µg / kg, country: UK
**curry powder hot, ***curry powder mild, ****mixes
→ spices

Curry paste may contain the following → mycotoxins:
→ aflatoxins (AFB_1, AFB_2, AFG_1, AFG_2)
incidence: 1/4, conc.: 1.2 µg / kg, country: UK
→ fumonisins (FB_1, FB_2)
incidence: 1/4, conc.: 56 µg / kg, country: UK
→ nivalenol
incidence: nc/4, conc. range: 5-16 µg / kg, country: UK
→ ochratoxin A
incidence: 4/4, conc. range: 0.6-15.5 µg / kg, country: UK
→ zearalenone
incidence: 4/4, conc. range: 3.1-4.2 µg / kg, country: UK
→ spices

Cyclopiazonic acid (Abbr.: CPA) is an indole-tetramic acid (6a,7,11a,11b-tetrahydro-10-(1-hydroxyethylidene)-7,7-dimethyl-6H-pyrrolol[1′,2′:2,3]isoindolo[4,5,6-cd]indole-9,11-(2H,1OH)-dione) that was first isolated from → Penicillium aurantiogriseum Dierckx in 1968 (see Figure Cyclopiazonic acid).

Chemical Data
molecular formula: $C_{20}H_{20}N_2O_3$, molecular weight: 336

Fungal Sources
e.g. *A. flavus*, → Aspergillus oryzae (Ahlburg) Cohn, → Aspergillus versicolor (Vuill.) Tiraboshi, → Aspergillus spp., *P. aurantiogriseum* (also produces cyclopiazonic acid imine and bissecodehydrocyclopiazonic acid), → Penicillium camembertii Thom (consistent producer), → Penicillium commune Thom, → Penicillium roquefortii Thom, → Penicillium spp. *P. aurantiogriseum* (*P. cyclopium*) was previously known to be the most important CPA producer of the genus *Penicillium*. Because all CPA-producing strains of *P. aurantiogriseum* have now been assigned to *P. commune* this *Penicillium* species is currently regarded as being the most prominent CPA producer on natural substrates. Aflatoxin synthesis of → Aspergillus flavus Link is often accompanied by similar production of CPA. The importance of *A. flavus* as CPA producer should therefore not be underestimated (→ turkey "X" disease).

Natural Occurrence
→ cheese, → cheese, Camembert, → cheese, Camembert & Brie, → cheese, Gouda & Cheddar, → kodo millet, → maize, → peanuts, → sunflower seeds. Co-contamination of peanuts and maize with aflatoxin has been reported.
CPA has been detected in the → milk of lactating ewes within one day after experimental application. Presence of CPA in the milk was obvious even several days after withdrawal of the mycotoxin. With the exception of manufacturing unsweetened condensed milk (reduction ca. 40%) storing (4 °C) and processing caused only a minor decrease in CPA levels. In addition, CPA proved to be quite stable in → fermented products.

Toxicity
Necrotic (liver, gastrointestinal tissue, kidneys, skeletal muscles), carcinogenic, neurotoxic, → mutagenic (Ames test).
In humans clinical symptoms such as tremors (→ tremorgenic mycotoxins), sleepiness and giddiness have been observed.
LD_{50} (po): 36 and 63 mg / kg bw male and female rats, respectively.

Detection
capillary electrophoresis, colorimetic and spectrophotometry technique, ELISA, GC, HPLC (normal- and reversed-phase, ligand exchange), TLC

Possible Mycotoxicosis
Besides the → aflatoxins, this mycotoxin is involved in the turkey "X" disease. An additive effect with aflatoxin has been demonstrated.
Implication of this mycotoxin in → Kodua poisoning, a human malady in India, caused by the ingestion of kodo → millet seeds invaded by *Aspergillus* has been suggested.

Further Comments
Production: CPA formation occurred in the range of 4 °C (refrigerator temperature), 13 °C (ripening temperature for cheeses), 25 °C (optimal).
The minimum → a_w for CPA production on maize was a_w 0.90 at 30 °C (a_w 0.85 yeast extrat agar*). Largest amounts were produced at a_w 0.98 at 20 °C (optimum a_w 0.996 yeast extract agar*) (→ Penicillium commune Thom*, → Aspergillus flavus Link).
Stability: Simulation of the heat-treatments used by the dairy industry caused no considerable degradation of CPA in the milk. → Yogurt processing of artificially contaminated milk resulted in a significant reduction (> 70%) of CPA concentration after the first day of storage.
Significant decomposition of CPA occured in acidic buffers. In basic environments it was less pronounced while a

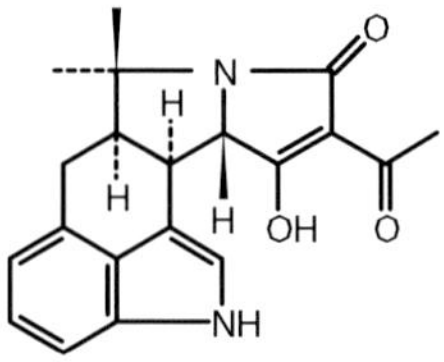

Cyclopiazonic acid

neutral pH caused minor rates of decomposition.
Assessment of possible health effects is difficult at this stage because analytical methods for the detection of CPA in foods are still being developed.

D

Dairy products Dairy products may be contaminated by → mycotoxins via two different routes. Indirect contamination occurs when contaminated feedstuffs are consumed by dairy → cattle. The cause of the direct contamination is the colonization of dairy products by mycotoxigenic molds, which may result in the contamination of the substrate. With respect to indirect contamination → aflatoxin M_1 the → milk metabolite of → aflatoxin B_1 is most important.
The insolubility of AFM_1 in the milk fat and absorption in the curd resulted in a specific pattern of distribution depending on the end-product, e.g. → butter, → cheese, → cream or whey (→ whey powder). About 10% of the original AFM_1 concentration in the milk is found in cream the remaining in the skimmed milk. Approximately 10% of the AFM_1 in the cream goes into the butter while up to 90% is retained in the buttermilk (→ milk-, butter). AFM_1 distribution in the single fractions is related to their content of non-fat milk solids, probably due to casein binding. The acidification during cottage cheese (→ cheese, cottage) production caused losses of AFM_1 concentration in the range of 20%, 30% is accumulated in the curd, 50% in the whey (see Figure Losses of AFM_1 during processing of milk).
Although → carry over of e.g. → ochratoxin A, → sterigmatocystin, → deoxynivalenol, → T-2 toxin and → zearalenone in milk has been reported, the rate of transmission and / or toxicity of the metabolites is low. Therefore, these mycotoxins do not represent a reasonable cause of concern.
Direct mycotoxin contamination may be due to starter cultures (e.g. → Penicillium roquefortii Thom and → Penicillium camembertii Thom) during cheese fermentation or accidental growth of molds on dairy products. Direct aflatoxin contamination is unlikely because → Aspergillus flavus Link and → Aspergillus parasiticus Speare do not belong to the frequent colonizers of these substrates. Cheeses, with their lower a_w-values promote fungal growth (and potential mycotoxin contamination), and therefore belong to the more susceptible dairy products.
Dairy products may contain the following mycotoxins:
→ aflatoxin B_1
incidence: 1/22, conc.: 6.4 µg / kg, country: Germany
incidence: 2/23*, conc. range: 10-20 µg / kg, ∅ conc.: 15 µg / kg, country: India, *indigenous
→ cheese

Decontamination Decontamination should be cheap and simple, ideally using the existing technology. The procedure should be effective against a variety of → mycotoxins and not lead to the toxic degradation metabolites. No reduction in the nutritional and palatable properties of → grains or grain products should occur. Detoxification processes may be divided into three categories: physical, chemical, and biological.
Physical methods include cleaning and washing, dehulling as well as → milling. Their effectiveness greatly depends on the relative distribution of mycotoxins throughout the grains and the degree of contamination. Because of additional cost for drying, washing is only suitable as a cleaning step prior to wet milling.
Separation of mycotoxin-contaminated grains due to differing physical properties is possible by fractionation (specific gravity table), density segregation (certain liquids) or fluorescence under ultraviolet light. The heat stability of most food-relevant mycotoxins reduces the

effectivity of cooking, baking, roasting and microwave heat.
Most of the **chemicals** used for decontamination have only a limited effect on the mycotoxins. Their effectivity is greatly influenced by the moisture content of the substrate and the processing temperature. Different chemicals like calcium hydroxide monomethylamine, hydrogen peroxide, sodium hypochlorite or sodium bisulfite have been used. Ascorbic acid essentially removed → patulin from contaminated → apple juice. For commercial decontamination of aflatoxin-containing cotton seeds, → maize and peanut cakes / meal (→ peanuts) ammonia is used in the US, France, Nigeria, etc. At present, ammonia decontamination is the most effective and economically feasible method.
Biological methods include e.g. the addition of mold inhibitors or potential mycotoxin-binding agents to the feed. In addition, various microorganisms have been tested for their detoxification potential. → Flavobacterium aurantiacum essentially removed → aflatoxin B_1 from different kinds of food while *Saccharomyces cerevisiae* detoxified → patulin in → apple juice during → cider production.

Deer → Roe deer

Deoxynivalenol (Syn.: DON, Rd-toxin, vomitoxin) belongs to the group of naturally-occurring → trichothecenes (3α,7α,15-trihydroxy-12,13-epoxytrichothec-9-en-8-one) and is produced by different species of the genus → Fusarium, with → Fusarium graminearum Schwabe being the most important (see Figure Deoxynivalenol). The first isolation of Rd-toxin (previous name) was reported in 1972 for Japanese *Fusarium*-damaged → barley, which showed a simultaneous contamination with → nivalenol. Subsequent isolations from *F. graminearum*-infected → maize, which caused vomiting in swine in the United States, led to the trivial name vomitoxin (1973).

CHEMICAL DATA
Empirical formula: $C_{15}H_{20}O_6$, molecular weight: 296

FUNGAL SOURCES
Fusarium acuminatum (?),→ Fusarium culmorum (W. G. Smith) Sacc. , → Fusarium graminearum Schwabe, → Fusarium nivale (Fr.) Ces., → Fusarium sporotrichioides Sherb.

NATURAL OCCURRENCE
→ baby cereals, → baby food, → barley, → barley flour, → barley grits, → barley malt, → beans, → beer, → beer, barley, → beer, wheat, → bran, → bread, → breakfast cereals, → buns, → cereal products, → chapatti, → chilli powder, → cookies, → coriander, → crackers, → croissant butter, → croissant fat, → figazzas, → flour, → foods, → garlic, → ginger, → grains, → job's-tears, → libritos, maize, → maize flour, → maize grits, → maize meal, → maize, brewers, → maize, brewers flaked, → maize, brewers grits, → maize, canned, → maize, fiber cereal, → maize, hominy, → maize, infant cereal, → maize, infant cream corn, → maize, popped, → maize, preharvest, → maize, puffed, → maize, quality-protein, → maize, shelled, → maize, sweet, → masa, → millet, → millet meal, → muesli ingredients, → noodles, → oats, → pop corn, potatoes, → rice, → rye, → rye bran, → rye flour, → snack food, → sorghum, → soybean, → spaghetti, → triticale, → wheat, → wheat grits, → wheat products
Cereals like wheat, barley and maize usually contain the highest DON-concentrations. DON is the most important mycotoxin in grains in several countries such as Austria, Canada, Italy, South Africa, Sweden, UK, USA. Because of its stability, DON survives processing

(→ milling) resulting in the contamination of cereal products (e.g. corn steep liqour, corn starch). Fractions which are used as animal feed (e.g. gluten meal and wet fibre) may show high DON-levels. Simultaneous occurrence of DON, → zearalenone and → aflatoxin B_1 in scabby wheat is possible.
Rate of transmission (→ Carry over) into cow → milk is extremely low (<4 μg / l). Because of rapid elimination low to medium DON-levels in the diet do not result in the accumulation of residues in swine. Transmission / residues of DON in(to) meat, milk or eggs is negligible.

Toxicity

Acute toxicity is characterized by intestinal disorders and emesis, especially in swine. However, the presence of DON limits feed consumption at concentrations > 1 μg / kg, so that acute toxicity seldom occurs. → Poultry (egg quality, weight reduction) and → cattle (reductions in feed intake, conception rate and milk production), possibly due to extensive degradation to secondary metabolites in the rumen, are more tolerant. Dermatological lesions, gastrointestinal disorders, hematological changes (→ hemorrhage) and → immunosuppressive, → teratogenic as well as nephrotoxic (?) effects in animals have also been reported.
Humans seem to be quite sensitive to DON.
LD_{50} (po): 46 mg / kg bw mice
A potently-synergistic toxic effect to laboratory animals occurred in combination with culmorin, dihydroxycalonectrin and sambucinol as metabolites of *F. graminearum*.
The co-contamination of grains with other mycotoxins may cause unanticipated interactions to the detriment of animals and humans. It seems possible that the carcinogenicity of AFB_1 is enhanced by the immunosuppressive acting → trichothecenes.

Detection

ELISA, GC, HPLC, MS, TLC

Possible Mycotoxicosis

Outbreaks of acute gastrointestinal illness in humans (China, India).

Further Comments

DON is probably the most common mycotoxin contaminating food and feed. DON is a contaminant virtually wherever cereals are grown. Co-occurrence with → zearalenone is common in grain worldwide. The contamination of cereals with DON may be eliminated by plant breeding. DON is often co-occurring with → nivalenol, → diacetoxyscirpenol, and → T-2 toxin.
Distribution in grains: DON is primarily located in the grain at the sites of fungal growth. Only little translocation occurred to other sites in the kernel. Low levels of fungal and mycotoxin contamination (50-1000 μg / kg) typically result in DON-accumulation near the exterior surface of the kernel. Here, most of the fungal mycelium is to be expected. The → flour of such wheat will contain relatively low mycotoxin levels with respect to the whole kernel. Higher concentrations (> 4000 μg DON / kg) may cause a more even distribution throughout the kernel due to a deeper penetration of the fungus. Mycotoxin levels of flours prepared from highly contaminated grains are comparable to those in the → bran and other outer portions of the kernels. However, in some cases this pattern of distribution is not related to high DON-concentrations in individual kernels. It seems possible that a correlation exists between the distribution of DON and the degree of fungal (*Fusarium*) contamination of the kernels.
Stability: Processing (e.g. cleaning, milling, → baking) of contaminated → cereals usually does not result in significant

losses of DON in the finished product. During milling of wheat, DON was detected throughout all of the milling fractions: bran shorts, reduction flour, break flour (in decreasing order). Several studies confirmed an accumulation of DON in the bran fraction whereas the lowest concentrations were found in the flour (reduction $\approx$ 50%).
About 50% of DON survived the baking process. An even dramatic increase in DON-concentration (180%) has been observed during doughnut preparation. This might be due to enzymatic conversions of DON-precursors already present in the used soft wheat.
DON is the mycotoxin which best survives the brewing process. An increase in amount during mashing may occur.

Deoxynivalenol monoacetate → 3-acetyl-deoxynivalenol

Deoxynivalenol toxicosis Between July through September 1987, human food poisonings occurred in the Kashmir Valley in northwestern India. Approximately 50,000 people were affected by this non-communicable disease irrespective of age or sex. The consumption of → bread made from certain consignments of → wheat led to different symptoms like abdominal pain, a feeling of fullness in the abdomen, throat irritation, diarrhoea, emesis, blood in the stool and allergic reactions 15 min to 1 h after ingestion. Unseasonal rains during the harvest season caused a considerable mold contamination of the wheat. Local millers bought the moldy wheat for a much lower price and mixed it with good wheat (50 : 50). The corresponding flour was sold to local bakers, who in turn sold it to consumers as flour or bread. The consistency of bread made from the moldy wheat resembled "chewing gum".
Mycological examination of the grains and the flours (24 brands) always revealed a → Fusarium contamination. → Aspergillus spp. and → Penicillium sp. occurred to a minor extent. In addition, several mycotoxins could be isolated: → deoxynivalenol (conc. 346-8380 µg / kg, 11 samples), → nivalenol (conc. 30-100 µg / kg, 2 sa), acetyldeoxynivalenol (conc. 600-2400 µg / kg 4 sa), → T-2 toxin (conc. 550-4000 µg / kg, 5 sa). While identification of different → trichothecenes failed, the detection of pesticide residues, → aflatoxins and → ergot alkaloids was negative.

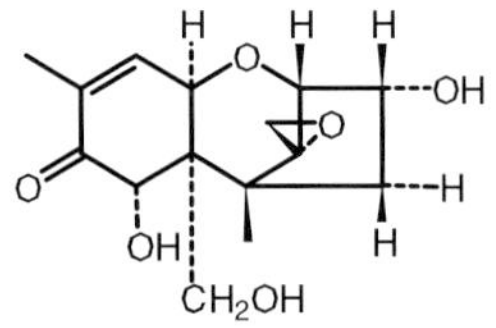

Deoxynivalenol

Diacetoxyscirpenol (Syn.: anguidine, DAS) belongs to the group of naturally occurring → trichothecenes (3α-hydroxy-4,15-diacetoxy-12,13-epoxytrichothec-9-ene), which is produced by different → Fusarium species, with *Fusarium sporotrichioides* Sherb. being the most important (see Figure Diacetoxyscirpenol). The first isolation was reported for → Fusarium equiseti (Corda) Sacc. sensu Gordon in 1961. Structure elucidation followed in 1965/1966.

CHEMICAL DATA
Empirical formula: $C_{19}H_{26}O_7$, molecular weight: 366

FUNGAL SOURCES
Fusarium acuminatum, *F. avenaceum* (?), *Fusarium equiseti* (Corda) Sacc. sensu Gordon, → Fusarium graminearum Schwabe, → Fusarium moniliforme Sheldon, → Fusarium oxysporum Schlecht. emend. Snyd. & Hansen, → Fusarium poae (Peck) Wollenw., → Fusarium sambucinum

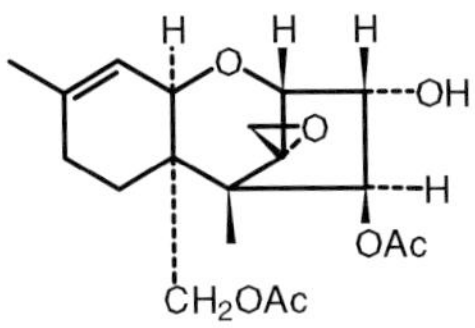

Diacetoxyscirpenol

Fuckel (good producer), *F. semitectum*, → Fusarium sporotrichioides Sherb.

NATURAL OCCURRENCE
→ barley, → beans, → beer, → chilli powder, → curry, → maize, → oats, → wheat

TOXICITY
cancerogenic, dermatoxic, hemorrhagic (→ hemorrhage) (enteritis), phytotoxic
LD_{50} (po): 7.3 mg / kg bw rats (21-day-old)

DETECTION
GC, MS, spectroscopy, TLC

POSSIBLE MYCOTOXICOSIS
Besides T-2 toxin DAS should also be involved in → alimentary toxic aleukia.

FURTHER COMMENTS
DAS often occurs naturally together with → deoxynivalenol.
The rapid and extensive metabolization of DAS in pigs has been reported. Although accumulation of this mycotoxin is not be expected in naturally exposed animals the toxicity and tissue distribution of unknown metabolites needs further clarification.

Dihydroalterperylenol (Syn.: altertoxin I, → altertoxin I-III)

DON → Deoxynivalenol

Dothideales → Ascomycota

Duck may contain the following → mycotoxins:
→ aflatoxin B_1
incidence: 33/41*, conc. range: 0.203-2.484 μg / kg, ∅ conc.: 0.84 μg / kg, country: Czechoslovakia, *wild duck, liver
incidence: 31/41*, conc. range: 0.3-3.605 μg / kg, ∅ conc.: 0.594 μg / kg, country: Czechoslovakia, *wild duck, kidney
→ ochratoxin A
incidence: 11/19, conc. range: ≤ 0.09 μg / kg, ∅ conc.: 0.02 μg / kg, country: Denmark
incidence: 4/7*, conc. range: ≤ 0.16 μg / kg, ∅ conc.: 0.06 μg / kg, country: Denmark, *liver
→ meat

Durum wheat → wheat

Dyspnea shortness of breath, difficult or labored breathing

E

Edema is characterized by the accumulation of an exessive amount of tissue fluid in intercellular spaces.

Egg products may contain the following → mycotoxins:
aflatoxin (no specification) (→ aflatoxins)
incidence: 1/112, conc.: 0.06 µg / kg, country: USA

ELEM → Equine leukoencephalomalacia, → fumonisins

Emericella → Trichocomaceae; anamorph → Aspergillus

Emu aran is a Nigerian indigenous beverage (palm juice) made from the sap of *Raphia vinifera* and *R. raphia.*
Emu aran may contain the following → mycotoxins:
aflatoxin B (→ aflatoxins)
incidence: 2/2, conc. range: 83-86 µg / kg, Ø conc.: 84.5 µg / kg, country: Nigeria

Encephalopathy and fatty degeneration of the viscera (Syn.: → Reye's syndrome)

Enchilada → Tortilla

Endemic Balkan nephropathy → Balkan Endemic Nephropathy

Endemic familiar arthritis of malnad This non-congenital disease is characterized by abnormal bone growth and occurred in the Malnad district in southern India from 1965-1975. In this area heavy rainfalls are common.
Members of 140 families belonging to the most impoverished castes were affected. Their diet mainly comprised → rice and various fauna like → fish and crabs. The victims were of both sexes and all ages although children younger than five years old did not show any symptoms. The bilateral, symmetrical lesions (osteoarthritis) primarily occur in the hip joint, pelvis, as well as vertebrae and may progress up to the knees. Other joints are rarely affected. Severe impairment of the patients mobility may result from the disease.
Although the consumed foodstuffs have not been investigated for mold and mycotoxin contamination, similarities (epidemiological, pathological) with other bone growth disorder diseases (→ Kashin-Beck disease = osteoarthritis, → Mseleni joint disease = lesions of the hip joint) in which → mycotoxins have been investigated as possible etiological agents are obvious. In addition, all three diseases occur in geographically isolated areas.

Endemic panmyelotoxicosis → Alimentary toxic aleukie

Enteritis is characterized by an inflammation of the intestines.

Epicladosporic acid → Cladosporium

Equine leukoencephalomalacia (Syn.: blindstaggers, foraging disease, corn stalk disease, leucoencephalitis, → moldy corn poisoning) (Abbr.: ELEM) is a fatal, disease which affects the co-ordination of horses and was first described in the late 1800s. This disease was associated with → Fusarium as early as 1904. Sporadic, seasonal, epidemic-like outbreaks have been reported in e.g. Argentina, Brazil, China, Egypt, South Africa and the United States. At present, two different forms have been reported: hepatotoxic and neurotoxic ELEM. The latter is more common.
The feeding of corn and other feeds highly infected with → Fusarium moniliforme Sheldon and contaminated with

→ fumonisins results in extensive damage to brain tissue. Lesions occur in form of none or more focal areas of liquefactive necrosis (= encephalomalacia) in the white matter (= leukoencephalomalacia) of the brain. One or both hemispheres may be affected. The encephalomalacic areas consist of large, irregular empty spaces. Random liquefactive (or malacic) lesions are characteristic for the subcortical white matter of the brain and the blood vessels show perivascular hemorrhages (→ hemorrhage) and → edema or a cuffing by infiltrating leukocytes.
→ esophageal cancer, → porcine pulmonary edema

Equsi meal is a Nigerian type of foodstuff from the plant *Cocumeropsis edulis.* Equsi meal may contain the following
→ mycotoxins
aflatoxin B (→ aflatoxins)
incidence: 1/1, conc.: 186 μg / kg, country: Nigeria

Ergot alkaloids In the view of toxicology and medicine, the alkaloids are the most important substances isolated from ergot. They particularly act on the nervous system. Based on their chemical structure the alkaloids are divided into lysergic acid, isolysergic acid and the clavine alkaloids (see Figure Ergot alkaloids). Lysergic acid derivatives are of the acid amide type and subdivided into the simple amides (e.g. ergometrine and ergine) and the peptide type comprising the ergotamine (e.g. ergotamine, ergosine), the ergotoxine (e.g. ergocristine, ergocornine, α-ergocryptine) and the ergoxine group. In the case of the clavine alkaloids, the carboxyl group, which is characteristic for the lysergic acid derivatives is reduced to a hydroxymethyl or a methyl group.
Ergot alkaloids are found in the sclerotia (→ ergots) of → Claviceps *purpurea.* Each sclerotium contains a total of over 100 compounds; ergocristine and ergotamine (lysergic acid dervatives) are generally the major components but alkaloid variation in individual sclerotia and throughout a contaminated field is high. The concentration and composition of alkaloids in ergot is influenced by different factors like strain and stage of maturity of the fungus, type of the host plant, climatic and geographic conditions.
Ergots of pearl → millet mainly contain alkaloids of the clavine type (*Claviceps fusiformis*), whereas ergot alkaloids of → rye and → wheat belong mainly to the ergotamine group (*C. purpurea*).

FUNGAL SOURCES
Claviceps spp., → Aspergillus spp. (e.g. → Aspergillus clavatus Desm., → Aspergillus fumigatus Fres.), → Emericella spp., → Penicillium spp. (e.g. *P. chermesinum, P. concavo-rugulosum*), → Rhizopus *nigricans*, and higher plants, e.g. *Ipomoea* spp.

NATURAL OCCURRENCE
→ baby cereals, → cereals, → rye flour, → triticale flour, → wheat
Hostplants like wheat, → barley, rye, → oats, → millet and Indian corn are found in the family of Graminae comprising the most important plants for human nutrition.
Wheat and rye flours usually contain only low alkaloid levels (< 100 μg / kg). Because of this situation there is almost no reason for concern.
No ergot alkaloids could be detected in → meat and → milk of livestock and → poultry after ingestion of contaminated feed which caused typical ergotism. Transmission of ergotism to breast-fed infants is not possible.

TOXICITY
Some ergot alkaloids are destroyed by ultraviolet light and there is much evidence to show that ergot sclerotia were more toxic when fresh than after storage.

Ingestion of higher alkaloid levels will result in neurological and / or gangrenous disorders. The nervous disorders include → ataxia, tremors, staggers, and → convulsions. The gangrenous form is characterized by vascoconstrictant effects (necrosis, sloughing of the extremities). Lower chronic levels are responsible for cardiac disorders.
Acute poisoning with gangrene occurred after the ingestion of between 5 and 72 mg ergotamine and 9 mg ergometrine. However, it was estimated that humans tolerate ca. 26 µg clavine alkaloids / kg bw without any toxic effects.

Detection
ELISA, densiometry, LC, spectrophotofluorometry, TLC

Mycotoxicosis
→ ergotism

Further Comments
Stability / Reduction: The lysergic acid derivatives are unstable to heat so significant losses occur during → bread processing. Baking caused a reduction in alkaloid concentration of up to 100% in whole wheat bread and up to 85% rye bread. During the making of triticale pancakes the losses amounted to 74%.
A reduction of ca. 90% in total alkaloid content was observed after treatment of wheat ergot sclerotia with chlorine.
During the normal cleaning and → milling process for grains, ergots are largely removed with the dockage. An accumulation of 70-80% of the ergot in the bran or shorts fractions was observed during milling. Therefore, these processing steps will usually result in a low alkaloid concentration in flour.

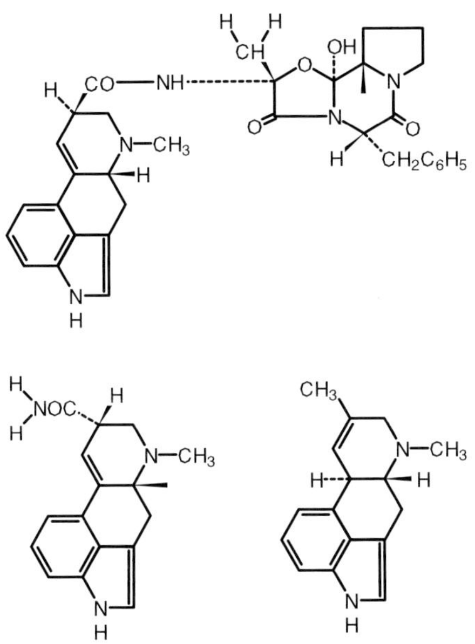

Ergot alkaloids. Ergotamine (lyserg acid), erginine (isolysergic acid), agroclavine (clavine alkaloid)

Ergotism Ergotism ("holy fire"), caused by → ergot alkaloids as derivatives of lysergic acid found in the sclerotia of → Claviceps, is probably the first recognized and best known → mycotoxicosis with respect to recorded effects on man. It is evident from history that ergotism has plagued humans and animals for centuries. In 430 B.C. an epidemic occurred among the Spartans that may have been due to ergot. In western and central Europe the use of contaminated → rye for → bread making led to large-scale epidemics in the Middle Ages. The first clear report of ergotism dates from 1582 but since 857 outbreaks of a disease resembling ergotism have been known in Central Europe. During the Middle Ages the disease was also called Saint Anthony´s fire because pilgrims suffering from it reported how they had been miraculously cured after paying homage at St. Anthony´s shrine in Dauphiné (France). The recovery of the patients was probably due to a change in diet made at the shrine.
Compared to the Middle Ages, human ergotism is now extremely rare, which is due to the change from rye to → wheat consumption and improvements in pre-

venting contaminated grain products from entering the food chain. Serious outbreaks sporadically occur in countries like India in 1975 (red millet / *Claviceps fusiformis*). In one outbreak in Ethiopia (1978), 93 people were gangrenous and 47 died after the consumption of wild → oats weeds contaminated with sclerotia of *C. purpurea*.
There are two types of ergotism, convulsive (neurological) and gangrenous (necrotic) ergotism. The latter form is due to the ingestion of sclerotia of *C. purpurea* and began with lassitude, sometimes accompanied by a prickling or an icy cold sensation in the limbs. Severe muscular pains, especially in the calf, followed. Although appetite and pulse remained constant at the beginning of the disease, the intellect was dulled. Swelling and inflammation of the limbs ensued. Similarly intensive burning pains with sensations of intense heat alternated with those of icy coldness. The pains sometimes ceased suddenly, leaving numbness. The skin was coverd with red to violet vesicles while the unaffected parts (face, white of the eyes) turned yellow indicating → jaundice. As gangrene set in, the toes and fingers became necrotic (black). In servere cases, the loss of fingers or toes, or even of all four limbs, occurred.
The impairment of the nervous system is characteristic for convulsive ergotism (→ convulsions) which is caused by *C. paspali*. The following symptoms are typical: sustained spasms, muscle cramps and twitching, numbness of the hands and feet, a tingling sensation under the skin, constriction of the blood vessels, followed by mortification of the limbs. Hallucination also occurs. Even in nonfatal cases full mental recovery was seldom. The mortality rate of ergotism ranges between 11 and 60%. Death may occur within several hours after ingestion of ergots but recovery is possible although not always completely. A higher susceptibility of previous victims of ergotism to recurrences has been reported. Ergotism occurred in Europe (particularly France and Germany), USA, Ethiopia as well as India and besides humans, → cattle, pigs, horses, sheep and → poultry are affected.

Ergots Ergot bodies are the → sclerotia of → Claviceps spp. which contain many toxic → ergot alkaloids (see Figure Ergots). About 50 species are known to infect many different grasses. The most widespread and common species is *C. purpurea* responsible for many cases of → ergotism in humans and animals.
The tightly-packed masses of fungal mycelium develop instead of kernels in grasses and → cereals (mainly → rye but also e.g. → wheat). The size and shape of the ergots may be roughly that of the kernels of the host plant but larger forms do exist. In general, not more than seven to eight ergots are found on a single spike of rye. Purple-black in colour they contain various pharmacologically active compounds, especially the → ergot alkaloids (conc. 0.1-0.8%). Low winter soil temperatures and wet springs stimulate the germination of the sclerotia. Infections of the host plants are enhanced by warm summers preceded by cold wet springs.

Ergots. Ergots in rye

Although large sclerotia are easily removed during grain cleaning, small and broken ones may pass through this processing step.
A maximum level of 0.05% and 0.3% ergot by weight has been suggested as an acceptable level for use in the production of → flour in Canada and other countries.

Erythema is characterized by redness of the skin due to congestion of the capillaries.

Esophageal cancer (Abbr.: EC) In certain parts of southern Africa, China, and northern Italy, the incidences of EC are extremely high with substantial varations in EC rates separated by only short geographical distances. In the high incidence areas very high fumonisin concentrations (FB_1, FB_2) have been detected in → maize and maize products intended for human consumption. In addition, → Fusarium moniliforme Sheldon strains isolated from Chinese maize (Linxian County) produced nitrosamines including *N*-methylbenzylnitrosamine, one of the most potent nitrosamines inducing esophageal cancer in experimental animals.
It has been concluded that the etiology of human esophageal cancer probably involves not one but several factors (e.g. vitamin and trace elements deficiencies in high risk populations in the Transkei). Although the experimental proof of a causative relationship between fumonisin contamination of corn-based staple diet and EC is still lacking, it is obvious that exposure to → fumonisins due to the ingestion of maize and maize products in the high EC areas of Transkei / South Africa, Linxian and Cixian Counties / northern China, northern Italy and southeastern United States is one etiological factor (of several) for human esophageal cancer.

Eumycota Kingdom of Eukaryota, the true → fungi

Eupenicillium → Trichocomaceae

Eurotiaceae (Syn.: → Trichocomaceae)

Eurotiales → Ascomycota

Eurotium → Trichocomaceae, anamorph: → Aspergillus
In marginally dried grain (→ a_w 0.65 0.70) *Eurotium* spp. besides → Aspergillus restrictus G. Sm. and *Eurotium halophilicum* belong to the earliest developing and most commonly encountered → storage fungi. However, in some case *Eurotium* spp. also occurs on → grains pre-harvest. They are a characteristically xerophilic group of fungi showing maximum growth rates at $a_w < 1.0$. Moisture contents in the range of 14.5-15% (→ cereals) enable their growth. Their metabolic water increases the a_w of the substrate contributing to the growth of mycotoxin producing fungi like *Aspergillus* spp. and → Penicillium spp. Important species are *E. amstelodami*, *E. chevalieri*, *E. herbariorum*, *E. rubrum* (see Figure Eurotium).
They are able to synthesize different mycotoxins like → ochratoxin A and → sterigmatocystin. However, accumulation probably does not reach dangerous concentrations. Some still unknown car-

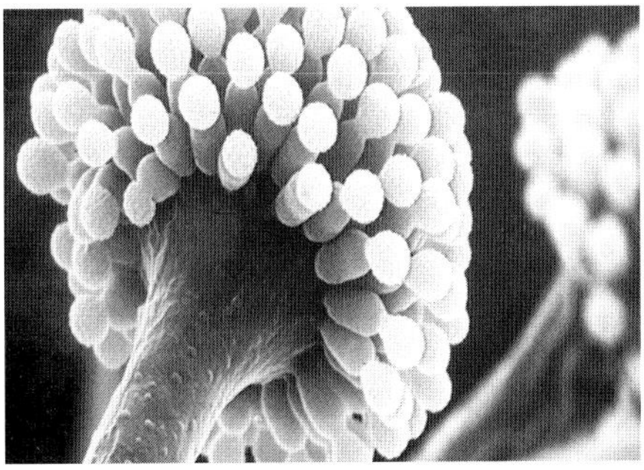

Eurotium. *Eurotium herbariorum*

cinogenic compounds should also be produced.

Expansin (Syn.: → patulin)

Extracellular mycotoxins like → aflatoxins, → citrinin, → kojic acid, → mycophenolic acid, → β-nitropropionic acid, → ochratoxins, → patulin, → penicillic acid, → PR-toxin, → rubratoxins, → T-2 toxin, and → zearalenone diffuse into the substrate. → Intracelullar mycotoxins, → mycotoxins

F

F-2 toxicosis (Syn.: estrogenic syndrome, hyperestrogenism, vulvo-vaginitis) → Zearalenone, mainly produced by → Fusarium graminearum Schwabe, and related metabolites (e.g. zearalenol) possess estrogenic activity. They may cause severe reproductive and infertility problems in domestic animals. Pigs are very susceptible, → cattle seem less susceptible, and chickens are apparently not affected. The effect of long term exposure of humans to low zearalenone levels in the diet is still unknown but this estrogene may cause hormone-dependent tumors in women.
Zearalenone production by *F. graminearum* is favored by both high moisture content and alternating moderate and low temperatures during → maize storage. In consequence, adequate drying of maize and storage at low moisture levels will reduce zearalenone contamination. The use of resistant varieties, as well as dilution of contaminated → cereals with sound cereals contribute to avoiding F-2 toxicosis. Clinical reports of hyperestrogenism in swine date as far back as the 1920s.

F-2 toxin → zearalenone

Fagicladosporic acid → Cladosporium

Fennel may contain the following → mycotoxins:
→ aflatoxin B_1
incidence: 2/10, conc. range: 11 µg / kg, country: India
incidence: 6/9, conc. range: 30-275 µg / kg, country: India
→ aflatoxin B_2
incidence: 1/10, conc.: 8 µg / kg, country: India
incidence: 6/9, conc. range: 28-173 µg / kg, country: India
→ aflatoxin G_1
incidence: 6/9, conc. range: 15-76 µg / kg, country: India
→ aflatoxin G_2
incidence: 6/9, conc. range: 9-69 µg / kg, country: India
→ aflatoxins (AFB_1, AFB_2, AFG_1, AFG_2)
incidence: 1/3, conc.: 1.2 µg / kg, country: UK
→ citrinin
incidence: 2/9, conc. range: 28-59 µg / kg, Ø conc.: 43.5 µg / kg, country: India
→ ochratoxin A
incidence: 1/9, conc.: nc, country: India
incidence: 3/3, conc. range: $\leq$ 0.2 µg / kg, country: UK
→ sterigmatocystin
incidence: 1/9, conc.: 142 µg / kg, country: India
→ zearalenone
incidence: 1/3, conc.: 7 µg / kg, country: UK
→ spices

Fenugreek may contain the following → mycotoxins:
→ aflatoxin B_1
incidence: 3/6, conc. range: 2-4.3 µg / kg, country: Egypt
→ aflatoxin B_2
incidence: 2/6, conc. range: 2.5-3 µg / kg, country: Egypt
→ aflatoxin G_1
incidence: 1/6, conc.: 1.8 µg / kg, country: Egypt
→ aflatoxins (AFB_1, AFB_2, AFG_1, AFG_2)
incidence: 1/41*, conc.: 2.5 µg / kg, country: UK, *miscellaneous → spices, imported

Fermented products may contain the following → mycotoxins:
→ zearalenone
incidence: 6/55, conc. range: 8-53 µg / kg, country: Swaziland
→ miso, → oriental fermentations

Fibrosis fibrous tissue formation

Field fungi The original source of these fungi is the field. They infect the developing and mature → grains while the plants are still growing in the field, or after the seeds are cut and swathed but before they are threshed. The most common field fungi which are present at the onset of storage like → Alternaria spp., → Cladosporiumspp., *Epicoccum* spp., → Fusarium spp. and *Drechslera* spp. are succeeded by → storage fungi with increasing storage time. Field fungi have high water requirements (90-100% relative humidities) which in → cereals amount to a moisture content of ≈ 20%. At lower moisture levels they do not compete well with the storage fungi and most of them die rapidly.
Some of the field fungi produce and cumulate → mycotoxins in kernels and chaff, e.g. → Alternaria mycotoxins and → Fusarium mycotoxins.

Figazzas may contain the following → mycotoxins:
→ deoxynivalenol
incidence: 8/8, conc. range: 212-2800 µg / kg, Ø conc.: 851 µg / kg, country: Argentina
→ wheat products

Fig paste may contain the following → mycotoxins:
→ aflatoxins (no specification)
incidence: 105/132, conc. range: 1-10 µg / kg (86 samples), 11-165 µg / kg (19 sa), country: Turkey

Figs Although → ochratoxin A and → kojc acid have been isolated from figs the → aflatoxins represent the main important → mycotoxins contaminating fig fruits.
During the ripening stage the → fruits become susceptible to aflatoxin contamination by → Aspergillus flavus Link while immature fig fruits do not support development and aflatoxin formation by *A. flavus*. Under experimental conditions those fruits taking longest to ripen contained up to 72,000 µg aflatoxin / kg when inoculated in the green stage. Maximum aflatoxin formation occurred on the fourth and sixth day and then decreased gradually. Only very little aflatoxin could be detected in fruits inoculated in the firm-ripe state if dried immediately. Aflatoxin amount increases with the extension of the drying time of the figs. During sun-drying on the tree, figs are very susceptible to fungal infection and development leading to fairly high levels of aflatoxins. Under natural conditions Turkish figs remain on the trees until they are shrivelled ripe. After falling to the ground (occasionally covered with cloths), they are dried in sunlight before the fruits are collected. These conditions seem to promote aflatoxin contamination of the fruits. Infection of the figs with *A. flavus* and → Aspergillus parasiticus Speare, due to gall wasps, during pollination will lead to potential aflatoxin contamination in a rather late stage during ripening. However, it is still being debated, whether aflatoxin contamination occurred only after the fruits had fallen from the trees and were lying on the ground for drying. The pattern of contamination in figs closely resembles that described for → peanuts: only some figs show an aflatoxin contamination but these individual fruits usually contain very high concentrations (5000 µg → aflatoxin B_1 / kg). It was estimated that the degree of contamination is in the range of ca. 1 in 100. The aflatoxin contamination is restricted to a great extent to that part of the fig showing surface fluorescence.
Although figs intended for retail sale are packed in very close contact with each other in small boxes, only slight cross-

contamination or none at all could be observed among packed fruits.
Figs may contain the following → mycotoxins:
aflatoxin B_1
incidence: 6/25*, conc. range: 0.1-3 µg / kg, country: Switzerland, *dried
incidence: 2/4*, conc. range: 2.5-11.8 µg / kg, country: Syria, *dried
incidence: 8/206* **, conc. range: 3.6-320 µg / kg, ∅ conc.: 112 µg / kg, country: Turkey, *dried, **lower grade figs
incidence: 94/386*, conc. range: 0.2-5 µg / kg (85 samples), 5-10 µg / kg (7 sa), 10-20 µg / kg (1 sa), 20-30 µg / kg (1 sa), country: Turkey, *dried, randomly selected
incidence: 37/52*, conc. range: 5-76,000 µg / kg, country: Turkey, *dried, selected, fluorescent
incidence: 8/16*, conc. range: 0.2-5 µg / kg (6 samples), 5-10 µg / kg (1 sa), 10-20 µg / kg (1 sa), country: Turkey, *dried, discolored
incidence: 52/62*, conc. range: 0.2-10 µg / kg (18 samples), 10-100 µg / kg (8 sa), 100-1.000 µg / kg (14 sa), 1000-10,000 µg / kg (11 sa), > 10,000 µg / kg (1 sa), country: Turkey, *fluorescent (BGY)
→ aflatoxin B_2
incidence: 4/206* **, conc. range: 23.5-71.8 µg / kg, ∅ conc.: 50.6 µg / kg, country: Turkey, *dried, **lower grade figs
→ aflatoxin G_1
incidence: 3/206* **, conc. range: 12.4-97.5 µg / kg, ∅ conc.: 61.4 µg / kg, country: Turkey, *dried, ** lower grade figs
incidence: 49/386*, conc. range: 0.2-5 µg / kg (45 samples), 5-10 µg / kg (2 sa), 20-30 µg / kg (2 sa), country: Turkey, *dried, randomly selected
incidence: 15/52*, conc. range: 5-180,000 µg / kg, country: Turkey, *dried, selected, fluorescent
incidence: 3/16*, 0.2-5 µg / kg (1 sample), 5-10 µg / kg (2 sa), country: Turkey, *dried, discolored
incidence: 21/62*, conc. range: 0.2-10 µg / kg (7 samples), 10-100 µg / kg (4 sa), 100-1000 µg / kg (4 sa), 1000-10,000 µg / kg (4 sa), > 10,000 µg / kg (2 sa), country: Turkey, fluorescent (BGY)
incidence: 1/106*, conc.: 10 µg / kg, country: USA, *dried, selected, fluorescent
→ aflatoxins (no specification)
incidence: 53/103, conc. range: 5-203 µg / kg, country: Sweden
incidence: 56/122*, conc. range: 1-10 µg / kg (43 samples), 12-96 µg / kg (13 sa), country: Turkey, *dried
incidence: 6/165, conc. range: 2-29 µg / kg, ∅ conc.: 13 µg / kg, country: USA
→ kojic acid
incidence: 52/52*, conc. range: 8-6,900,000 µg / kg, country: Turkey, *dried, selected, fluorescent
→ ochratoxin A
incidence: 1/39*, conc.: ≤ 0.6 µg / kg, country: Germany
incidence: 3/30*, conc.: ≤ 3.3 µg / kg, country: Germany
incidence: 1/9*, conc.: 160 µg / kg, country: Switzerland, *dried
incidence: 12/52*, conc. range: 5-12,000 µg / kg, country: Turkey, *dried, selected, fluorescent
→ fruits

Filberts → hazelnuts

Fish may contain the following → mycotoxins:
→ aflatoxin B_1
incidence: 1/1* **, conc.: 679 µg / kg, country: Thailand, *total: 795 µg AFB_1, AFB_2, AFG_1, AFG_2, / kg prepared food, **plaa tuu = Mackerel like, sun dried
→ aflatoxins
incidence: 7*/139**, ∅ conc.: 166 µg / kg, country: Thailand, **dried, and shrimps *total: ∅ conc.: 722 µg / kg AFB_1, AFB_2, AFG_1, AFG_2

→ ochratoxin A
incidence: 14/20, conc. range: 1000-2000 µg / kg (sqd), country: Sierra Leone
According to Frisvad (1988) potential mycotoxins in dried fish may be → ochratoxin A and → citreoviridin.

Flavobacterium aurantiacum removes → aflatoxins from fluid and solid foods such as cow and peanut milk, vegetable oil, → peanuts and → peanut butter as well as → maize. Rapid conversion of AFB_1 led to water-soluble degradation products. Release of CO_2 by the living cells of the bacterium contributes to the assumption that AFB_1 is at least in part metabolized.

Flavomycelin → Luteoskyrin

Flour (cereals, no specification)
Cereal flours mainly show contamination with species of the genera → Aspergillus and → Penicillium. The degree of contamination varies from sample to sample and probably reflects different sanitation standards in the mills.
Although mycotoxin-producers may contribute to the mold flora of → flour to a small extent, their detection is important because transmission into food products of which flour is an ingredient is possible. Improper processing of these food products may result in growth of the fungi and subsequent mycotoxin formation.
Flour may contain the following → mycotoxins:
→ citrinin
incidence: 11/21, conc. range: 0.2-1.0 µg / kg, ∅ conc.: 0.55 µg / kg, country: Switzerland
→ deoxynivalenol
incidence: 13/56, conc. range: 350-8380 µg / kg, country: India
incidence: 36/36, conc. range: 2-240 µg / kg, country: Japan
incidence: 2/2*, conc. range: 23-720 µg / kg, ∅ conc.: 372 µg / kg, country: Papua New Guinea, *imported, wholemeal self-raising flour
incidence: 2/2*, conc. range: 91-1460 µg / kg, ∅ conc.: 776 µg / kg, country: Papua New Guinea, *imported, wholemeal plain flour
→ nivalenol
incidence: 6 products analysed, conc. range: 37-190 µg / kg, country: Japan
incidence: 1/2*, conc.: 13 µg / kg, country: Papua New Guinea, *imported, wholemeal self-raising flour
incidence: 1/2*, conc.: 1375 µg / kg, country: Papua New Guinea, *imported, wholemeal plain flour
→ ochratoxin A
incidence: 3/80, conc. range: 0.4 µg / kg, country: Germany
incidence: 17/93, ∅ conc.: 2.2 µg / kg, country: Germany
incidence: 26/52, conc. range: 0.1-0.49 µg / kg (11 samples), 0.5-1.49 µg / kg (13 sa), 1.5-9.99 µg / kg (2 sa), country: Germany
incidence: 11/11, conc. range: < 2.5-20 µg / kg, country: Japan
incidence: 48/215, ∅ conc.: 4370 µg / kg, country: Poland
incidence: 2/7*, conc. range: 490-2900 µg / kg, country: UK, *moldy
incidence: 28/57, conc. range: ≤ 2.0 µg / kg, country: UK
incidence: 49/57, conc. range: ≤ 1.6 µg / kg, country: UK
incidence: 48/61, conc. range: ≤ 3.2 µg / kg, country: UK
incidence: 21/31, conc. range: ≤ 1.0 µg / kg, country: UK
→ zearalenone
incidence: 2/2*, conc. range: 1450-2150 µg / kg, ∅ conc.: 1800 µg / kg, country: Papua New Guinea, *imported, wholemeal self-raising flour
incidence: 2/2*, conc. range: 1400-2570 µg / kg, ∅ conc.: 1985 µg / kg, country:

Papua New Guinea, *imported, wholemeal plain flour
→ cereals, → barley flour, → buckwheat flour, → maize flour, → rye flour, → soybean flour, → wheat flour

Food Estimations of the FAO stated that ca. 25% of the worldwide produced foodstuff contains mycotoxin(s) at detectable levels. The contamination of food (and feeds) mainly depends on the prevailing environmental conditions that favor mold growth and subsequent mytotoxin formation. As a consequence of the import/export of food (and feeds) the problem of → mycotoxicosis is not limited to any one geographical area but represents a real or potential problem in all areas of the world where food (and feeds) are consumed. It is evident that nearly all staple food products consumed anywhere in the world are prone to mycotoxin (→ mycotoxins) contamination.

Foods (canned, no specification)
may contain the following → mycotoxins:
→ aflatoxins (AFB_1, AFB_2, AFG_1, AFG_2)
incidence: nc/4, conc. range: 0.2-1.4 µg/kg, country: UK
→ deoxynivalenol
incidence: nc/4, conc. range: 4-9 µg/kg, country: UK
→ fusarenon X
incidence: 1/4, conc.: 15 µg/kg, country: UK
→ nivalenol
incidence: 1/4, conc.: 18 µg/kg, country: UK
→ ochratoxin A
incidence: nc/4, conc. range: 0.1-0.3 µg/kg, country: UK
→ zearalenone
incidence: 1/4, conc.: 6.1 µg/kg, country: UK

Fresh cheese → cheese, fresh

Frontoethmoidal encephalomenigocele
(Abbr.: FEEM) Teratogens (→ teratogenic), produced by fungi in grain staples, may be the cause of FEEM in Myanmar (formerly Burma). The disease occurs in countries like Australia, England, Germany, India, South Africa and United States but it is less common than in Myanmar, Russia and Thailand.
As a neural tube defect the disease is characterized by a tumor protusion between the eyes or at the base of the nose. The protusion diameter is in the range of ca. 1.5 cm to ca. 8 cm. Although hypertelorism is common and smell as well as vision can be affected, the disease does not lead to serious debilitation or dislablement.
It is speculated that a teratogen consumed with fungus-contaminated → rice during a critical period of pregnancy may interfere with the development of the embryo.
As yet there is only one supposed case - consumption of blight-affected → potatoes by pregnant women - but correct mycotoxicological justification is lacking.

Fruit juices (no specification)
may contain the following → mycotoxins:
→ patulin
incidence: 2/3, conc. range: $\leq$ 50 µg/kg, country: Germany
incidence: 12/58, conc. range: 5-15 µg/kg, country: Italy

Fruit products (no specification)
may contain the following → mycotoxins:
→ patulin
incidence: 18/52, conc. range: 5-32 µg/kg, country: Australia

Fruits (no specification)
The high → a_w and nutrient content of ripe fruits and → vegetables make them highly susceptible to the infection by toxigenic molds. At full maturity fruits

are easily injured and predisposed to fungal attack.
The most important mycotoxigenic fungus on fruits is → Penicillium expansum Link. The growth of this fungus leads to → patulin contamination, especially in apples.
The contamination with → Alternaria spp. one of the most common microorganisms responsible for the spoilage of fruits and vegetables may result in the production of copious amounts of → mycotoxins. Mycotoxin formation is favored by the high moisture content of fruits so that all three groups of → Alternaria mycotoxins are found. The incorporation of contaminated fruits into processed products, e.g. juices, preserves, sauces, due to faulty sorting procedures or neglect, is a potential health hazard.
According to Frisvad (1988), the following mycotoxins may be found in fruits and fruit products: *Alternaria* mycotoxins, → Fusarium mycotoxins, patulin.
Fruits may contain the following → mycotoxins:
→ aflatoxin (no specification)
incidence: 6/157*, conc. range: 2-20 μg / kg, country: Uruguay, *dried
patulin
incidence: 4/74*, conc. range: nc, country: India, *dried
incidence: 1/1, conc.: 10 μg / kg, country: UK
→ zearalenone
incidence: 1/99*, conc.: > 200 μg / kg, country: Uruguay, *dried
→ apples, → bananas, → blueberries, → cherries, → cranberries, → lingonberries, → mandarin fruits, → mango, → oranges, → peaches

Fumonisin B_1 (Syn.: macrofusin) is a 2-amino-12,16-dimethyl-3,5,10-trihydroxy-14,15-propane-1,2,3-tricarboxy icosane (→ mycotoxins) which was first isolated from → Fusarium moniliforme Sheldon in 1988 (see Figure Fumonisin B_1).

Chemical Data
Empirical formula: $C_{34}H_{59}NO_{15}$, molecular weight: 721

Fungal Sources
see → fumonisins

Natural Occurrence
→ Asparagus, → beans, → beer, → bread, → breakfast cereals, → corn flakes, → maize, → maize-based thickeners, → maize bran, → maize chips, → maize flakes, → maize flour, → maize grits, → maize meal, → maize products, → maize snacks, → maize, canned, → maize, fiber cereal, → maize, hominy, → maize, popped, → maize, puffed, → maize, quality protein, → maize, sweet, → masa, → milk, → muffin mix, → pop corn, → rice, → sorghum meal, → starch, → tortillas, → tortilla chips
For further information see → fumonisins

Toxicity
see → fumonisins

Detection
see → fumonisins

Possible Mycotoxicosis
see → fumonisins

Further Comments
Stability / reduction: At atmospheric pressure chemical ammonia was ineffective for the detoxification of FB_1-contaminated maize. However, at high pressure this treatment caused losses of FB_1 to almost 80%. Treatment of fumonisin contaminated maize with 2% ammonia at low pressure for 4 days, a process that successfully decontaminates aflatoxin-contaminated maize, did not result in complete destruction of the mycotoxin. Calcium hydroxide was highly effective in removing FB_1 from contaminated maize while potassium hydroxide and hydrochloric acid hydrolyze FB_1 to HFB_1. FB_1 was destroyed by using sodium hypochlorite. The effect of ammoniation on

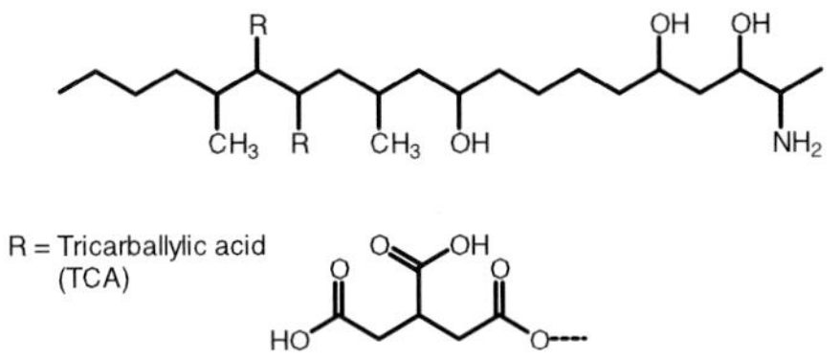

Fumonisin B_1

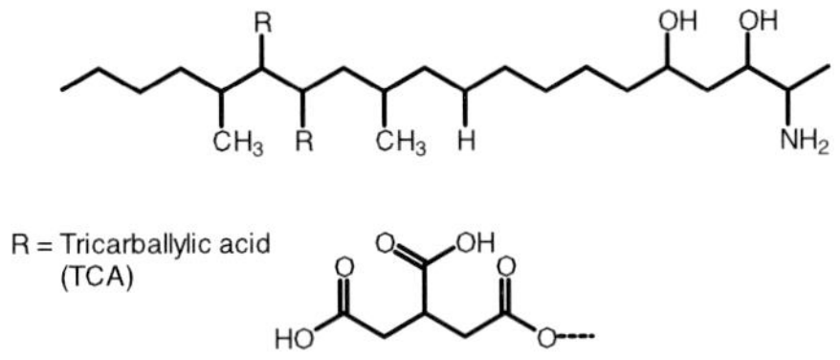

Fumonisin B_2

FB_1 reduction varies with experimental conditions. Potassium hydroxide and hydrochloric acid caused hydrolyzation of fumonisins to tricarballylic acid and a C_{22} aminopolyol.
FB_1 losses during **baking** may be related to the nonenzymatic browning reaction. As the heat increases, more FB_1 (the primary amine group) reacts with free aldehyde or ketone groups in reducing sugars (i.e. glucose and fructose). However, commercial drying and baking temperatures in general are not sufficient to significantly reduce the FB_1 concentration in corn muffins whereas the fumonisin content of maize bread was lowered to almost 50%.
Neither **drying** nor normal **food processing** and **cooking** are effective in the destruction of FB_1.

Fumonisin B_2 is a 2-amino-12,16-dimethyl-3,5-dihydroxy-14,15-propane-1,2,3-tricarboxy icosane (→ mycotoxins) which was first isolated from → Fusarium moniliforme Sheldon in 1988 (see Figure Fumonisin B_2).

CHEMICAL DATA
Empirical formula: $C_{34}H_{59}NO_{14}$, molecular weight: 705

FUNGAL SOURCES
see → fumonisins

NATURAL OCCURRENCE
→ Asparagus, → beer, → breakfast cereals, → corn flakes, → incarpina, → maize,→ maize-based thickeners, → maize flour, → maize grits, → maize meal, → maize products, → maize snacks, → maize, fiber cereal,→ maize, hominy, → maize, puffed, → maize, quality protein, → maize, sweet, → masa, → muffin mix, → pop corn, → rice, → starch, → tortillas, → tortilla chips
For further information see → fumonisins

TOXICITY
see → fumonisins

DETECTION
see → fumonisins

POSSIBLE MYCOTOXICOSIS
see → fumonisins

Fumonisin B_3 is a 2-amino-12,16-dimethyl-3,10-dihydroxy-14,15-propane-1,2,3-tricarboxy icosane (mycotoxins) which was first isolated from → Fusarium moniliforme Sheldon in 1988 (see Figure Fumonisin B_3).

CHEMICAL DATA
Empirical formula: $C_{34}H_{59}NO_{14}$, molecular weight: 705

FUNGAL SOURCES
see → fumonisins

NATURAL OCCURRENCE
→ breakfast cereals, → maize, → maize flour, → maize meal, → maize, quality protein, → rice, → tortillas
For further information see → fumonisins.

TOXICITY
see → fumonisins

DETECTION
see → fumonisins

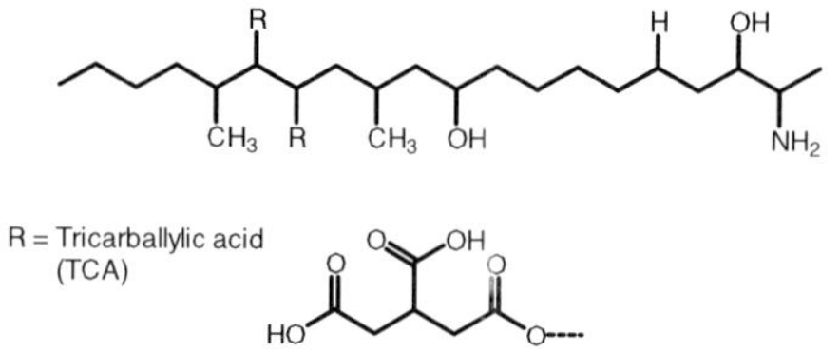

Fumonisin B_3

POSSIBLE MYCOTOXICOSIS
see → fumonisins

Fumonisins are long-chain polyhydroxyl alkylamines containing two propane tricarboxyclic acid moieties which are esterified to hydroxyl groups on adjacent carbon atoms.→ Fumonisin B_1 is the most important of the fumonisins. → Fumonisin B_2 and → fumonisin B_3 are homologs but FB_2 lacks the hydroxyl at C-10 while FB_3 lacks the hydroxyl group at C-5. These three fumonisins account for most of the fumonisins that are both found in naturally contaminated → maize as well as under cultivated conditions.
At least 13 fumonisins, four B´s (B_1, B_2, B_3, B_4) having a free amine and three A´s (A_1, A_2, A_3) which are amides, fumonisin C_1, C_3 and FC_4 (analogs of FB_1, FB_3 and FB_4, respectively) as well as fumonisin P_1, P_2, P_3 have been isolated from *F. moniliforme*. FA_1, FA_2 and FA_3 are the N-acetyl derivatives of FB_1, FB_2 and FB_3, respectively. Within each series differing hydroxyl substitution results in different fumonisins. FC_1, FC_3 and FC_4 lacking the C-1 terminal methyl group which is characteristic for the other fumonisins. In comparison to FC_1 the hydroxylated FC_1 (OH-FC_1) has one more hydroxy group at the C-3 position.
The fumonisins were first reported in South Africa (1988) and belong to the most recently described → Fusarium mycotoxins. High rates of → esophageal cancer in the rural population of South Africa and the death of many horses due to → equine leukoencephalomalacia, mainly in New Caledonia, led to their detection. The involved feed was highly infested with → Fusarium moniliforme Sheldon. Because researches were unable to find toxic substances in the usual organic extracts of *F. moniliforme* cultures, they concentrated on the aqueous fractions. The isolation of fumonisin B_1 succeeded in South Africa while independently fumonisin B_1 was isolated under the name → macrofusin from culture material of *F. moniliforme*, which is responsible for equine leukoencephalomalacia in New Caledonia in 1989 (see Figure Fumonisins).

FUNGAL SOURCES
F. moniliforme, → Fusarium proliferatum (Matsushima) Nirenberg (the main producers); *F. anthophilum*, *F. dlamini*, *F. globosum*, *F. napiforme*, *F. nygamai*, and *F. subglutinans* seems to be a non-consistent producer. *A. alternata* f. sp. *lycopersici* is also known for FB_1 production.

NATURAL OCCURRENCE
→ beer, → bread, → breakfast cereals, → chilli pickles, → corn flakes, → curry, → curry paste, → maize muffin, → maize pops cereals, → maize starch, → maize, infant cereal, → maize, infant cream corn, → noodles, → spices, → tandoori,
For further information see fumonisin B_1, fumonisin B_2, fumonisin B_3
These so-called "aflatoxins of the nineties" are widespread in maize and maize-based products in numerous countries of the world. FB_1, FB_2 and FB_3 are the major compounds produced in nature (→ food and feed). FB_1 is the predominating fumonisin in naturally-contaminated maize kernels with a ratio of 3:1 (FB_1:FB_2) and 12:1 (FB_1:FB_3) which corresponds to ca. 70% of the total fumonisin concentration detected. However, *in vitro* there are some isolates of *F. moniliforme* producing more FB_2 than FB_1. FB_4, FC_1 and FA_1-FA_3 are synthezised in relatively

minor quantities while the three latter ones do not occur naturally. In contrast, FC_1, FC_3 and FC_4 as well as FB_4 have been detected in Korean moldy maize samples intended for animal consumption.

Fumonisin contamination of maize may be related to dry weather at or just prior to pollination. The contamination may occur world-wide but higher levels in food and feedstuff may be present in countries with a warm, dry climate. In countries having a cool, damp climate only low fumonisin levels are expected. However, some studies indicate that the contamination levels of maize and maize products are similar from country to country.

Whole kernel maize, grits and flour that undergo the mildest forms of processing are most frequently affected, usually showing medium (grits, flour) to high (kernels) fumonisin concentrations while maize → bran is also affected. Low contamination or none at all occurred in highly processed maize-based products-such as corn flakes, maize chips, corn pop cereals, tortillas and tortilla chips but in part recovery problems may be responsible.

This pattern of fumonisin distribution in maize and maize-based products agrees with the growth characteristics of *F. moniliforme*, which colonizes the tip and germ area of the kernel, just beneath the pericarp. In dry milled fractions the bran and germ fractions were highly contaminated with the fungus and fumonisins while the flour and flaking grit fractions contain low to medium mycotoxin concentrations. Almost no contamination was found in maize and tortilla chips, which may be explained in part by recovery problems, while hominy corn, tortillas and popcorn showed low contamination levels. It is under discussion whether the (apparent) loss of fumonisins by heating is due to degradation and loss of toxicity or may result from ineffective detectability by current methods of extraction and analysis. It seems that corn starch as a product of the wet milling process is usually free of fumonisins. The widespread occurrence of fumonisins in Eastern and Southern Africa is documented by a positive of 92.5% of the maize samples analyzed. In these countries the daily maize intakes amounts to more than 200 g / person / day (FAO 1992) with peaks in Malawi (468.8 g), Zambia (418.6 g), and Zimbabwe (330.9 g). An daily average intake of 245 µg fumonisins was calculated for Zimbabwe´s population on the basis of maize consumption. The high natural contamination of maize with fumonisins is of concern, particularly with respect to the much lower levels of other mycotoxins, like aflatoxin, T-2 toxin or zearalenone, present in food and feedstuffs (although these mycotoxins possess a relatively higher toxicity). Serious health implications may arise, taking into account that 10 and 100 µg fumonisin / g are dangerous to horses and pigs, respectively.

In north-eastern Italy an increasing risk of developing human esophageal cancer with increasing consumption of maize (→ polenta) was observed.

In animal tissues, so far, only trace amounts of fumonisins have been found. It seems that residues in → meat, → milk and eggs are not a problem.

Toxicity

FB_1 causes severe animal diseases like leukoencephalomalacia (LEM, "hole in the head syndrome") in horses (ca. 10,000 µg FB_1 + FB_2 / kg bw), pulmonary → edema syndrome (PES) in pigs (ca. 100,000 µg FB_1 + FB_2 / kg bw), and liver cancer in rats (15,000 µg / kg bw). In the last case, FB_1 acts as a cancer initiator and promoter. Cattle seem to be less susceptible than pigs which are less susceptible than horses. Besides hepatotoxicity

FB_1 caused nephrotoxicity and diverse effects on the immunsystem in rodents. Toxic reactions also occurred in the case of turkey → poultry and broiler chickens but laying hens seem to be not sensitive to low levels of fumonisin. Therefore, the FDA (Center for Veterinary Medicine) recommended that products with fumonisin levels greater than 1, 10, 30, and 50 mg / kg should not be fed to horses, pigs, beef → cattle, and poultry, respectively. FB_2 and FB_3 showed hepatotoxic effects similar to FB_1 and similar, although weak, cancer-initiating potential.
In addition, esophageal cancer (EC) in humans has been observed in distinct areas of the world (Transkei / South Africa, Linxian and Cixian Counties / northern China, northern Italy and south-eastern United States) where extremely high levels of fumonisins occurred in moldy home-grown maize and maize-based food products.
Since FB_1 inhibits the uptake of folate, it is also under discussion whether the fumonisins are involved in malformations of the central nervous system, e.g. neural tube defects. Such birth defects may be related to dietary exposure to fumonisin. In contrast to AFB_1, FB_1 is not → mutagenic or → genotoxic, whereas the cytotoxicity is low.
The fumonisins bear a remarkable structural similarity to the long-chain base sphingosine as a component of the long-chain backbone of various sphingolipids. These are highly active components of cell membranes. The disruption of their metabolism may result in serious effects on cell behaviour, differentiation and growth. FB_1 and FB_2 were the first naturally occurring specific inhibitors of sphingolipid synthesis to be discovered. Fumonisins inhibit ceramide synthetase (sphingosine and sphinganine *N*-acyltransferase) resulting in an alteration in sphingolipid base ratios (sphinganine). This alteration causes massive liquefactive necrosis of the cerebral hemisphere. Neurological manifestations occur in horses, such as abnormal movements, aimless circling, lameness, etc. Interference with sphingolipid biosynthesis is the main cause of their toxicity in horses and probably also in pigs, as well as their tumor-promoting effects.
In chicks ($\approx$ 14 days old) fumonisins, perhaps together with other metabolites, may cause "spiking mortality syndrome" involving several neurological signs, reduced growth and mortality.
The nixamalization product, the hydrolyzed fumonisin B_1 (HFB_1), resulting from cleavage of the tricarballylic side chains at C-14 and C-15, appears to be more toxic to rats than FB_1 itself since 50 mg / kg of FB_1 or 10 mg / kg of HFB_1 (maize, canned) possessed almost equal toxicity in rat feeding studies. However, the fact that HBF_1 did not initiate cancer in liver may be due to lack of absorption.

Detection

ELISA, GC-MS, HPLC, LC, TLC
Besides chromatographic, mass spectrometric and immunochemical methods liquid chromatography is most commonly used in analysing food extracts for fumonisins.

Further Comments

The fumonisins are unusual → mycotoxins in that they do not contain cyclic or ring groups. They are not unique in nature since structural similarities have been observed with the AAL-toxins, sphingofungins (antifungal agents isolated from → Aspergillus fumigatus Fres.) and sphingosine (see above).
In addition, they are relatively water soluble (more soluble in acetonitrile-water or methanol, insoluble in organic solvents) but are as heat-stable as many other mycotoxins (see below).
Unfortunately visual assessment and subsequent separation of the ears into good and moldy lots is not sufficient to pre-

vent fumonisin intake by humans because mycotoxins may also be present in visibly undamaged homegrown Transkeien maize.

Commercial maize hybrids in the US differ in their degree of fumonisin accumulation. Higher concentrations were detected in hybrids grown outside their adapted range.

20 °C was the best temperature for the production of FB_1 on corn.

Suitable storage conditions (e.g. low oxygen tension, kernel moisture content < 22%), reduce or prevent toxin production in stored maize.

Within 24 h, FB_1 is eliminated to more than 99% in the unmetabolized form in the faeces of rat. Traces are found only in the liver, kidney, urine and red blood cells. It is assumed that the adsorption of FB_1 is poor or there is a rapid elimination by biliary excretion.

Stability / Reduction: Fumonisins are appreciably stable during beer **fermentation.** If contaminated maize grits are used as brewing adjuncts only small decreases in FB_1 and FB_2 concentrations ($\approx$ 20-30%) occurred during the fermentation. Fumonisin uptake by yeast was negligible. Although the distilled ethanol was free of FB_1 all the other fermentation products contained FB_1.

Like other mycotoxins, fumonisins are heat stable. Minor losses occurred after **heating** aqueous solutions of FB_1 and FB_2 at temperatures < 150 °C. Only higher temperatures (150 °C) were effective. Temperatures of $\approx$ 200 °C (60 min) are necessary to cause substantial fumonisin reduction in dry or moist corn meal. A partial reduction of the fumonisin concentration was detected in muffins that had been baked at 220 °C for 25 min.

No reduction in FB_1 and FB_2 levels could be detected in whole milk heated for 30 min at 62 °C.

Canning (121 °C for various times) of different maize products did not result in significant losses of fumonisins ($\leq$ 15%). However, the apparent loss of fumonisin content in thermally processed foods may be due to matrix-related problems of recovery and detection by analytical methods.

During the **nixtamalization** process for manufacturing masa or tortilla flour at 100 °C, calcium hydroxide (0.01 M) causes the loss of the two propane-1,2,3-tricarboxylic acid (tricarballylic acid) moieties of FB_1, leading to hydrolyzed FB_1 (HFB_1). This amino pentol chain is found in commercial masa, tortilla chips and canned sweet corn, formed as a result of alkaline conditions and heating during processing. The toxicity of both HBF_1 and HBF_2 was higher than that of FB_1 and FB_2 when mammalian cell cultures and jimsonweed leaf bioassays were used.

Removal of the corn fines (or screenings) from bulk shipments of corn by **sieving** reduced the total fumonisin levels down to almost 30%. This is due to the accumulation of fumonisins in the outer pericarp layers of broken kernels in these screenings (61,000-268,000 µg FB_1 / kg) and (19,000-86,000 µg FB_2 / kg). The removal of fine particulate matter from bulk shipments of maize, prior to processing, might be an effective procedure for the preliminary decontamination of affected maize.

A further reduction is achieved by the **milling** process. Increasing refinement of maize meal means that combined fumonisin levels could be lowered by as much 95% in fine maize meal compared to maize screenings. Dry milling caused an accumulation of fumonisins in the bran, germ and fines fractions that are widely used in the production of animal feed, although bran is sometimes also used in certain breakfast cereals. Flaking grits, widely used in breakfast cereals and → snack foods were relatively free of con-

tamination. However, decreasing grit size led to an increase in fumonisin concentration. During wet milling most of fumonisin was found in steep water, gluten fiber, and germ, whereas no detectable levels occurred in the starch fractions.
Steeping naturally contaminated corn in water or solutions of sodium bisulfite may reduce fumonisin levels.
Regulations: Based on their toxicological potential (carcinogenic in experimental animals) *F. moniliforme* toxins, including → Fusarin C, have been classified as potential carcinogens for humans (class 2B carcinogens) by the IARC.
In the EU there is still no legislation on fumonisin B_1 levels. Switzerland is the only country with a maximum tolerated level for fumonisins in maize produced for human consumption (sum of FB_1 and $FB_2 \leq 1000\ \mu g/kg$). Since maize constitutes only a small component in the diet of the first world population considerably higher tolerance levels may be adequate for sufficient for protecting of the population against these mycotoxins. However, in areas where maize is a staple food levels even lower than 1000 μg / kg seem to be necessary.

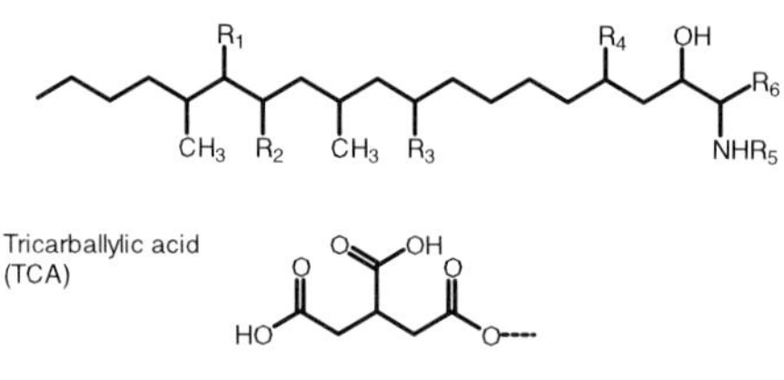

	R_1	R_2	R_3	R_4	R_5	R_6
FB_4	TCA	TCA	H	H	H	CH_3
FA_1	TCA	TCA	OH	OH	$COCH_3$	CH_3
FA_2	TCA	TCA	H	OH	$COCH_3$	CH_3
FA_3	TCA	TCA	OH	H	$COCH_3$	CH_3
FC_1	TCA	TCA	OH	OH	H	H
$HHFB_{1a}$	TCA	OH	OH	OH	H	CH_3
$HHFB_{1b}$	OH	TCA	OH	OH	H	CH_3
AP_1	OH	OH	OH	OH	H	CH_3

Fumonisins. Structure and substituents of fumonisins

Fungi Kingdom of Eukaryota, the true fungi

Fusaproliferin is a bicyclic sesterterpene (3-[2-(acetyloxy)-1-methylethyl]-4,7,8,9,12,13,16,16a-octahydro-2,7-dihydroxy-6,10,14,16a-tetramethyl-1(3aH)-cyclopentacyclopentadecenone), characterized from a toxigenic strain maize culture of → Fusarium proliferatum (Matsushima) Nirenberg in 1993/1995 (see Figure Fusaproliferin).

CHEMICAL DATA
Empirical formula: $C_{27}H_{40}O_5$, molecular weight: 444

FUNGAL SOURCES
F. proliferatum

NATURAL OCCURRENCE
→ maize

TOXICITY
→ teratogenic, toxic to *Artemia salina* and mammalian cells

DETECTION
HPTLC, TLC

Fusarenon X (Syn.: 4-acetylnivalenol, nivalenolmonoacetate, fusarenon) belongs to the group of naturally-occurring B → trichothecenes (3α,7α,15-trihydroxy-

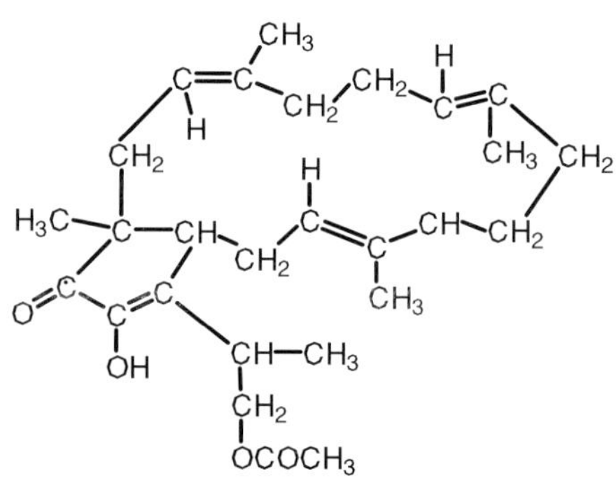

Fusaproliferin

4β-acetoxy-12,13-epoxythrichothec-9-en-8-one). Fusarenon X was first isolated in 1967 and is produced by different species of the genus → Fusarium (see Figure Fusarenon X).

CHEMICAL DATA
Empirical formula: $C_{17}H_{22}O_8$, molecular weight: 354

FUNGAL SOURCES
→ Fusarium equiseti (Corda) Sacc. sensu Gordon, → Fusarium graminearum Schwabe, → Fusarium oxysporum Schlecht., → Fusarium semitectum Berk. & Rav., → Fusarium sporotrichioides Sherb., → Fusarium sambucinum Fuckel (= *F. sulphureum*),

NATURAL OCCURRENCE
→ foods, → garlic, → maize, → oats, → wheat

TOXICITY
LD_{50} (po): 4.4 mg / kg bw rat
→ immunosuppressive, carcinogenic, cytotoxic, emetic, causes diarrhea, → hypothermia, decreased respiratory rate (experimental animals)

DETECTION
GC, MS, spectroscopy, TLC

FURTHER COMMENTS
Fusarenon X might occur more often in the warmer and subtropical parts of the world.

Fusarin C consists of a polyene chromophore with all the olefinic bonds in the *trans* configuration, linked in position C_{13} to a 2-pyrrolidone moiety and with a C_{13}-C_{14} epoxide group (2-ethylidene-11-[4-hydroxy-4-(2-hydroxyethyl)-2-oxo-6-oxa-3-azabicyclo[3.1.0]-hex-1-yl]-4,6,10-trimethyl-11-oxo-3,5,7,9-undecatetraenoic acid; methyl ester). It is the most important mycotoxin (→ mycotoxins) in the group of fusarins which include Fusarin A, D, E, F, X, Z. Fusarin C was first described in 1981, isolated from → Fusarium moniliforme Sheldon (see Figure Fusarin C).

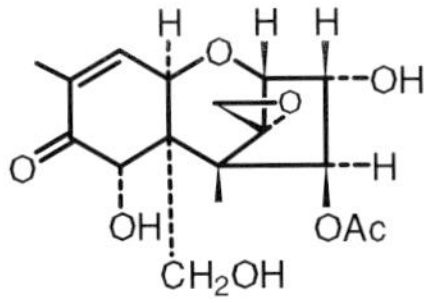

Fusarenon X

CHEMICAL DATA
Empirical formula: $C_{23}H_{29}NO_7$, molecular weight: 431

FUNGAL SOURCES
Different → Fusarium species (13) such as → Fusarium avenaceum (Fr.) Sacc., *F. crookwellense*, → Fusarium culmorum (Wm. G. Smith) Sacc., → Fusarium graminearum Schwabe, *F. moniliforme*, → Fusarium sambucinum Fuckel, → Fusarium sporotrichioides Sherb.

NATURAL OCCURRENCE
→ maize
Visibly *Fusarium*-infected as well as healthy looking corn kernels in South Africa were affected. It was also found in maize from Linxian county, China.

TOXICITY
mutagenic as → aflatoxin B_1 and → sterigmatocystin, genotoxic, → immunosuppressive, production and functioning of macrophages are inhibited
Fusarin A and D are two less-toxic and non-mutagenic forms.
Although the biological activity of fusarin E is unknown (first described in 1991), its chemical structure may impart a comparable activity to that of fusarin C.

DETECTION
HPLC

FURTHER COMMENTS
Fusarin C is one of the most unstable mycotoxins and therefore the significance of this fungal metabolite to the etiology of human illness is questionable.
Although a moderate stability was established after storing contaminated ground

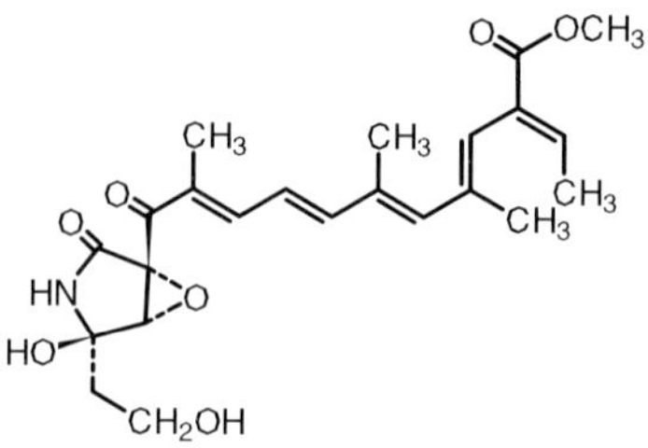

Fusarin C

maize at room temperature for 7 days in the dark, the high thermal instability led to an almost complete loss in → maize meal as well as → wheat flour during cooking / heating. No fusarin C was detected in maize meal muffins (230 °C) (→ maize muffin) made from contaminated maize kernels. Because stability of fusarin C decreases with increasing pH, thorough cooking at a slightly basic pH will effectively destroy most of this mycotoxin.
Fusarin C may be produced on soybeans and other cereals.

Fusariogenins → Fusarins

Fusariotoxicoses (in China)
Since 1961 the consumption of moldy → wheat and → maize in China has been linked with 35 outbreaks of toxicosis in man. Symptons like nausea, diarrhea, dizziness, and headache were accompanied by fever and disturbances of the nervous system 5 to 30 min after ingesting the incriminated → grains. 26 outbreaks definitely occurred in the time between March and July and it seems that → deoxynivalenol contaminated grain (wheat $\leq$ 40,000 µg DON / kg, maize $\leq$ 92,800 µg DON / kg) was the causal agent.

Fusarium anamorphic → Hypocreaceae, teleomorphs → Gibberella, → Nectria
Fusaria are adapted to a wide range of environmental conditions, which explains their ubiquitous distribution in diverse soil and organic substrates. They belong to the (ecological) group of → field fungi requiring minimum → a_w values of 0.88-0.91 for growth in soil. However, they occasionally develop in stored → cereals (especially in Scandinavia) when the moisture content is high (22-33%) and temperature is low. During wet summers, resulting in late harvest cereal, → grains are primarily affected by Fusaria infections. The invaded kernels are characterized by shriveled, discolored kernels called scab, tombstone, or head blight. Due to climatic conditions and inadequate drying techniques in Scandinavian grains, Fusaria are quite common in stored cereals.
Toxigenic species often synthesize more than one mycotoxin, e.g. trichothecenes, with one or two of them being dominant. Strains from cold areas are usually more toxic than those from tropical and subtropical regions. Higher yields of toxic metabolites are produced after periods of low temperature. In the temperate countries of northern Europe, Canada and the northern regions of the USA, trichothecenes and zearalenone are more common than → aflatoxins.
With respect to human and / or animal health problems → Fusarium graminearum Schwabe, → Fusarium moniliforme Sheldon, and → Fusarium sporotrichioides Sherb. are the most important but more than 20 problematic species are known. The Fusaria produce more than 100 biologically active secondary metabolites which belong to different structural groups. Grains damaged by *Fusarium* spp. are considered as significantly toxic. Grain toxicity may be measured by the percentage of such damaged kernels in a given lot. From the standpoint of human exposure, mycotoxins such as → trichothecenes (e.g. → deoxynivalenol, → nivalenol, → T-2 toxin), → zearalenone, as

well as the → fumonisins, have attracted the most attention.
The enhanced production of trichothecenes at low temperatures led to the wrong assumption that these mycotoxins served as an agent of war ("yellow rain" = bee faeces containing a mixture of trichothecenes) in Southeast Asia. However, these trichothecenes as well as zearalenone have been isolated from grains grown in this tropical part of the world.

Fusarium avenaceum (Fr.) Sacc. teleomorph: *Gibberella avenacea* Cook is of worldwide distribution and possesses a very broad host range such as → cereals, broad bean (→ beans), → potatoes. This species may produce → mycotoxins such as antibiotic Y, → moniliformin, → zearalenone.

Fusarium crookwellense Burgess, Nelson & Toussoun may produce → mycotoxins such as acetylnivalenol, → fusarin C, → nivalenol, zearalenols, → zearalenone.

Fusarium culmorum (W. G. Smith) Sacc. teleomorph: unknown
is a pathogen of → wheat, → rye, → barley, → oats and → maize, often co-occurring with → Fusarium graminearum Schwabe. *F. culmorum* is of worldwide distribution but prefers cooler climatic regions such as northern Europe and southern Australia (see Figure *Fusarium culmorum* (W.G. Smith) Sacc.). It co-occurs with *F. graminearum* as a causal agent of *Fusarium* head blight, invading cereal heads at the time of flowering.
→ Deoxynivalenol and → zearalenone are the primarily produced → mycotoxins but some more toxic metabolites such as → butenolide, culmorin, → diacetoxyscirpenol, → fusarenon X, → HT-2 toxin, → neosolaniol, → nivalenol, → T-2 toxin may be synthesized.

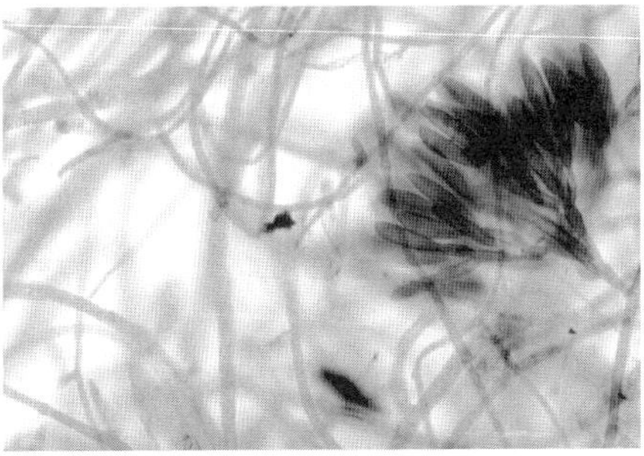

Fusarium culmorum (W.G. Smith) Sacc.

Fusarium equiseti (Corda) Sacc. sensu Gordon teleomorph: *Gibberella intricans* Wollenw. as a weak pathogen of → cereals, → vegetables, legumes, and → fruits prefers subtropical and tropical climates.
→ Grains harvested and stored with high moisture contents are likely to be invaded. Several animal diseases like degnala disease, fescue foot, → bean hulls poisoning, and tibial dyschondroplasia are probably due to the → mycotoxins of *F. equiseti*. Leukemia in man may be another disease caused by this fungus. *F. equiseti* may produce mycotoxins such as → 15-acetylscirpentriol, → butenolide, → diacetoxyscirpenol, equisetin, → fusarenon X, → fusarochromanone, → HT-2 toxin, → neosolaniol, → nivalenol, → scirpentriol, → T-2 toxin, → zearalenone.

Fusarium graminearum Schwabe teleomorph: *Giberella zeae* (Schw.) Petch.
F. graminearum produces several dozen metabolites, four or five accumulate in quantity most often in → cereals grown in warmer climates (see Figure *Fusarium graminearum* Schwabe). This species is divided into Groups I and II. Strains of Group II commonly infest cereal → grains especially → wheat and → maize (scab or head blight) and produce significant amounts of B → trichothecenes and → zearalenone (up to 60,000,000 μg / kg). They have been separated into the "NIV-chemotype" (→ nivalenol and → fusarenon X producers) and

Fusarium graminearum Schwabe. Macroconidia of *Fusarium graminearum*

the "DON-chemotype" which is further separated into chemotype IA (→ deoxynivalenol and → 3-acetyldeoxynivalenol) and "chemotype IB" (DON and → 15-acetyldeoxynivalenol producers). → T-2 toxin production occurs at a low optimal temperature of 6-12 °C.
The following further → mycotoxins such as 4-acetamido-2-butenoic acid, → butenolide, → diacetoxyscirpenol, 3,15-dihydroxy-12,13-epoxythrichothec-9-ene-8-one, → HT-2 toxin, → monoacetoxyscirpenol, → neosolaniol, → nivalenol, and T-2 toxin may be produced.
F. graminearum is involved in the following → mycotoxicosis:
feed refusal and emetic syndromes, → F-2 toxicosis, → red mold disease

Fusarium moniliforme Sheldon teleomorph: *Gibberella fujikuroi* (Swada) Ito in Ito & K. Kimura
is widespread in humid and subhumid temperate zones. It is found also in subtropical and tropical zones, but is uncommon in cooler temperate zones.
It is one of the most prevalent seedborne fungi of → maize in most dry and warm (corn growing) areas of the world, but crops like → peanuts, → rice, → sorghum, → soybeans, sugar-cane, → bananas etc. are also attacked.
The most important → mycotoxins, the → fumonisins, appear to be less common in → maize grown in cooler climates, e.g. northern Europe and Canada, but are of general significance in maize of warm and dry climatic regions, e.g. South Africa, China, Italy. Mating population A of *F. moniliforme* as well as the D mating population of *G. fujikuroi* (*F. proliferatum*) are much better FB_1 producers than the F population of *F. moniliforme*. Several animal diseases like ELEM (horses), PES (swine), hepatocarcinogenicity in rats etc. are caused by these mycotoxins, while EC (→ esophageal cancer) in man is probably due to these toxic fungal metabolites.
The following mycotoxins such as → diacetoxyscirpenol, fusaric acid, fusarins (→ fusarin C), → moniliformin, → T-2 toxin, → zearalenone may also be produced.

Fusarium mycotoxicosis → akakabi-byo disease, → alimentary toxic aleukie, → Kashin-Beck disease, → moldy corn toxicosis, → onyalai, → pellagra, → premature thelarche. These diseases are predominantly found in the temperate regions of the world due to → Fusarium mycotoxins. Temperatures of 8 °C and grain humidities between 20-25 °C, especially in cold rainy summers contribute to the occurrence of these → mycotoxicosis.

Fusarium mycotoxins → Fusarium spp. are well known producers of the → thrichothecenes, as well as the estrogenic mycotoxin, → zearalenone. Food-relevant *Fusarium* → mycotoxins are e.g. → 3-acetyldeoxynivalenol, → 15-acetyldeoxynivalenol, → 15-acetylscirpentriol, → butenolide, → deoxynivalenol, → diacetoxyscirpenol, → fumonisins, → fusarenon X, → HT-2 toxin, → moniliformin, → neosolaniol, → nivalenol, → T-2 toxin, → zearalenone

Fusarium nivale (Fr.) Ces. teleomorph: unknown
is a (seedborne) pathogen of cereal → grains, particularly under snow cover, prefering colder to temperate climates as found in e.g. Asia, Australia, Europe, and North America. This "snow mold" may produce → mycotoxins such as → deoxynivalenol, → 3-acetyldeoxynivalenol, → zearalenone.

Fusarium oxysporum Schlecht. emend. Snyd. & Hansen teleomorph: unknown
is the most economically important member of the genus *Fusarium* with a cosmopolitan distribution. Pathogenic strains are involved in damping-off diseases and cause vascular wilts in different crop plants (e.g. → cereals). This fungus also plays a role in the → Moldy sweet potato toxicosis. *F. oxysporum* may produce → mycotoxins such as → diacetoxyscirpenol, diacetylnivalenol, → 7α,8α-dihydroxydiacetoxyscirpenol, → fusarenon X, enniatins, fusaric acid, 7-hydroxydiacetoxyscirpenol, → moniliformin, → neosolaniol, → T-2 toxin?, → zearalenone.

Fusarium poae (Peck) Wollenw. teleomorph: unknown
This species is of wide geographical distribution (predominantly temperate regions) often co-occurring with → Fusarium sporotrichioides Sherb.. It has numerous hosts (cereal → grains) and is a weak parasite or saprophyte after the death of cereal host plants (see Figure *Fusarium poae* (Peck) Wollenw.) → T-2 toxin and other → trichothecenes are produced at low optimal temperatures (6-12 °C), especially during freezing and thawing conditions in overwintering unharvested crops or during storage. Probably due to the production of type A trichothecenes it might be involved in → Alimentary toxic aleukia, → Moldy corn toxicosis and → Kashin-Beck disease (Urov Disease).

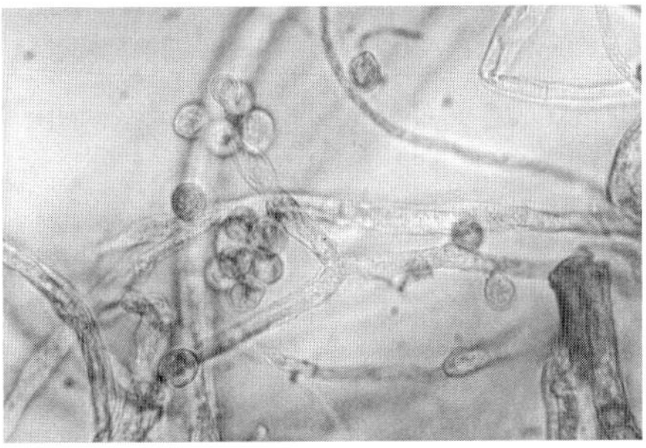

Fusarium poae (Peck) Wollenw.

F. poae may produce → mycotoxins such as → butenolide, → diacetoxyscirpenol, → HT-2 toxin, → neosolaniol, "poin" (water soluble substance, no structure elucidation, contamination with trichothecenes), → T-2 toxin, T-2 tetraol.

Fusarium proliferatum (Matsushima) Nirenberg teleomorph: unknown
This taxon was distinguished only recently (1976) from what may now be considered the → Fusarium moniliforme Sheldon complex. In consequence, there are similarities with that fungus concerning hosts, pathogenic associations with → maize, fumonisin production and toxicity on → maize.
This species, often misidentified as *F. moniliforme*, is cosmopolitan but predominant in tropical and subtropical countries, as well as in greenhouses in temperate zones and in a wide range of host plants (e.g. → rice, → fruits). *F. proliferatum* may produce → mycotoxins such as → fumonisins, fusaric acid, → fusarin C, → moniliformin, naphthoquinone pigments.

Fusarium sambucinum Fuckel (Syn.: *Fusarium sulphureum*) teleomorph: *Gibberela pulicaris* (Fr.) Sacc.
This ubiquitous species, which is more common in the northern but less frequently in the southern hemisphere has a wide host range, including stored

→ fruits and → potatoes. The involvement in human → esophageal cancer is discussed. The following → mycotoxins such as 4-→ acetoxyscirpenol, 4-acetoxyscirpenediol, 8-acetylneosolaniol, → butenolide?, → diacetoxyscirpenol, → fusarenon X, → monoacetoxyscirpenol, → nivalenol?, → sambutoxin, triacetoxyscirpenol, → zearalenone may be produced.

Fusarium sporotrichioides Sherb. teleomorph: unknown
This species is almost exclusively found in temperate to cold areas of the world on a wide variety of host plants, e.g. → cereals and their products, stone → fruits. It often co-occurs with → Fusarium poae (Peck) Wollenw. in overwintered cereals. *F. sporotrichioides* is the principal agent of → Alimentary toxic aleukia (ATA) and involved in → Moldy corn toxicosis, fescue foot, → Akakabi byo disease, → Bean hulls poisoning. Mycotoxin production occurs at low temperatures, between 4 and 1.5 °C but the optimum temperature seems to be 1.5-4 °C. The following → mycotoxins such as acetyl T-2 toxin, → butenolide, → deoxynivalenol, → diacetoxyscirpenol, → diacetylnivalenol, → fusarenon X, → HT-2 toxin, → neosolaniol, → nivalenol, NT-1 toxin (= T-1 toxin: 4β, 8α-diacetoxy-3α,15-dihydroxy-12,13-epoxytrichothec-9-ene), NT-2 toxin (4β-acetoxy-3α, 8α,15-trihydroxy-12,13-epoxytrichothec-9-ene), → T-2 toxin, T-2 tetraol, → zearalenone may be produced.

Fusarium sulphureum → Fusarium sambucinum Fuckel

Fusarochromanone (Syn.: TDP-1) is a water-soluble chromone derivative containing an amino group at C-5 and a side chain at C-6 (5-amino-6-(3-amino-4-hydroxy-1-oxobutyl)-2,3-dihydro-2,2-dimethyl-4H-1-benzopyran-4-one). As a metabolite of → Fusarium equiseti (Corda) Sacc. Sensu Gordon, it was first isolated and described in 1986 (see Figure Fusarochromanone).

CHEMICAL DATA
Empirical formula: $C_{15}H_{20}N_2O_4$, molecular weight: 292

FUNGAL SOURCES
F. equiseti

NATURAL OCCURRENCE
It should be present in → cereals, viz. → maize and → wheat.

TOXICITY
tibial dyschondroplasia in cattle, chickens, dogs, horses, pigs, and turkeys; hatching reduction of fertile eggs (experimental conditions)

DETECTION
fluorescence detection, HPLC

POSSIBLE MYCOTOXICOSIS
It may be involved in the → Kashin-Beck disease.

FURTHER COMMENTS
There are two derivatives: TDP-2 the C-3'-N-acetyl derivative, TDP-6 containing a hydroxyl group on C-3' and a methoxyl group on C-4'.

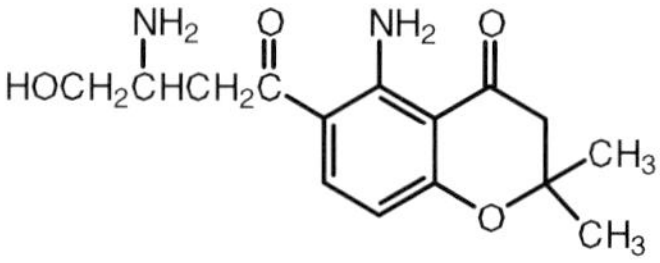

Fusarochromanone

G

Gabi → tubers

Galgant (*Alpinia officinarum* Hance) is a ginger-like spice.
Glagant may contain the following → mycotoxins:
→ aflatoxin B_1
incidence: 1/4, conc.: $\leq$ 5 µg / kg, country: Germany
→ spices

Garlic may contain the following → mycotoxins:
→ aflatoxin B_1
incidence: 1/6, conc.: 12 µg / kg, country: India
→ aflatoxin B_2
incidence: 1/6, conc.: 15 µg / kg, country: India
→ aflatoxin G_1
incidence: 1/6, conc.: 10 µg / kg, country: India
→ deoxynivalenol
incidence: 1/4 conc.: 14 µg / kg, country: UK
→ fusarenon X
incidence: 1/4, conc.: 5 µg / kg, country: UK
→ nivalenol
incidence: 1/4, conc.: 21 µg / kg, country: UK
→ spices

Garlic onions may contain the following → mycotoxins:
→ aflatoxin B_1
incidence: 2*/58, ∅ conc.: 67 µg / kg, country: Thailand, *total ∅ conc.: 60 µg / kg AFB_1, AFB_2, AFG_1, AFG_2
→ spices

Garlic pickle may contain the following → mycotoxins:
→ aflatoxins (AFB_1, AFB_2, AFG_1, AFG_2)
incidence: nc/4, conc. range: 0.2-0.6 µg / kg, country: UK
→ ochratoxin A
incidence: 3/4, conc. range: 0.9-2.5 µg / kg, country: UK
→ zearalenone
incidence: 1/4, conc.: 3.8 µg / kg, country: UK
→ spices

Garlic powder may contain the following → mycotoxins:
→ aflatoxins (AFB_1, AFB_2, AFG_1, AFG_2)
incidence: 1/41*, conc.: 3.3 µg / kg, country: UK, *imported, miscellaneous → spices

genotoxic changes the genom

Gigantic acid (Syn.: → patulin)

Ginger is a dried rhizome of tropical origin. Ways of mycotoxin contamination have not yet been elucidated.
Ginger may contain the following → mycotoxins:
→ aflatoxin B_1
incidence: 8/15, conc. range: < 2.5-25 µg / kg, country: India
incidence: 3/5, conc. range: 1.4-6.5 µg / kg, ∅ conc.: 4.03 µg / kg, country: USA
→ aflatoxin B_2
incidence: 1/5, conc.: 0.2 µg / kg, country: USA
→ aflatoxin G_1
incidence: 1/5, conc.: 2.5 µg / kg, country: USA
→ aflatoxin G_2
incidence: 1/5, conc.: 0.2 µg / kg, country: USA
→ aflatoxins (AFB_1, AFB_2, AFG_1, AFG_2)
incidence: 3/41*, conc. range: 1.3-8.4 µg / kg, ∅ conc.: 3.9 µg / kg, country: UK, *imported, miscellaneous → spices
incidence: nc/4, conc. range: 4.2-13.5 µg / kg, country: UK

incidence: 2/3, conc. range: $\leq$ 2 μg / kg, country: USA
aflatoxins (no specification)
incidence: 1/3*, conc.: 2 μg / kg, country: USA, *imported
→ deoxynivalenol
incidence: 1/4, conc.: 9 μg / kg, country: UK
→ neosolaniol
incidence: 1/4, conc.: 23 μg / kg, country: UK
→ nivalenol
incidence: 1/4, conc.: 34 μg / kg, country: UK
→ ochratoxin A
incidence: 2/4, conc. range: 2.1-7.5 μg / kg, country: UK
→ T-2 toxin
incidence: 1/4, conc.: 18 μg / kg, country: UK
→ spices

Goose may contain the following
→ mycotoxins:
→ ochratoxin A
incidence: 5/12, conc. range: $\leq$ 0.1 μg / kg, ∅ conc.: 0.03 μg / kg, country: Denmark
incidence: 4/12*, conc. range: $\leq$ 0.06 μg / kg, ∅ conc.: 0.02 μg / kg, country: Denmark, *goose liver
→ meat

Grains (no specification)
Grinding destroys the protective outer testa of → cereals and thus enables the rich nutrients inside to be colonized by mycotoxin-producing fungi. Therefore, ground → grains are often more contaminated than intact grains. Ca. 25% of the strains of → Aspergillus and → Penicillium isolated from grain are able to produce → ochratoxin A besides other → mycotoxins. OTA levels seem to be a good indicator of proper storage of grain. Grains may contain the following
→ mycotoxins:
→ aflatoxins
incidence: 19/3489, ∅ conc.: 5 μg / kg, country: USA
→ deoxynivalenol
incidence: 2/17, conc. range: 20-130 μg / kg, country: Germany
incidence: 4/11, conc. range: 420-520 μg / kg, ∅ conc.: 470 μg / kg, country: Sweden
→ nivalenol
incidence: 57/190*, conc. range: 20-290 μg / kg, country: Germany, *moldy
→ ochratoxin A
incidence: 2/49, conc.: 18-22 μg / kg, country: Germany
→ T-2 toxin
incidence: 9/230, conc. range: 10-50 μg / kg, country: Finland
→ zearalenone
incidence: 9/114, conc. range: 5-30 μg / kg, country: Austria
incidence: 18/51, conc. range: 10-500 μg / kg, country: Germany
incidence: 3/584, conc. range: 200-1200 μg / kg, country: Poland
incidence: 26/1417, conc. range: > 20 μg / kg, country: UK
→ barley, → buckwheat, → cereals, → cereal products, → maize, → millet, → oats, → rice, → rye, → sorghum, → triticale, → wheat

Grape juice may contain the following
→ mycotoxins:
→ ochratoxin A
incidence: 1/6*, conc.: 0.73 μg / kg, country: Germany, *red
incidence: 12/14*, conc.: $\leq$ 4.7 μg / kg, country: Germany, *white
incidence: 6/7*, ∅ conc.: 0.218 μg / l, country: Switzerland, *red, imported
incidence: 2/3*, ∅ conc.: ca. 0.004 μg / kg, country: Switzerland, *white, partly imported
incidence: 6/18*, conc. range: < 0.005-0.11 μg / l, country: Switzerland, *white, red, rosé

→ patulin
incidence: 8/8*, conc. range: 360-4200 µg / kg, ∅ conc.: 1500 µg / kg, country: Canada, *moldy
incidence: 21/55, conc. range: 1-230 µg / l, country: Germany
incidence: 8/16, conc. range: 1-8 µg / l, country: UK
→ apple juice, → breakfast drinks, → fruit juice, → fruits, → soft drinks

Groundnut toffee is an Indian peanut based snack. It consists of the crashed kotyledons of the → peanuts, without seedcoat, which are cooked mild in hot concentrated jaggery syrup. The aflatoxin contamination may be lower than that of → bondakaledkai. This may result from a certain degree of cleaning of the infested seeds.

Groundnut toffees may contain the following → mycotoxins:
→ aflatoxin B_1
incidence: 19/67, conc. range: 2-400 µg / kg, country: India
→ aflatoxin B_2
incidence: 3/67, conc. range: 3-120 µg / kg, country: India
→ congressbele

Groundnuts → peanuts

Gushing It could be shown that commercial beers (→ beer) suspected of gushing, had significantly higher concentrations of → deoxynivalenol compared with non-gushing beers.

H

Ham may contain the following
→ mycotoxins:
→ aflatoxin B_1
incidence: 1/1, conc.: 100 µg / kg, country: Germany
→ ochratoxin A
incidence: 60/206*, conc. range: 40-70 µg / kg, country: Yugoslavia, *total of smoked → meat

Hare (wild)
may contain the following → mycotoxins:
→ aflatoxin B_1 →
incidence: 89/168*, conc. range: 0.3-1.421 µg / kg, ∅ conc.: 0.407 µg / kg, country: Czechoslovakia, *liver
incidence: 94/168*, conc. range: 0.3-3.21 µg / kg, ∅ conc.: 0.658 µg / kg, country: Czechoslovakia, *kidney
→ meat

Hazelnuts (no specification)
may contain the following → mycotoxins:
→ aflatoxin B_1
incidence: 1/199*, conc.: 325 µg / kg, country: Finland, *imported
incidence: 18/29*, conc. range: 5-50,000 µg / kg, country: Germany, *moldy
incidence: 1/3*, conc.: 0.6 µg / kg, country: UK, *shelled
incidence: 11/142*, conc. range: 2-100 µg / kg, ∅ conc.: 33 µg / kg, country: USA, *imported
→ aflatoxin B_2
incidence: 1/199*, conc.: 29 µg / kg, country: Finland, *imported
→ aflatoxins (AFB_1, AFB_2, AFG_1, AFG_2)
incidence: 3/35*, conc. range: 6-10 µg / kg, country: Sweden *imported; edible, possibly edible and inedible
→ nuts
incidence: 2/18*, conc. range: 0.5-5 µg / kg with a maximum of 0.7 µg / kg, country. UK, *in-shell
→ aflatoxins (no specification)
incidence: 18/20, conc. range: 25-175 µg / kg, country: Egypt
→ ochratoxin A
incidence: 3/57, conc. range: ≤ 4.7 µg / kg, country: Germany
incidence: 2/11, conc. range: ≤ 1.49 µg / kg, country: Germany
nuts

Hematuria blood in the urin

Hemorrhage bleeding, escape of blood

Hemorrhagic aleukia (Syn.: → Alimentary toxic aleukia)

Hemorrhagic syndrome → Alimentary toxic aleukia, → Moldy corn toxicosis

Hens may contain the following
→ mycotoxins:
→ ochratoxin A
Levels up to 29 µg / kg were found in the muscle of hens and chickens collected at a slaughterhouse. The birds had been rejected because of → nephropathy.
→ meat

hepatic pertaining to the liver

Hepatitis inflammation of the liver

hiptagenic acid → β-nitropropionic acid

Holy fire → Ergotism

Hot dog The → aflatoxins detected in hot dogs derive from the use of mycotoxin-contaminated → spices and / or the incorporation of aflatoxin producers.
Hot dog may contain the following
→ mycotoxins:
→ aflatoxin B_1
incidence: 1/25, conc.: 5 µg / kg, country: Egypt
→ aflatoxin B_2
incidence: 1/25, conc.: 2 µg / kg, country: Egypt
→ meat

HT-2 toxin is a 3α,4β-dihydroxy-4,15-diacetoxy-8α-(3-methylbutyryloxy)-12,13-epoxytrichothec-9-ene which belongs to the trichothecene (→ trichothecenes) → mycotoxins as a metabolite of → Fusarium spp. (see Figure HT-2 toxin).

CHEMICAL DATA
Empirical formula: $C_{22}H_{23}O_8$, molecular weight: 424

FUNGAL SOURCES
Fusarium acuminatum, → Fusarium graminearum Schwabe, → Fusarium poae (Peck) Wollenw., → Fusarium sporotrichioides Sherb.

NATURAL OCCURRENCE
→ barley, → chilli powder, → curry, → maize, → oats, → rye, → soybean, → wheat

TOXICITY
dermatoxic (similar to → T-2 toxin)
inhibition of the initiation step in protein synthesis
LD_{50} (ip): 9 mg / kg bw mice

DETECTION
GC, MS, spectroscopy, TLC

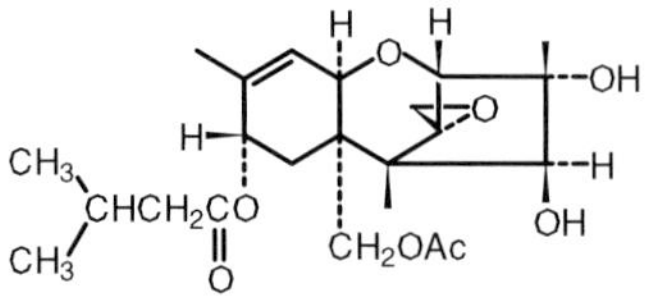

HT-2 toxin

Human breast milk The ingestion of aflatoxin-contaminated (→ aflatoxins) foods by humans will result in the elimination of variable levels of the toxin in body fluids or the accumulation in the tissue. This is currently a considerable problem for people living in tropical and subtropical countries because → aflatoxin B_1 and the corresponding metabolites in human blood and breast → milk represent a serious health hazard to the mother, to the fetus, and to newborn infants. The → aflatoxin M_1 contamination of breast milk is mainly caused by the consumption of food of plant origin, e.g. → peanuts, → maize.
EU data indicate an ochratoxin A contamination between 0.007-0.58 µg / l human milk. Breast-fed infants may ingest (very) high levels of OTA.
Human breast milk may contain the following → mycotoxins:
→ aflatoxicol
incidence: 3/264, conc. range: 0.64-0.27 µg / l, country: Ghana, Nigeria
aflatoxin B_1
incidence: 17/264, conc. range: 0.13-8.218 µg / l, country: Ghana, Nigeria
→ aflatoxin B_2
incidence: 2/264, conc. range: 0.04-0.05 µg / l, country: Ghana, Nigeria
aflatoxin M_1
incidence: 2/2, conc. range: 0.17-0.79 µg / l, Ø conc.: 0.48 µg / l, country: Algeria
incidence: 11/73, conc. range: 0.028-1.031 µg / l, country: Australia
incidence: 1/1, conc.: 0.158 µg / l, country: Bahrain
incidence: 6/6, conc. range: 0.006-0.174 µg / l, Ø conc.: 0.061 µg / l, country: Bangladesh
incidence: 48/48, conc. range: 0.004-0.72 µg / l, Ø conc.: 0.092 µg / kg, country: Egypt
incidence: 90/264, conc. range: 0.02-1.816 µg / l, country: Ghana
incidence: 163/510, conc. range: 0.005-1.379 µg / l, country: Ghana
incidence: 48/48, conc. range: 0.004-0.6 µg / l, Ø conc.: 0.099 µg / l, country: India
incidence: 2/2, conc. range: 0.003-0.051 µg / l, Ø conc.: 0.027 µg / l, country: Indonesia
incidence: 3/3, conc. range: 0.051-1.6 µg / l, Ø conc.: 0.58 µg / l, country: Iran
incidence: 2/2, conc. range: 0.008-0.014 µg / l, Ø conc.: 0.011 µg / l, country: Iraq

incidence: 42/42 conc. range: 0.002-0.88 µg / l, ∅ conc.: 0.122 µg / kg, country: Jordan
incidence: 53/191, conc. range: 0.005-1.379 µg / l, country: Kenya
incidence: 15/15, conc. range: 0.014-1.0 µg / l, ∅ conc.: 0.181 µg / l, country: Lebanon
incidence: 3/3, conc. range: 0.007-0.15 µg / l, ∅ conc.: 0.056 µg / l, country: Morocco
incidence: 6/6, conc. range: 0.07-0.978 µg / l, ∅ conc.: 0.35 µg / l, country: Oman
incidence: 44/44, conc. range: 0.002-1.1 µg / l, ∅ conc.: 0.178 µg / l, country: Pakistan
incidence: 54/55, conc. range: 0-0.84 µg / l, ∅ conc.: 0.115 µg / l, country: Palestine
incidence: 2/2, conc. range: 0.25-0.58 µg / l, ∅ conc.: 0.415 µg / l, country: Philippines
incidence: 3/7, conc.: nc, country: Philippines
incidence: 2/2, conc. range: 0.058-0.395 µg / l, ∅ conc.: 0.227 µg / l, country: Saudi Arabia
incidence: 18/18, conc. range: 0.002-1.0 µg / l, ∅ conc.: 0.217 µg / l, country: Somalia
incidence: 37/99, conc. range: 0.005-1.379 µg / l, country: Sudan
incidence: 44/44, conc. range: 0.003-2.1 µg / l, ∅ conc.: 0.285 µg / l, country: Sudan
incidence: 13/99, conc. range: 0.005-0.064 µg / l, country: Sudan
incidence: 36/36, conc. range: 0.003-0.8 µg / l, ∅ conc.: 0.204 µg / l, country: Syria
incidence: 10/64, conc. range: 0.3-1.3 µg / l, country: UAE
incidence: 37/37, conc. range: 0.009-3.0 µg / l, ∅ conc.: 0.412 µg / l, country: UEA
incidence: 5/11, conc. range: 0.039-1.736 µg / l, country: Thailand
incidence: 1/1, conc.: 0.02 µg / l, country: The Netherlands
incidence: 27/28, conc. range: 0-1.6 µg / l, ∅ conc.: 0.17 µg / l, country: Yemen
incidence: 6/64, conc. range: ≤ 0.05 µg / l, country: Zimbabwe
→ aflatoxin M_2
incidence: 18/264, conc. range: 0.016-2.075 µg / l, Ghana, Nigeria
incidence: 11/99, conc. range: 0.003-0.020 µg / l, country: Sudan
aflatoxin M_1 & M_2
incidence: 13/99, conc. range: 0.003-0.084 µg / l, country: Sudan
→ ochratoxin A
incidence: 4/36, conc. range: 0.017-0.03 µg / l, ∅ conc.: 0.024 µg / l, country: Germany
incidence: 9/50, conc. range: 1.7-6.6 µg / l, country: Italy
incidence: 22/111, conc. range: 0.1-12 µg / l, country: Italy
incidence: 38/115, conc. range: 0.001-0.13 µg/l, country: Norway
incidence: 23/40, conc. range: 0.01-0.04 µg / l, country: Sweden
ochratoxin A methyl ester
incidence: 4/40, conc. range: 0.01-0.04 µg / l, country: Sweden
→ dairy products

Human hepacellular carcinoma → Aflatoxin B_1 as an extremely potent hepatocarcinogen, is distributed in human foodstuffs especially in sub-Saharan African countries (e.g. Kenya, Mozambique, Swaziland) and southeast Asia (Thailand), where a high incidence of liver cancer can be found. Epidemiological studies showed a highly significant positive correlation between the liver-cancer rate and the level of dietary aflatoxin intake. However, infection with hepatitis B virus may predispose people for primary hepatocellular carcinoma. Although some other agents may also be involved in the development of this disease, interaction between the hepatitis B virus and aflatoxin appears the most plausible explanation available.

Human milk → human breast milk

Human serum About 50% of the European human sera investigated showed contamination with → ochratoxin A (< 0.1-57 ng OTA / ml serum), with a slightly higher incidence in rural areas. OTA positive human blood sera also occurred in Canada. There are three main causes for this high contamination rate:

- long biological half-time of OTA which is bound to serumalbumines,
- intake of OTA-contaminated foodstuff,
- inhalation of OTA-contaminated conidia.

Hydnocarpus laurifolia (medicinal seeds) may contain the following → mycotoxins:
→ aflatoxin B_1
incidence: nc/nc, conc. range: 20-650 µg / kg, country: India
→ citrinin
incidence: nc/nc, conc. range: 10-490 µg / kg, country: India

Hydrolyzed fumonisin B_1 (Abbr.: HFB_1, → fumonisins)

Hydroxydihydroaflatoxin B_1 → Aflatoxin B_{2a}

4-Hydroxyochratoxin A Although this mycotoxin is produced by → Penicillium viridicatum Westling this compound primarily seems to be a detoxification product in animals (e.g. rats) dosed with → ochratoxin A.

3´-Hydroxy HT-2 toxin is a metabolite of → HT-2 toxin and a contaminant of → milk, plasma and the excreta of cows. There is no accumulation in any organ.

3´-Hydroxy T-2 toxin is a metabolite of → T-2 toxin and a contaminant of → milk, plasma and the excreta of cows. There is no accumulation in any organ.

Hyperemia engorgement of blood

Hyperestrogenism → F-2 toxicoses

Hyperplasia an abnormal increase in the number of cells.

Hypocreaceae → Hypocreales

Hypocreales → Ascomycota

Hypothermia an unusually low body temperature

I

Ice cream Manufacturing of naturally contaminated → milk will result in → aflatoxin M_1 contamination of ice → cream because no toxin destruction occurred after 8 month of frozen storage.
→ coconut ice

Icterus → Jaundice

immunosuppressive increased susceptibility to diseases caused by bacteria, viruses and fungi

Incarpina is a product consisting of → maize plus cottonseed → flour.
Incarpina may contain the following → mycotoxins:
→ fumonisin B_2
incidence: 1/1, conc.: 140 µg / kg, country: Guatemala

Indian cassia (*Cinnamomum tamala* (Bush.-Ham.)
may contain the following → mycotoxins:
→ aflatoxin B_1
incidence: 1/6, conc.: 13 µg / kg, country: India
→ aflatoxin B_2
incidence: 1/6, conc.: 11 µg / kg, country: India
→ aflatoxin G_1
incidence: 1/6, conc.: 8 µg / kg, country: India
→ aflatoxin G_2
incidence: 1/6, conc.: 4 µg / kg, country: India
→ spices

Indian childhood cirrhosis This disease caused vague gastrointestinal symptoms and → anorexia. The subsequent hepatomegaly often resulted in → icterus, → ascites and → hepatic coma, mainly in children with a peak incidence at 3 years, in certain areas of India. The detection of aflatoxin-like fluorescent substances succeeded in the mother´s breast milk (→ Human breast milk), the urine of affected children, parboiled → rice and the → peanut oil used for frying most foods. However, because no chemical confirmation of the identity of these compounds was carried out, the etiology of this lethal disease is unresolved.

Ingwer may contain the following → mycotoxins:
→ aflatoxin B_1
incidence: 8/15, conc. range: < 2.5-12.5 µg / kg, country: Canada
incidence: 1/12, conc.: ≤ 5 µg / kg, country: Germany
incidence: 3/5, conc. range: 1.4-6.5 µg / kg, country: USA
→ spices

Intracellular mycotoxins like → penitrem A, → roquefortine C, → sterigmatocystin, verrrucosidin are mainly intracellular.
→ Extracellular mycotoxins, → mycotoxins

Islanditoxin is composed of L-serine, L-ß-phenyl-ß-aminopropionic acid, L-α-amino-n-butyric acid, and L-dichloroproline in the mole ratio of 2:1:1:1 (→ mycotoxins). This cyclic, water soluble, colorless chloropeptide was first isolated in 1955 and structurally elucidated in 1959 (see Figure Islanditoxin).

CHEMICAL DATA
Empirical formula: $C_{24}H_{31}O_7N_5Cl_2$, molecular weight: 571

FUNGAL SOURCES
Pencillium islandicum Sopp

NATURAL OCCURRENCE
It may be a contaminant of "yellow rice".

TOXICITY
LD_{50} (po): 6.55 mg / kg bw mice
Clinical signs include respiratory and circulatory disturbances, low body temperature, decrease of muscle and skin tension,

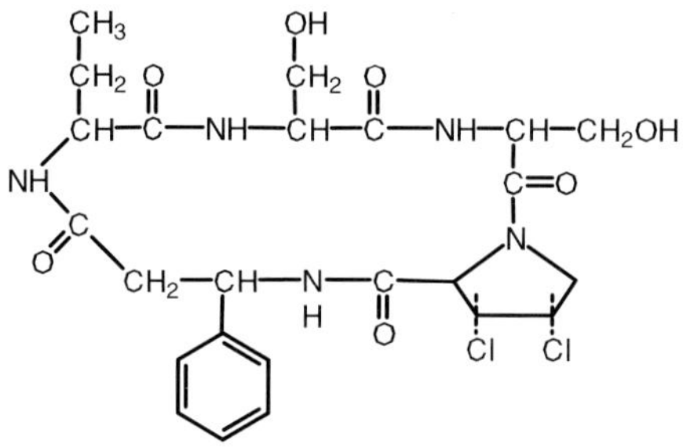

Islanditoxin

enlargement of the liver (significant decline in → hepatic glycogen content, concomitant decrease in hepatic glycogen synthetase activity), hemorrhagic changes (→ hemorrhage) in the small intestines.

DETECTION
TLC

POSSIBLE MYCOTOXICOSIS
In combination with → luteoskyrin, islanditoxin should be responsible for the → Yellow rice disease

FURTHER COMMENTS
Compared to luteoskyrin it is more toxic. If the chlorine atoms are removed, the toxicity of islanditoxin is significantly reduced.

Isofumigaclavine A, B (Syn.: → roquefortine A & B)

J

Jam (no specification)
may contain the following → mycotoxins:
→ patulin
incidence: 10/20, conc. range: 5-50 µg/kg, country: Italy
incidence: 15/35, conc. range: 2-20 µg/kg, country: Germany
→ fruits

Jaundice is characterized by very high levels of bilirubin in the blood while bile pigment is deposited in the skin and mucous membrane, causing to a yellow appearance.

Job's-tears is an oriental kind of seed. Job's-tears may contain the following → mycotoxins:
→ aflatoxin B_1
incidence: 34/144, conc. range: 0.1-14.9 µg/kg, country: Japan
→ aflatoxin B_2
incidence: 34/144, conc. range: tr-1.8 µg/kg, country: Japan
→ aflatoxin G_1
incidence: 34/144, conc. range: 0.3-0.7 µg/kg, country: Japan
→ deoxynivalenol
incidence: 2/12, conc. range: 48-496 µg/kg, country: Japan
→ nivalenol
incidence: 11/12, conc. range: 3-920 µg/kg, country: Japan
→ zearalenone
incidence: 7/7, conc. range: 10-440 µg/kg, country: Japan

K

Kashin-Beck disease (Syn.: Urov disease) is neither heritable nor congenital. Its etiology is still unknown KBD is endemic to northern China, North-Korea, Russia (Siberia) and was first described along the Urov river in Russia in 1861. Especially in Russia certain climatic conditions, like significant temperature changes during the day, major rainfall during late summer and / or early fall at grain maturing and harvesting in connection with this disease have been reported. More recently KBD (named after the two Russian scientists Kashin and Beck who studied the disease from the 1860s) has also been detected in Taiwan, Japan, Sweden, and Holland. In China about two million people are affected, predominantly peasants in rural areas. The consumption of → maize and → wheat infected with → Fusarium spp. may be responsible (→ Fusarium equiseti (Corda) Sacc. sensu Gordon / → fusarochromanone). Besides the possible involvement of different *Fusarium* species (*F. equiseti*, → Fusarium oxysporum Schlecht. emend. Snyd. & Hansen, → Fusarium poae (Peck) Wollenw.), selenium deficiency as well as trace metal toxicity have been discussed as possible etiological agents.
Pre-adolescent and adolescent children are primarily affected. They show bone and joint deformation, typically in the elbows, knees, and ankles, which leads to impaired mobility; disproportionate dwarfism may also occur. At an early stage, reversibility of the disease is possible if the patients leave the endemic area. The decline of KBD in some areas may be the result of improved hygienic conditions, together with the import of grain from non-endemic regions. Further studies are needed to elucidate the causative agents of this widespread, crippling disease.

Kodua poisoning occurs in India and may be due to the ingestion of kodo → millet seeds (*Paspalum* spp.) by cattle and humans that are contaminated with → Aspergillus spp. / → cyclopiazonic acid. Cattle show symptoms of nervousness, lack of muscular coordination, depression and spasms, death sometimes occurs. The accidental consumption of the contaminated, dehusked grains cooked like → rice or used in → bread baking caused tremors (→ tremorgenic mycotoxins), giddiness, and sleepiness.

Kojic acid is a 2-hydroxymethyl-5-hydroxy-2-γ-pyrone (→ mycotoxins) produced by several → Aspergillus and → Penicillium species (see Figure Kojic acid).

CHEMICAL DATA
Empirical formula: $C_6H_6O_4$, molecular weight: 142

FUNGAL SOURCES
Aspergillus spp. (e.g. → Aspergillus candidus Link, → Aspergillus flavus Link, → Aspergillus oryzae (Ahlburg) Cohn, → Aspergillus parasiticus Speare, *A. tamarii* group, *A. wentii* group), *Penicillium* spp. (→ Penicillium citrinum Thom, *P. lanosum*, *P. rubrum*) and *Verticillium dahliae*.

NATURAL OCCURRENCE
→ figs, → maize

TOXICITY
convulsive (→ convulsions), → mutagenic insecticidal
LD_{50} (ip): 30 mg / mice
Large amounts are necessary to produce server intoxication or death in animals. Up to now, no natural cases of kojic acid

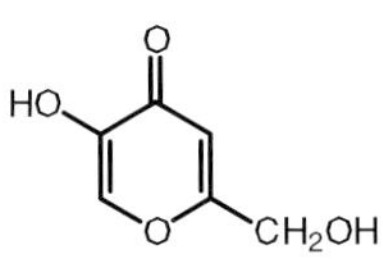

Kojic acid

toxicosis have appeared in animals or humans.

DETECTION
TLC

Koshk → yoghurt

Kubeba is an Egyptian meat product. Detection of → aflatoxins in kubeba results from the use of mycotoxin contaminated → spices and / or the incorporation of aflatoxin producers.
Kubeba may contain the following → mycotoxins:
→ aflatoxin B_1
incidence: 1/25, conc.: 150 µg / kg, country: Egypt
→ aflatoxin B_2
incidence: 1/25, conc.: 25 µg / kg, country: Egypt
→ meat

Kulen is a Yugoslavian → meat speciality and may contain the following → mycotoxins:
→ ochratoxin A
incidence: 27/206* conc. range: 10-460 µg / kg, country: Yugoslavia, *total of smoked → meat

Kwashiorkor is common in tropical and subtropical countries / areas like parts of Brazil, Central America, southern India, Mexico, South Africa, Uganda, and parts of Zaire. The local main staple foods, e.g. → maize, → rice and / or plantains (high in starch, low in protein), are often contaminated with → aflatoxins. Accumulation of these → mycotoxins in the body fluids and tissues of very young children suffering from kwashiorkor has been reported. The symptoms shown by these children may in part also be due to protein malnutrition.
The disease is characterized by several clinical signs like hypoalbuminaemia, → edema, immunosuppression (→ immunosuppressive), and fatty liver. These symptons are also caused by aflatoxins in experimental animals (guinea pigs). Although an association between aflatoxin and kwashiorkor has been established, conclusive evidence is still lacking.

L

Lasiosphaeriaceae → Sordariales

Lemons (pickled in salt)
may contain the following → mycotoxins:
→ aflatoxin B_1
incidence: 19/40*, Ø conc.: 195 µg / kg, country: India
incidence: 3/18**, conc. range: 20-60 µg / kg, country: Germany, **moldy
→ aflatoxin B_2
incidence: 19/40*, Ø conc.: 42 µg / kg, country: India
→ aflatoxin G_1
incidence: 19/40*, Ø conc.: 110 µg / kg, country: India
→ aflatoxin G_2
incidence: 19/40*, Ø conc.: 25 µg / kg, country: India
*stored in polythene bags
→ fruits

Lentils may contain the following → mycotoxins:
→ aflatoxin B_1
incidence: 1/6, conc.: 3.1 µg / kg, country: Egypt
incidence: 1/4, conc.: 72 µg / kg, country: Egypt
incidence: 1/20, conc.: 8-10 µg / kg, country: Italy
incidence: 1/2*, conc.: 1.8 µg / kg, country: Syria, *ground
→ aflatoxins
incidence: 1/20, conc.: 20 µg / kg, country: Egypt
→ ochratoxin A
incidence: 2/10, conc. range: 0.1-0.19 µg / kg, country: Germany
→ beans, → cabbage, → cowpeas, → pigeon peas, → peas, → soybeans, → vegetables

Leucopin (Syn.: → patulin)

Leukocytosis transient increase in the amount of the white blood cells in the blood.

Lewia → Pleosporaceae

Libritos may contain the following → mycotoxins:
→ deoxynivalenol
incidence: 4/4, conc. range: 210-1023 µg / kg, Ø conc.: 581 µg / kg, country: Argentina
→ wheat products

Lima beans → Beans, lima

Lineseed oil may contain the following → mycotoxins:
→ aflatoxin B_1
incidence: 1/10, conc.: 1.2 µg / kg, country: Germany

Lineseeds may contain the following → mycotoxins:
→ aflatoxin B_1
incidence: 1*/6, conc.: 1.1 µg / kg, country: Germany
→ aflatoxin G_1
incidence: 1*/6, conc.: 0.9 µg / kg, country: Germany
*soaked for 36 h

Lingonberries may contain the following → mycotoxins:
→ patulin
incidence: 1/2, conc.: 265 µg / kg, country: Sweden
→ fruits

Liver → Cattle liver
→ Pig liver
→ Sausage

Losses Worldwide losses in the export market due to mycotoxin contamination in only five crops (→ barley, cottonseed, → maize, → peanuts, and → rice) were estimated for 1985 at about 1.5 billion dollars. These losses did not consider human costs due to mycotoxin (→ mycotoxins) contamination, losses in pigs / abortions, dairy → cattle / → milk production and → poultry / egg production, losses in domestic animals due to reduced weight gain because of mycotoxin-contaminated animal feeds, losses due to lower prices for lower quality → grains etc. In view of all these costs, a major research effort in mycotoxin prevention and control is necessary.

Low water activity foods Direct mycotoxin (→ mycotoxins) contamination in foods characterized by low water activities is most unlikely. The minimum → a_w for mycotoxin production is a_w 0.80 (*Aspergillus ochraceus* / → penicillic acid).

Luteoskyrin (Syn.: flavomycelin) is a 2,2´,4,4´,5,5´,8,8´-octahydroxy-2,2´,3,3´-tetrahydro-7,7´-dimethyl-1,1´-bianthraquinone (→ mycotoxins). This yellow anthraquinone-like pigment is produced by → Penicillium islandicum Sopp which was first isolated in 1912 from skyr, an Iceland kind of → yogurt (see Figure Luteoskyrin). Due to pyrolysis, this bisanthraquinone decomposes into cateniarin and → islanditoxin in a molar ratio of 1:1.

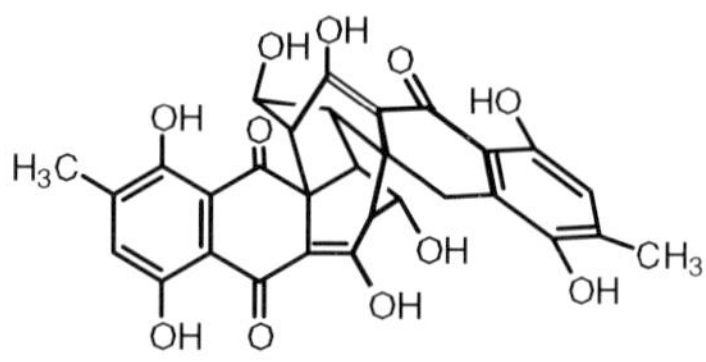

Luteoskyrin

Chemical Data
Empirical formula: $C_{30}H_{22}O_{12}$, molecular weight: 574

Fungal Sources
P. islandicum

Natural Occurrence
P. islandicum is promoted in tropical and subtropical climates especially in the → rice-growing areas of Asia and Africa where high temperatures and humid conditions are common. This yellow rice may be contaminated with luteoskyrin. A high incidences of diseases such as liver cirrhosis and carcinoma occurs in such areas.
P. islandicum grows on contaminated rice, → maize, and other → cereals. Because of the lipophilic nature of luteoskyrin, → oil processed from contaminated rice husks might be a high risk foodstuff. In Europe animal feed is mainly affected by luteoskyrin contamination, while food contamination is rare.

Toxicity
hepatotoxic: the liver shows yellow discoloration, centrilobular necrosis, fatty degeneration, liver tumors (mice); carcinogenic
LD_{50} (po): 221 mg / kg bw mice, significant toxicological variation depending on the route of administration. Chemically luteoskyrin is very similar to → rugulosin which caused the same clinical signs.

Detection
TLC

Possible Mycotoxicosis
→ Yellow rice disease

Further Comments
→ Apples and → grape juice are very good substrates for luteoskyrin production.

Lymphocytosis exessive increase in the number of lymph cells

M

Macrofusin (Syn.: fumonisin B_1 → fumonisins)

Maize Among cereal → grains maize as a staple food is considered as a high risk crop for mycotoxin production. The toxin levels found are generally higher than those of other small-grain → cereals.
→ Fusarium graminearum Schwabe, → Fusarium proliferatum (Matsushima) Nirenberg and → Fusarium moniliforme Sheldon are extremely common on maize ears and their → mycotoxins represent a serious health hazard to man and domestic animals. In addition, *F. subglutinans* is also very common on maize.
F. graminearum and *F. moniliforme* colonize maize seeds on the cob or in the ear. The developing corn is invaded by *F. graminearum* at the silking stage, especially in periods of heavy rainfall. Wet or insufficiently dried stored kernels promote mycelial growth, while low temperatures ($\leq 15\ ^\circ C$) are essential for → zearalenone production. Ears stored in cribs are most frequently affected by zearalenone contamination.
F. moniliforme is primarily an internally seed-borne fungus but it is also soil-borne and survives in plant residues. *F. moniliforme* requires a minimum moisture content of 18.4% (maize) for vegetative growth. The fungus invades the seed through the pedicle to colonize the internal section of the kernel, including the embryo. In consequence, fumonisin contamination may occur. Since fumonisin production is favored by high seed moisture contents, maximum formation probably occurs during harvest or before drying and storage. In the field → fumonisin B_1 formation in maize succeeded FB_2 and FB_3 production while most of the fumonisin is located in the pericarp layer of maize kernels.
In Argentinian maize a good agreement between fungal contamination (*F. moniliforme, F. proliferatum*) at the medium and late stages of maturity and fumonisin contamination was found.
Conventional grading of corn for human consumption is not effective to lower fumonisin levels because "first-grade" maize may also contain high fumonisin concentrations. In general, processing does not remove or inactivate fumonisin B_1 in maize and maize products. So, these foodstuffs are regarded as the main sources for human and animal FB_1 intake. However, it could be shown that → milling caused the accumulation of → fumonisins in maize screenings and → bran. Therefore, increasing the level of refinement of corn meal may cause a decrease in fumonisin levels.
→ Fusarium mycotoxins occur in different parts of the maize plant, e.g. stalk, leaf axis, peduncle, rachis and kernels but they are unevenly distributed. It seems that the kernels are one of the least contaminated parts of the plant. Therefore, mycotoxin contamination of maize grain may be minimized by prompt, undelayed, harvesting.
In addition, maize may be prone to field infection with → Aspergillus flavus Link and → Aspergillus parasiticus Speare resulting in significant aflatoxin contamination before harvesting. It is suggested that insect damage and inoculum spread is the major cause of maize infections in the USA. Colonization of the base of the kernels is due to the inoculum present on the infected maize silks. The high carbohydrate and low nitrogen content of the seeds favor aflatoxin production but regional differences in contamination (higher levels: e.g. southeastern States USA, western and northern regions India) do occur.
The aflatoxigenic fungi are primarily surface-borne but occasionally internally seed-borne. The maturing kernels are

highly susceptible during the late milk and dough stage, but infection is also possible at a later stage before harvesting. Integrity of the kernel in general prevents aflatoxin contamination but direct colonization of intact kernels through the silk scars by the fungus without harming the process of fertilization and growth of the developing embryo is possible. However, breaks in the pericarp tissue due to insects (European corn borer = *Ostrinia nubialis*, corn earworm = *Heliothis zea* and the rice weevil = *Sitophilus oryzae*) which also transmit spores of *A. flavus*, plant stress (e.g. drought, low soil fertility, weed competition) and / or mechanical damage major contribute to infection of maize kernels.

During warm weather at harvest there is a high risk of aflatoxin contamination of maize. Temperatures of 24 °C or a moisture content of 17.5% are necessary for aflatoxin B_1 production in stored maize. → Aflatoxins as well as zearalenone formation occurs during the development of "hot spots".

Single kernels or pieces of kernels of a maize sample may contain very high levels of → aflatoxin B_1 (88,500-101,000 µg / kg). In naturally contaminated maize → aflatoxin G_1 is always found to a lower extent than AFB_1 and never occurs in the absence of AFB_1. Aflatoxins may also be present in all corn fractions including sound kernels, damaged and discolored kernels, fluorescing kernels, kernels with visible fluorescence beneath the seed coat, broken corn-foreign material. Aflatoxin-containing particles can be removed to a different degree by cleaning processes, e.g. blowers and sieves, because the contaminated fragments shatter easily. Electronic sorting devices are also helpful.

Wet- or dry-milling of maize will result in the accumulation of aflatoxins mainly in the feed fractions. This phenomenon is even more pronounced in the wet milling process.

Besides the simultaneous contamination of maize with different *Fusarium* mycotoxins co-contamination with aflatoxin and fumonisin B_1 has been detected. Under favorable conditions the growth and mycotoxin production of aflatoxigenic fungi as well as *F. moniliforme* and / or *F. proliferatum* is possible but negative relationships between these fungi and mycotoxin production have also been reported. It is assumed that fumonisins are more evenly distributed in maize kernels compared to the distribution of aflatoxins.

The excellent mycotoxin-promoting nature of maize is confirmed by the fact that maize kernels contained nearly 10 times more moniliformin than *Fusarium* damaged wheat kernels (Ø ca. 16,000 µg / kg).

Maize may contain the following → mycotoxins:

→ 3-acetyldeoxynivalenol
incidence: 1/1, conc.: 100 µg / kg, country: Austria
incidence: 5/24, conc. range: 30-185 µg / kg, Ø conc.: 113 µg / kg, country: China
incidence: 12/36*, conc. range: 20-1500 µg / kg, Ø conc.: 200 µg / kg, country: Korea, *moldy
incidence: 2/35*, conc. range: 50-200 µg / kg, Ø conc.: 200 µg / kg, country: Korea, *healthy
incidence: 1/9*, conc.: 300 µg / kg, country: Poland, *healthy and damaged kernels

→ 15-acetyldeoxynivalenol
incidence: 5/24, conc. range: 160-1435 µg / kg, Ø conc.: 495 µg / kg, country: China
incidence: 30/36*, conc. range: 20-4600 µg / kg, Ø conc.: 900 µg / kg, country: Korea, *moldy

incidence: 6/35*, conc. range: 2-100 µg / kg, ∅ conc.: 40 µg / kg, country: Korea, *healthy
incidence: 4/9*, conc. range: 2800-7700 µg / kg, ∅ conc.: 4725 µg / kg, country: Poland, *healthy and damaged kernels
incidence: 7/20*, conc. range: 900-7900 µg / kg, ∅ conc.: 1800 µg / kg, country: USA
*moldy
→ 4-acetylnivalenol
incidence: 14/36*, conc. range: 2-2200 µg / kg, ∅ conc.: 400 µg / kg, country: Korea, *moldy
incidence: 3/35*, conc. range: 4-30 µg / kg, ∅ conc.: 10 µg / kg, country: Korea, *healthy
aflatoxicol I
incidence: 2/2, conc. range: 12.9-25.4 µg / kg, ∅ conc.: 19.15 µg / kg, country: Thailand
aflatoxicol II
incidence: 2/2, conc. range: 7.9-15.7 µg / kg, ∅ conc.: 11.8 µg / kg, country: Thailand
→ aflatoxin B_1
incidence: 445/2271, conc. range: $\leq$ 560 µg / kg, ∅ conc.: 11.6 µg / kg, country: Argentina
incidence: 5/150, conc. range: 10-50 µg / kg, ∅ conc.: 24 µg / kg, country: Argentina
incidence: 3/174, conc. range: 1-3 µg / kg, country: Australia
incidence: 1/1, conc.: 131 µg / kg, country: Burma
incidence: 30/36, conc. range: 0.54-76.32 µg / kg, ∅ conc.: 15.8 µg / kg, country: Costa Rica
incidence: 1/6, conc.: 2.6 µg / kg, country: Egypt
incidence: 2/75, conc. range: 10 µg / kg, ∅ conc.: 10 µg / kg, country: France
incidence: 1*/3, conc.: 25 µg / kg, country: Germany, *moldy
incidence: 975/2074, conc. range: > 5-666 µg / kg, country: India
incidence: 6/6, conc. range: tr-15,600 µg / kg, country: India
incidence: 11/16, conc. range: 4-428 µg / kg, ∅ conc.: 102 µg / kg, country: Indonesia
incidence: 10/12, conc. range: 1-3300 µg / kg, ∅ conc.: 352 µg / kg, country: Indonesia
incidence: 50/111, conc. range: 0.02-1.2 µg / kg, country: Italy
incidence: 2/161, conc.: 0.1 µg / kg, country: Japan
incidence: 3/3, conc. range: 8.8-37.5 µg / kg, ∅ conc.: 21.8 µg / kg, country: Nepal
incidence: 44/50, conc. range: 1-430 µg / kg, ∅ conc.: 49 µg / kg, country: Philippines
incidence: 39/155, conc. range: < 5-1500 µg / kg, country: South Africa
incidence: 2/2, conc. range: 131-340 µg / kg, ∅ conc.: 236 µg / kg, country: Thailand
incidence: 158/162, conc. range: 500-1200 µg / kg, country: Thailand
incidence: 17/27, conc. range: 1-606 µg / kg, ∅ conc.: 63 µg / kg, country: Thailand
incidence: 22*/62, ∅ conc.: 400 µg / kg, country: Thailand, *total: ∅ conc.: 2730 µg / kg AFB_1, AFB_2, AFG_1, AFG_2
incidence: 3/38, conc. range: 48-62 µg / kg, country: Tunisia
incidence: 27/167, conc. range: 2-73.9 µg / kg, country: Turkey
incidence: 24/29, conc. range: < 5 µg / kg (21 samples), 6-10 µg / kg (2 sa), 11-15 µg / kg (1 sa), country: UK
incidence: 16/567, conc. range: 20-350 µg / kg, country: USA
incidence: 6/283, conc. range: 6-25 µg / kg, ∅ conc.: 15 µg / kg, country: USA
incidence: 21/60, conc. range: 4-308 µg / kg, ∅ conc.: 58.6 µg / kg, country: USA
incidence: 8/293, conc. range: < 6-25 µg / kg, country: USA

incidence: 27/28, conc. range: 0-321 µg / kg, ∅ conc.: 73 µg / kg, country: USA
incidence: 11*/34, conc. range: 0.7-47 µg / kg, country: USA, *single damaged kernels contained 88,500-101,000 µg AFB_1 / kg
incidence: 25/353, conc. range: 3-19 µg / kg, ∅ conc.: 10.4 µg / kg, country: USA
incidence: 10/918, conc. range: 3-19 µg / kg, ∅ conc.: 9 µg / kg, country: USA
→ aflatoxin B_2
incidence: 92/2271, conc. range: 130 µg / kg, ∅ conc.: 28.15 µg / kg, country: Argentina
incidence: 1/174, conc.: 50 µg / kg, country: Australia
incidence: 1/1, conc.: 18 µg / kg, country: Burma
incidence: 18/36, conc. range.: 0.16-5.82 µg / kg, ∅ conc.: 1.9 µg / kg, country: Costa Rica
incidence: 1/6, conc.: 3.7 µg / kg, country: Egypt
incidence: 9/16, conc. range: 1-160 µg / kg, ∅ conc.: 9 µg / kg, country: Indonesia
incidence: 8/12, conc. range: 1-680 µg / kg, ∅ conc.: 90 µg / kg, country: Indonesia
incidence: 34/50, conc. range: 1-78 µg / kg, ∅ conc.: 14 µg / kg, country: Indonesia
incidence: 3/3, conc. range: 2.3-5 µg / kg, ∅ conc.: 4.1 µg / kg, country: Nepal
incidence: 2/2, conc. range: 17-47 µg / kg, ∅ conc.: 32 µg / kg, country: Thailand
incidence: 135/162, conc. range: 49-260 µg / kg, country: Thailand
incidence: 11/27, conc. range: 1-73 µg / kg, ∅ conc.: 14 µg / kg, country: Thailand
incidence: 8/167, conc. range: 1.5-6 µg / kg, country: Turkey
incidence: 4/567, conc. range: 52-129 µg / kg, country: USA
incidence: 15/60, conc. range: tr-40 µg / kg, country: USA
→ aflatoxin G_1
incidence: 2/150, conc. range: 10-25 µg / kg, ∅ conc.: 17.5 µg / kg, country: Argentina
incidence: 1/174, conc.: 2 µg / kg, country: Australia
incidence: 1/3, conc.: 57.6 µg / kg, country: Nepal
incidence: 2/50, conc. range: 40-78 µg / kg, ∅ conc.: 59 µg / kg, country: Philippines
incidence: 17/162, conc. range: 50-250 µg / kg, country: Thailand
incidence: 3/27, conc. range: 2-7 µg / kg, ∅ conc.: 5 µg / kg, country: Thailand
incidence: 3/38, conc. range: 8-22 µg / kg, country: Tunisia
incidence: 3/167, conc. range: 2-5.4 µg / kg, country: Turkey
incidence: 2/283, conc. range: tr-12 µg / kg, country: USA
incidence: 5/60, conc. range: tr-10 µg / kg, country: USA
incidence: 3/353, conc. range: 3-8 µg / kg, ∅ conc.: 5.7 µg / kg, country: USA
incidence: 3/918, conc. range: tr-3 µg / kg, country: USA
→ aflatoxin G_2
incidence: 5/16, conc. range: tr-8 µg / kg, country: Indonesia
incidence: 2/50, conc. range: 3-33 µg / kg, ∅ conc.: 18 µg / kg, country: Indonesia
incidence: 1/3, conc.: 9.7 µg / kg, country: Nepal
incidence: 2/162, conc. range: 49-110 µg / kg, country: Thailand
incidence: 2/167, conc. range: 2-3 µg / kg, country: Turkey
incidence: 2/60, conc. range: tr-1 µg / kg, country: USA
aflatoxin (no specification)
incidence: 1/71*, conc.: 2-20 µg / kg, country: Uruguay, *and by-products
→ aflatoxins (no specification)
incidence: 1*/36, conc.: < 25 µg / kg, country: Canada, *AFB_1, AFB_2, AFG_1, AFG_2

incidence: 9*/10, conc. range: 2-35 µg / kg, Ø conc.: 9.7 µg / kg, country: Gambia, *AFB_1, AFB_2, AFG_1, AFG_2
incidence: 304/364, conc. range: nc, country: Germany
incidence: 7/22, conc. range: 12-160 µg AFB_1 / kg, 25-90 µg AFB_2 / kg, 10-95 µg AFG_1 / kg, 65 µg AFG_2 / kg, country: India
incidence: 2/8, conc. range: nc, country: Hong Kong
incidence: 2/52, conc. range: nc, country: Mocambique
incidence: 22*/49, conc. range: 1-100 µg / kg (13 samples), 100-1000 µg / kg (9 sa), country: Uganda, * 19 samples contained AFB_1, 11 AFB_2, 14 AFG_1, 4 AFG_2
incidence: 39/45, conc. range: 1-2300 µg / kg, Ø conc. 252 µg / kg, country: USA
incidence: 717/4651, conc. range: 20-100 µg / kg, country: USA
incidence: 40/1594, conc. range: 3-37 µg / kg, Ø conc.: 9 µg / kg, country: USA
incidence: 21/60, conc. range: 6-348 µg / kg, Ø conc.: 66 µg / kg, country: USA
incidence: 235/2866, conc. range: 15 µg / kg, country: USA
incidence: 281/743, Ø conc.: 135 µg / kg, country: USA
incidence: 46/123*, Ø conc.: 130 µg / kg, country: USA
incidence: 49/101*, Ø conc.: 187 µg / kg, country: USA
incidence: 36/99*, Ø conc.: 58 µg / kg, country: USA
incidence: 33/114*, Ø conc.: 118 µg / kg, country: USA
incidence: 81/99*, Ø conc.: 167 µg / kg, country: USA
incidence: 11/90*, Ø conc.: 110 µg / kg, country: USA
incidence: 24/117*, Ø conc.: 176 µg / kg, country: USA
*dent maize
incidence: 49/109, conc. range: ≤ 123 µg / kg, Ø conc.: 30 µg / kg, country: USA
incidence: 12/28, conc. range: ≤ 98 µg / kg, Ø conc.: 20 µg / kg, country: USA
incidence: 63/197, conc. range: ≤ 1019 µg / kg, Ø conc.: 77 µg / kg, country: USA
incidence: 57/315, conc. range: tr-845 µg / kg, country: USA
incidence: 27/28, conc. range: 0-321 µg / kg, Ø conc.: 73 µg / kg, country: USA
incidence: 218/1669, conc. range: 20-99 µg / kg (167 samples), 100 µg / kg (51 sa), country: USA
→ beauvericin
incidence: 6/22*, conc. range: tr-520,000 µg / kg, Ø conc.: 102,833 µg / kg, country: Italy, *visibly infected
→ citrinin
incidence: 1/1, conc.: 212 µg / kg, country: Burma
incidence: 2/2, conc. range: 174-1390 µg / kg, Ø conc.: 782 µg / kg, country: Thailand
incidence: 1/1, conc.: 450 µg / kg, country: UK
→ cyclopiazonic acid
incidence: 23/45, conc. range: < 25-2800 µg / kg, Ø conc.: 467 µg / kg, country: USA
→ deoxynivalenol
incidence: 2/20, Ø conc.: 111 µg / kg, country: Argentina
incidence: 33/100, conc. range: tr-200 µg / kg, country: Argentina
incidence: 1/1*, conc.: 1450 µg / kg, country: Argentina, *flint maize
incidence: 14/58, conc. range: 200-400 µg / kg, country: Argentina
incidence: 77/78, conc. range: ≤ 6200 µg / kg, Ø conc.: 790 µg / kg, country: Austria
incidence: 1/1, conc.: 90,000 µg / kg, country: Austria
incidence: 46/51, conc. range: 40-3700 µg / kg, Ø conc.: 730 µg / kg, country: Austria
incidence: 3/6*, conc. range: 550-50,500 µg / kg, Ø conc.: 17,400 µg / kg, country: Austria, *visibly moldy (*Fusarium* spp.)
incidence: 3/3, conc. range: 1300-7900 µg / kg, country: Austria

incidence: 77/78, conc. range: ≤ 6200 µg / kg, ∅ conc.: 780 µg / kg, country: Austria
incidence: 1/1, conc.: 960 µg / kg, country: Canada
incidence: 2/2*, conc.: 130-700 µg / kg, ∅ conc.: 415 µg / kg, country: Canada, *No. 2
incidence: 243/283, conc. range: 20-4090 µg / kg, ∅ conc.: 610 µg / kg, country: Canada
incidence: 28/28, conc. range: ≤ 4500 µg / kg, ∅ conc.: 1960 µg / kg, country: China
incidence: 24/24, conc. range: 360-12,670 µg / kg, ∅ conc.: 5376 µg / kg, country: China
incidence: 4/4, conc. range: 20-100 µg / kg, country: France
incidence: 1/1, conc.: 40 µg / kg, country: France
incidence: 3/3, conc. range: 20-60 µg / kg, ∅ conc.: 40 µg / kg, country: France
incidence: 9/23, conc. range: 10-1800 µg / kg, ∅ conc.: 900 µg / kg, country: Germany
incidence: 10/35, conc. range: 30-2000 µg / kg, country: Germany
incidence: 2/11, conc. range: 200-1300 µg / kg, country: Germany
incidence: 2/4*, conc. range: 280-640 µg / kg, ∅ conc.: 460 µg / kg, country: Germany, *organic produce
incidence: 2/16, conc. range: 21-32 µg / kg, ∅ conc.: 27 µg / kg, country: Indonesia
incidence: 2/3, ∅ conc.: 402 µg / kg, country: Italy
incidence: nc/6, conc. range: 20-670 µg / kg, country: Italy
incidence: 1/1, conc.: 67,000 µg / kg, country: Italy
incidence: 1/1*, conc.: 20,000 µg / kg, country: Italy, *visible moldy (*Fusarium* spp.)
incidence: 2/2, conc. range: 101-500 µg / kg, country: Italy
incidence: 14/15, conc. range: 22-442 µg / kg, ∅ conc.: 145 µg / kg, country: Korea
incidence: 34/36*, conc. range: 6-15,200 µg / kg, ∅ conc.: 4000 µg / kg, country: Korea, *moldy
incidence: 8/35*, conc. range: 10-100 µg / kg, ∅ conc.: 40 µg / kg, country: Korea, *healthy
incidence: 1/3, conc.: 352 µg / kg, country: Korea
incidence: 3/9, ∅ conc: 541 µg / kg, country: Nepal
incidence: 11/20, conc. range: ≤ 300 µg / kg, ∅ conc.: 100 µg / kg, country: New Zealand
incidence: 73/91, conc. range: 3500 µg / kg, country: New Zealand
incidence: 8/9*, conc. range: 1400-132,000 µg / kg, ∅ conc.: 49,350 µg / kg, country: Poland, *healthy and damaged kernels
incidence: 14/36, conc. range: tr-820 µg / kg, country: South Africa
incidence: 1/5, conc.: 140 µg / kg, country: South Africa
incidence: 24/24, conc. range: 50-12,100 µg / kg, ∅ conc.: 2900 µg / kg*, 300 µg / kg**, country: South Africa, *low-prevalence EC area, **high-prevalence EC area
incidence: 2*/2, conc. range: 420-2500 µg / kg, ∅ conc.: 1460 µg / kg, country: South Africa, *moldy
incidence: 7/10, conc. range: 20-100 µg / kg, country: South Africa
incidence: 50/50, conc. range: 7-7400 µg / kg, country: South Africa
incidence: 43/72, conc. range: 10-15,800 µg / kg, country: South Africa
incidence: 2/2*, conc. range: 120-180 µg / kg, ∅ conc.: 150 µg / kg, country: USA, *yellow maize No. 3
incidence: 7/100, conc. range: 95-312 µg / kg, country: USA
incidence: 24/52, conc. range: 500-10,000 µg / kg, ∅ conc.: 5000 µg / kg, country: USA

incidence: 93/198, conc. range: ≤ 2470 µg / kg, ∅ conc.: 400 µg / kg, country: USA
incidence: 44/52, conc. range: 500-10,700 µg / kg, country: USA
incidence: 17/20*, conc. range: 400-65,800 µg / kg, ∅ conc.: 19,700 µg / kg, country: USA, *moldy
incidence: 33/33, conc. range: 20-100 µg / kg (2 samples), 101-500 µg / kg (17 sa), > 500 µg / kg (14 sa), country: USA
incidence: 1/1*, conc.: 100 µg / kg, country: USA, *dent maize No. 2
incidence: 1/1*, conc.: 550 µg / kg, country: USA, *waxy maize
incidence: 19/19*, conc. range: 69,960-722,450 µg / kg, ∅ conc.: 445,790 µg / kg, country: USA, *moldy, tip section of sweet maize ears
incidence: 1/12, ∅ conc.: 6 µg / kg, country: Yemen
→ diacetoxyscirpenol
incidence: 2/100, conc. range: 400-450 µg / kg, ∅ conc.: 425 µg / kg, country: Argentina
incidence: 1/6*, conc.: 400 µg / kg, country: Austria, **Fusarium* infected
incidence: 1/77, conc.: 31,500 µg / kg, country: Germany
incidence: 6/11, conc. range: 500-2100 µg / kg, country: Germany
incidence: 1*/nc, conc.: 14,000 µg / kg, country: India, *moldy
incidence: 5/100, conc. range: nc, country: Italy
incidence: 6/20, conc. range: ≤ 900 µg / kg, ∅ conc.: 350 µg / kg, country: New Zealand
incidence: 8/100, conc. range: nc, country: Yugoslavia
→ fumonisin B_1
incidence: 1/1, conc.: 900 µg / kg, country: Argentina
incidence: 17/17*, conc. range: 1110-6695 µg / kg, ∅ conc.: 2877 µg / kg, country: Argentina, *field-trial corn
incidence: nc/547*, conc. range: ≤ 4330 µg / kg, ∅ conc.: 290 µg / kg, country: Argentina, *export corn for South Africa
incidence: 47/47*, conc. range: 50-720 µg / kg, ∅ conc.: 300 µg / kg, country: Argentina, *export corn for South Africa
incidence: 8/8, conc. range: 85-8791 µg / kg, ∅ conc.: 2131 µg / kg, country: Argentina
incidence: 1/1*, conc.: 250 µg / kg, country: Bahrain, *imported from The Netherlands
incidence: 9/11*, conc. range: 20-2630 µg / kg, ∅ conc.: 506 µg / kg, country: Benin, *corn genotypes
incidence: 2/2, conc. range: 165-350 µg / kg, ∅ conc.: 258 µg / kg, country: Botswana
incidence: 48/48, conc. range: 600-18,520 µg / kg, ∅ conc.: 5080 µg / kg, country: Brazil
incidence: 6/6, conc. range: 12,200-75,200 µg / kg, country: Burundi
incidence: 1/3*, conc.: 120 µg / kg, country: Canada, *fresh maize
incidence: 16/48, conc. range: 160-2300 µg / kg, ∅ conc.: 760 µg / kg, country: Canada
incidence: 2/5, conc. range: 5300-8400 µg / kg, ∅ conc.: 6800 µg / kg, country: China
incidence: 16/19*, conc. range: 18,000-155,000 µg / kg, ∅ conc.: 74,000 µg / kg, country: China, *moldy corn
incidence: 15/15*, conc. range: 20,000-60,000 µg / kg, ∅ conc.: 35,300 µg / kg, country: China, *fine corn
incidence: 13/27*, conc. range: 186-2964 µg / kg, ∅ conc.: 872 µg / kg, country: China, *high-EC area
incidence: 5/20*, conc. range: 197-1732 µg / kg, ∅ conc.: 890 µg / kg, country: China, *low-EC area
incidence: 7/7, conc. range: 365-3276 µg / kg, ∅ conc.: 1428 µg / kg, country: China

incidence: 8/8, conc. range: 1700-4780 µg / kg, ∅ conc.: 2803 µg / kg, country: Costa Rica
incidence: 11/19*, conc. range: 10-60 µg / kg, ∅ conc.: 19.1 µg / kg, country: Croatia, *corn genotypes
incidence: 25/25*, conc. range: tr (< 25 µg / kg) -3350 µg / kg, ∅ conc.: 868 µg / kg, country: France, *imported from The Netherlands
incidence: 3/3*, conc. range: 100-560 µg / kg, ∅ conc.: 277 µg / kg, country: Greece, *imported from The Netherlands
incidence: 16/16, conc. range: 51-2440 µg / kg, ∅ conc.: 788 µg / kg, country: Indonesia
incidence: 7/12, conc. range: 226-1780 µg / kg, ∅ conc.: 843 µg / kg, country: Indonesia
incidence: 26/26*, conc. range: 10-2330 µg / kg, ∅ conc.: 382 µg / kg, country: Italy, *corn genotypes
incidence: 7/7, conc. range: 100-5310 µg / kg, ∅ conc.: 2807 µg / kg, country: Italy
incidence: 20/22*, conc. range: tr-300,000 µg / kg, ∅ conc.: 74,500 µg / kg, country: Italy, *visibly infected
incidence: 1/1, conc.: 130 µg / kg, country: Kenya
incidence: 93/197, conc. range: 110-12,000 µg / kg, ∅ conc.: 670 µg / kg, country: Kenya
incidence: 33/36*, conc. range: 100-168,800 µg / kg, ∅ conc.: 23,200 µg / kg, country: Korea, *moldy
incidence: 10/35*, conc. range: 90-12,500 µg / kg, ∅ conc.: 3200 µg / kg, country: Korea, *healthy
incidence: 7/8, conc. range: 20-115 µg / kg, ∅ conc.: 67.1 µg / kg, country: Malawi
incidence: 3/3, conc. range: 240-295 µg / kg, ∅ conc.: 260 µg / kg, country: Mozambique
incidence: 12/24, conc. range: 50-4600 µg / kg, ∅ conc.: 600 µg / kg, country: Nepal
incidence: 26/50, conc. range: 57-1820 µg / kg, ∅ conc.: 419 µg / kg, country: Philippines
incidence: 2/7*, conc. range: 10-20 µg / kg, ∅ conc.: 15 µg / kg, country: Poland, *corn genotypes
incidence: 9/9*, conc. range: 90-2300 µg / kg, ∅ conc.: 1031 µg / kg, country: Portugal, *corn genotypes
incidence: 3/6*, conc. range: 10-20 µg / kg, ∅ conc.: 13.3 µg / kg, country: Romania, *corn genotypes
incidence: 2/12*, conc. range: ≤ 550 µg / kg, ∅ conc.: 375 µg / kg, country: South Africa, *good corn, low-EC area
incidence: 12/12*, conc. range: 50-7900 µg / kg, ∅ conc.: 1600 µg / kg, country: South Africa, *good corn, high-EC area
incidence: 11/11*, conc. range: 450-18,900 µg / kg, ∅ conc.: 6520 µg / kg, country: South Africa, *moldy corn intended for beer brewing or animal feed, low-EC area
incidence: 12/12*, conc. range: 3450-46,900 µg / kg, ∅ conc.: 23,900 µg / kg, country: South Africa, *moldy corn intended for beer brewing or animal feed, high-EC area
incidence: 5/6*, conc. range: 210-5380 µg / kg, ∅ conc.: 1840 µg / kg, country: South Africa, *good corn, high-EC area
incidence: 6/8*, conc. range: ≤ 3310 µg / kg, ∅ conc.: 667 µg / kg, country: South Africa, *good corn, low-EC area
incidence: 7/7*, conc. range: 110-11,340 µg / kg, ∅ conc.: 4050 µg / kg, country: South Africa, *moldy corn, low-EC area
incidence: 6/6*, conc. range: 3020-117,520 µg / kg, ∅ conc.: 53,740 µg / kg, country: South Africa, *moldy corn, high-EC area
incidence: 1/1, conc.: 600 µg / kg, country: South Africa
incidence: 50/68**, conc. range: < 50-5420 µg / kg, ∅ conc.: 570 µg / kg (all samples), country: South Africa
incidence: 55/66**, conc. range: < 20-5030 µg / kg, ∅ conc.: 380 µg / kg (all samples), country: South Africa

incidence: nc/77**, conc. range: ≤ 3050 µg / kg, ∅ conc.: 320 µg / kg (all samples), country: South Africa
incidence: nc/71**, conc. range: ≤ 1810 µg / kg, ∅ conc.: 340 µg / kg (all samples), country: South Africa
incidence: nc/113**, conc. range: ≤ 5640 µg / kg, ∅ conc.: 320 µg / kg (all samples), country: South Africa
**white corn
incidence: 31/53***, conc. range: < 50-1120 µg / kg, ∅ conc.: 180 µg / kg (all samples), country: South Africa
incidence: 50/62***, conc. range: < 20-1060 µg / kg, ∅ conc.: 180 µg / kg (all samples), country: South Africa
incidence: nc/82***, conc. range: ≤ 1840 µg / kg, ∅ conc.: 190 µg / kg (all samples), country: South Africa
incidence: nc/76***, conc. range: ≤ 740 µg / kg, ∅ conc.: 170 µg / kg (all samples), country: South Africa
incidence: nc/117***, conc. range: ≤ 11,700 µg / kg, ∅ conc.: 680 µg / kg (all samples), country: South Africa
***yellow corn
incidence: 24/68*, conc. range: < 50-865 µg / kg, ∅ conc.: 280 µg / kg, country: South Africa, *export corn for Taiwan
incidence: 3/3, conc. range: 400-4440 µg / kg, ∅ conc.: 2447 µg / kg, country: South Africa
incidence: 8/9, conc. range: 25-165 µg / kg, ∅ conc.: 79.4 µg / kg, country: Tanzania
incidence: 16/18, conc. range: 63-18,800 µg / kg, ∅ conc.: 1790 µg / kg, country: Thailand
incidence: 19/27, conc. range: 63-18,800 µg / kg, ∅ conc.: 1580 µg / kg, country: Thailand
incidence: 9/19*, conc. range: 8-380 µg / kg, ∅ conc.: 209 µg / kg, country: The Netherlands, *intended for bread production
incidence: 2/10*, conc. range: 8-110 µg / kg, country: The Netherlands, *intended for popcorn production
incidence: 1/1, conc.: 605 µg / kg, country: Uganda
incidence: 7/7, conc. range: 105-1915 µg / kg, ∅ conc.: 635 µg / kg, country: USA
incidence: 6/7, conc. range: 1100-2600 µg / kg, ∅ conc.: 2083 µg / kg, country: USA
incidence: nc/175, conc. range: ≤ 37,900 µg / kg, ∅ conc.: 2984 µg / kg (all samples), country: USA
incidence: 24/28, conc. range: ≤ 1820 µg / kg, ∅ conc.: 870 µg / kg, country: USA
incidence: nc/80, conc. range: ≤ 1600 µg / kg, ∅ conc.: 50 µg / kg (all samples), country: USA
incidence: nc/91, conc. range: ≤ 8400 µg / kg, ∅ conc.: 370 µg / kg, country: USA
incidence: 284/886*, conc. range: 1-10 µg / kg (276 samples), > 10 µg / kg (8 sa), country: USA, *field-trial corn
incidence: 13/99, conc. range: 1200-3200 µg / kg, ∅ conc.: 2400 µg / kg, country: USA
incidence: 5/6*, conc. range: < 50-4100 µg / kg, ∅ conc.: 2220 µg / kg, country: USA, *export corn for Japan
incidence: nc/846, conc. range: ≤ 7470 µg / kg, ∅ conc.: 950 µg / kg (all samples), country: USA, *export corn for South Africa
incidence: nc/836, conc. range: ≤ 7600 µg / kg, ∅ conc.: 960 µg / kg (all samples), country: USA, *export corn for South Africa
incidence: 79/79*, conc. range: 890-3860 µg / kg, ∅ conc.: 2350 µg / kg, country: USA, *export corn for South Africa
incidence: 5/5*, conc. range: 300-3400 µg / kg, ∅ conc.: 2400 µg / kg, country: USA, *Indian maize
incidence: 7/7*, conc. range: 80-16,310 µg / kg, ∅ conc.: 2883 µg / kg, country: USA, *including 1 white maize sample

incidence: 7/7, conc. range: 280-33,450 µg / kg, ∅ conc.: 6617 µg / kg, country: USA
incidence: 20/20, conc. range: 20-1420 µg / kg, ∅ conc.: 180 µg / kg, country: Zambia
incidence: 1/2, conc.: 125 µg / kg, country: Zimbabwe
incidence: 32/33*, conc. range: 30-1240 µg / kg, ∅ conc.: 488 µg / kg, country: unknown origin, *imported from The Netherlands
→ fumonisin B_2
incidence: 1/1, conc.: 800 µg / kg, country: Argentina
incidence: 17/17*, conc. range: 325-2680 µg / kg, ∅ conc.: 1137 µg / kg, country: Argentina, *field-trial corn
incidence: nc/547*, conc. range: $\leq$ 1250 µg / kg, ∅ conc.: 20 µg / kg, country: Argentina, *export corn for South Africa
incidence: 41/47*, conc. range: 50-500 µg / kg, ∅ conc.: 110 µg / kg, country: Argentina, *export corn for South Africa
incidence: 7/8, conc. range: 78-2267 µg / kg, ∅ conc.: 583 µg / kg, country: Argentina
incidence: 7/11*, conc. range: 20-680 µg / kg, ∅ conc.: 147 µg / kg, country: Benin, *corn genotypes
incidence: 2/2, conc. range: 50-105 µg / kg, ∅ conc. 77.5 µg / kg, country: Botswana
incidence: 48/48, conc. range: 1200-19,130 µg / kg, ∅ conc.: 4213 µg / kg, country: Brazil
incidence: 2/5, conc. range: 2300-4300 µg / kg, ∅ conc.: 3300 µg / kg, country: China
incidence: 3/27*, conc. range: 298-550 µg / kg, ∅ conc.: 448 µg / kg, country: China, *high-EC area
incidence: 2/20*, conc. range: 213-447 µg / kg, ∅ conc.: 330 µg / kg, country: China, *low-EC area
incidence: 4/7, conc. range: 96-2834 µg / kg, ∅ conc.: 1223 µg / kg, country: China
incidence: 11/19*, conc. range: 10 µg / kg, ∅ conc.: 10 µg / kg, country: Croatia, *corn genotypes
incidence: 8/16, conc. range: tr-376 µg / kg, ∅ conc.: 182 µg / kg, country: Indonesia
incidence: 3/12, conc. range: 231-556 µg / kg, ∅ conc.: 442 µg / kg, country. Indonesia
incidence: 7/7, conc. range: 30-1480 µg / kg, ∅ conc.: 839 µg / kg, country: Italy
incidence: 13/26, conc. range: 20-520 µg / kg, ∅ conc.: 143 µg / kg, country: Italy
incidence: 1/1, conc.: 275 µg / kg, country: Kenya
incidence: 31/36*, conc. range: 70-48,400 µg / kg, ∅ conc.: 7500 µg / kg, country: Korea, *moldy
incidence: 8/35*, conc. range: 100-5400 µg / kg, ∅ conc.: 1100 µg / kg, country: Korea, *healthy
incidence: 1/8, conc.: 30 µg / kg, country: Malawi
incidence: 3/3, conc. range: 75-110 µg / kg, ∅ conc.: 90 µg / kg, country: Mozambique
incidence: 7/24, conc. range: 100-5500 µg / kg, ∅ conc.: 1600 µg / kg, country: Nepal
incidence: 1/7*, conc.: 10 µg / kg, country: Poland, *corn genotypes
incidence: 6/50, conc. range: 58-1210 µg / kg, ∅ conc.: 286 µg / kg, country: Philippines
incidence: 8/9*, conc. range: 250-4450 µg / kg, ∅ conc.: 1211 µg / kg, country: Portugal, *corn genotypes
incidence: 1/6*, conc.: 10 µg / kg, country: Romania, *corn genotypes
incidence: 1/1, conc.: 300 µg / kg, country: South Africa
incidence: 3/12*, conc. range: 0-150 µg / kg, ∅ conc.: 83 µg / kg, country: South Africa, *good corn, low-EC area

incidence: 10/12*, conc. range: ≤ 2250 µg / kg, Ø conc.: 610 µg / kg, country: South Africa, *good corn, high-EC area
incidence: 11/11*, conc. range: 150-6750 µg / kg, Ø conc.: 2500 µg / kg, country: South Africa, *moldy corn intended for beer brewing or animal feed, low-EC area
incidence: 12/12*, conc. range: 900-16,300 µg / kg, Ø conc.: 7550 µg / kg, country: South Africa, *moldy corn intended for beer brewing or animal feed, high-EC area
incidence: 2/8*, conc. range: ≤ 970 µg / kg, Ø conc.: 515 µg / kg, country: South Africa, *good corn, low-EC area
incidence: 5/6*, conc. range: 150-1320 µg / kg, Ø conc.: 508 µg / kg, country: South Africa, *good corn, high-EC area
incidence: 6/7*, conc. range: ≤ 3700 µg / kg, Ø conc.: 1277 µg / kg, country: South Africa, *moldy corn, low-EC area
incidence: 6/6*, conc. range: 750-22,960 µg / kg, Ø conc.: 13,680 µg / kg, country: South Africa, *moldy corn, high-EC area
incidence: 50/68**, conc. range: < 50-1600 µg / kg, Ø conc.: 190 µg / kg (all samples), country: South Africa
incidence: 55/66**, conc. range: < 20-1670 µg / kg, Ø conc.: 140 µg / kg (all samples), country: South Africa
incidence: nc/77**, conc. range: ≤ 270 µg / kg, Ø conc.: 30 µg / kg (all samples), country: South Africa
incidence: nc/71**, conc. range: ≤ 740 µg / kg, Ø conc.: 50 µg / kg (all samples), country: South Africa
incidence: nc/113**, conc. range: ≤ 1430 µg / kg, Ø conc.: 80 µg / kg (all samples), country: South Africa
**white corn
incidence: 31/53***, conc. range: < 50-700 µg / kg, Ø conc.: 50 µg / kg (all samples), country: South Africa
incidence: 50/62***, conc. range: < 20-320 µg / kg, Ø conc.: 70 µg / kg (all samples), country: South Africa
incidence: nc/82***, conc. range: ≤ 690 µg / kg, Ø conc.: 30 µg / kg (all samples), country: South Africa
incidence: nc/76, conc. range: ≤ 540 µg / kg, Ø conc.: 30 µg / kg (all samples), country: South Africa
incidence: nc/117***, conc. range: ≤ 5690 µg / kg, Ø conc.: 220 µg / kg (all samples), country: South Africa
***yellow corn
incidence: 24/68*, conc. range: < 50-250 µg / kg, Ø conc.: 130 µg / kg, country: South Africa, *export corn for Taiwan
incidence: 3/3, conc. range: 150-1300 µg / kg, Ø conc.: 833 µg / kg, country: South Africa
incidence: 1/9, conc.: 60 µg / kg, country: Tanzania
incidence: 12/18, conc. range: 50-1400 µg / kg, Ø conc.: 251 µg / kg, country: Thailand
incidence: 12/27, conc. range: 50-1400 µg / kg, Ø conc.: 251 µg / kg, country: Thailand
incidence: 1/1, conc.: 155 µg / kg, Uganda
incidence: 6/7, conc. range: 70-460 µg / kg, Ø conc.: 182 µg / kg, country: USA
incidence: 6/7, conc. range: 600-10,200 µg / kg, Ø conc.: 2867 µg / kg, country: USA
incidence: nc/175, conc. range: ≤ 12,300 µg / kg, Ø conc.: 821 µg / kg, country: USA
incidence: 5/6*, conc. range: < 100-10,200 µg / kg, Ø conc.: 3120 µg / kg, country: USA
incidence: nc/846*, conc. range: ≤ 2470 µg / kg, Ø conc.: 120 µg / kg (all samples), country: USA
incidence: nc/836*, conc. range: ≤ 3120 µg / kg, Ø conc.: 140 µg / kg (all samples), country: USA
incidence: 79/79*, conc. range: 260-1120 µg / kg, Ø conc.: 670 µg / kg, country: USA, *export corn for South Africa

incidence: 7/7*, conc. range: 30-4020 μg / kg, ∅ conc.: 811 μg / kg, country: USA, *including 1 white maize sample
incidence: 5/5, conc. range: 32-4200 μg / kg, ∅ conc.: 1187 μg / kg, country: USA
incidence: 15/20, conc. range: 10-290 μg / kg, ∅ conc.: 50.7 μg / kg, country: Zambia
incidence: 1/2, conc.: 40 μg / kg, country: Zimbabwe
→ fumonisin B_3
incidence: 17/17*, conc. range: ≤ 110-855 μg / kg, ∅ conc.: 372 μg / kg, country: Argentina, *field-trial corn
incidence: 28/47*, conc. range: 50-500 μg / kg, ∅ conc.: 80 μg / kg, country: Argentina, *export corn for South Africa
incidence: 6/8, conc. range: 50-980 μg / kg, ∅ conc.: 348 μg / kg, country: Argentina
incidence: 2/2, conc. range: 40-70 μg / kg, ∅ conc.: 55 μg / kg, country: Botswana
incidence: 2/7, conc. range: 230-545 μg / kg, ∅ conc.: 388 μg / kg, country: China
incidence: 4/16, conc. range: 57-222 μg / kg, ∅ conc.: 108 μg / kg, country: Indonesia
incidence: 1/1, conc.: 780 μg / kg, country: Kenya
incidence: 31/36*, conc. range: 50-10,600 μg / kg, ∅ conc.: 6300 μg / kg, country: Korea, *moldy
incidence: 7/35*, conc. range: 50-500 μg / kg, ∅ conc.: 300 μg / kg, country: Korea, *healthy
incidence: 55/66**, conc. range: < 20-400 μg / kg, ∅ conc.: 40 μg / kg (all samples), country: South Africa
incidence: nc/77**, conc. range: ≤ 340 μg / kg, ∅ conc.: 10 μg / kg (all samples), country: South Africa
incidence: nc/71**, conc. range: ≤ 180 μg / kg, ∅ conc.: 10 μg / kg (all samples), country: South Africa
incidence: nc/113**, conc. range: ≤ 400 μg / kg, ∅ conc.: 30 μg / kg (all samples), country: South Africa
**white corn
incidence: 50/62***, conc. range: < 20-200 μg / kg, ∅ conc.: 20 μg / kg (all samples), country: South Africa
incidence: nc/82***, conc. range: ≤ 120 μg / kg, ∅ conc.: < 10 μg / kg (all samples), country: South Africa
incidence: nc/76, conc. range: ≤ 330 μg / kg, ∅ conc.: 10 μg / kg (all samples), country: South Africa
incidence: nc/117, conc. range: ≤ 1960 μg / kg, ∅ conc.: 110 μg / kg (all samples), country: South Africa
***yellow corn
incidence: 1/1, conc.: 85 μg / kg, country: Uganda
incidence: nc/175, conc. range: 2800 μg / kg, ∅ conc.: 290 μg / kg, country: USA
incidence: 79/79*, conc. range: 80-550 μg / kg, ∅ conc.: 250 μg / kg, country: USA, *export corn for South Africa
fumonisin (no specification) (→ fumonisins)
incidence: 17/17*, conc. range: 500-48,500 μg / kg, ∅ conc.: 17,864 μg / kg, country: USA, *hybrid maize
incidence: 4/4*, conc. range: 37-1400 μg / kg, ∅ conc.: 411 μg / kg, country: Germany,
*organic produce
incidence: 3/3, conc. range: 25-50 μg / kg, ∅ conc.: 40 μg / kg, country: Mozambique
→ fusaproliferin
incidence: 9/22*, conc. range: 600-500,000 μg / kg, ∅ conc.: 72,222 μg / kg, country: Italy, *visibly infected
→ fusarenon X
incidence: 2/2*, conc. range: 400-900 μg / kg, ∅ conc.: 650 μg / kg, country: Austria,
*visibly moldy (*Fusarium* spp.)

incidence: 5/15, conc. range: 15-72 µg / kg, ∅ conc.: 27 µg / kg, country: Korea
→ fusarin C
incidence: 2/2, conc. range: 20-280 µg / kg, ∅ conc.: 150 µg / kg, country: South Africa
→ HT-2 toxin
incidence: 2/52, conc. range: 500-800 µg / kg, ∅ conc.: 650 µg / kg, country: Germany
incidence: 1/56, conc.: 600 µg / kg, country: Germany
incidence: 2/11, conc. range: 500-700 µg / kg, country: Hungary
incidence: 3*/162, conc. range: 53,000-645,000 µg / kg, ∅ conc.: 294,333 µg / kg, country: Poland, *heavily damaged kernels
→ kojic acid
incidence: 3/155, conc.: nc, country: South Africa
→ moniliformin
incidence: 2/12, conc. range: 60-200 µg / kg, ∅ conc.: 130 µg / kg, country: Canada
incidence: 23/58, conc. range: 80-650 µg / kg, country: Germany
incidence: 15/29, conc. range: ≤ 280 µg / kg, country: New Zealand
incidence: 20/20*, conc. range: 4200-399,300 µg / kg, ∅ conc.: 97,710 µg / kg, country: Poland
incidence: 57/57*, conc. range: 16,800-425,000 µg / kg, ∅ conc.: 172,000 µg / kg, country: Poland
*hand selected, visible fungal damage
incidence: 15/36, conc. range: tr-12,000 µg / kg, country: South Africa
incidence: 24/24, conc. range: 350-11,570 µg / kg, ∅ conc.: 3500 µg / kg*, 800 µg / kg**, country: South Africa, *low-prevalence EC area, **high-prevalence EC area
incidence: 2*/2, conc. range: 16,000-25,000 µg / kg, ∅ conc.: 20,500 µg / kg, country: South Africa, *moldy
incidence: 64/64, conc. range: < 50-3160 µg / kg, country: different countries, mainly Africa
→ neosolaniol
incidence: 1/100, conc.: traces, country: Argentina
incidence: 2*/162, conc. range: 19,400-27,200 µg / kg, ∅ conc.: 23,300 µg / kg, country: Poland, *heavily damaged kernels
→ nivalenol
incidence: 5/100, conc. range: tr-500 µg / kg, country: Argentina
incidence: 2/2, conc. range: 700-2200 µg / kg, ∅ conc.: 1450 µg / kg, country: Austria
incidence: 2/2*, conc. range: 500-1800 µg / kg, country: Austria, *visibly moldy (*Fusarium* spp.)
incidence: 1/1, conc.: 12 µg / kg, country: Canada
incidence: 28/28, conc. range: ≤ 4050 µg / kg, ∅ conc.: 1960 µg / kg, country: China
incidence: 24/24, conc. range: 54-2760 µg / kg, ∅ conc.: 757 µg / kg, country: China
incidence: 2/16, conc. range: 49-169 µg / kg, ∅ conc.: 109 µg / kg, country: Indonesia
incidence: 8/15, conc. range: 26-332 µg / kg, ∅ conc.: 168 µg / kg, country: Korea
incidence: 32/36*, conc. range: 6-15,600 µg / kg, ∅ conc.: 1700 µg / kg, country: Korea, *moldy
incidence: 6/35*, conc. range: 20-200 µg / kg, ∅ conc.: 80 µg / kg, country: Korea, *healthy
incidence: 1/3, conc.: 624 µg / kg, country: Korea
incidence: 6/9, ∅ conc.: 892 µg / kg, country: Nepal
incidence: 73/91, conc. range: ≤ 3600 µg / kg, country: New Zealand
incidence: 7/50, conc. range: 18-102 µg / kg, ∅ conc.: 43 µg / kg, country: Philippines

incidence: 6/36, conc. range: tr-240 µg / kg, country: South Africa
incidence: 24/24, conc. range: 880-15,200 µg / kg, Ø conc.: 4600 µg / kg*, 1800 µg / kg**, country: South Africa *low-prevalence EC area, **high-prevalence EC area
→ ochratoxin A
incidence: 3/27, conc. range: 5-100 µg / kg, country: Austria
incidence: 1/12, conc.: 32 µg / kg, country: Brazil
incidence: 12/52, conc. range: 25-35 µg / kg, country: Bulgaria
incidence: 87/151*, conc. range: 0.2-1418 µg / kg, country: Bulgaria, *area with endemic nephropathy
incidence; 30/113, conc. range: 0.2-235 µg / kg, country: Bulgaria
incidence: 1/28, conc.: 55 µg / kg, country: Chile
incidence: 1/3, conc.: 12 µg / kg, country: Egypt
incidence: 18/924, conc. range: 15-200 µg / kg, country: France
incidence: 2/75, conc.: 10 µg / kg, Ø conc.: 10 µg / kg, country: France
incidence: 2/49*, conc. range: 18-22 µg / kg, Ø conc.: 20 µg / kg, country: Germany, *moldy
incidence: 3/40, conc. range: 1.7-82 µg / kg, Ø conc.: 80.3 µg / kg, country: Germany
incidence: 1/7, conc.: 0.1 µg / kg, country. Germany
incidence: 38/112, conc. range: ≤ 0.7 µg / kg, country: Italy
incidence: 14/90, conc. range: ≤ 2.0 µg / kg, country: Italy
incidence: 39/111, conc. range: 0.1-1.02 µg / kg, country: Italy
incidence: 1/22, conc.: nc, country: India
incidence: 1/26, conc.: 3 µg / kg, country: Indonesia
incidence: 2/123, conc. range: 25-400 µg / kg, Ø conc.: 213 µg / kg, country: Poland
incidence: 1/30, conc.: 2.5 µg / kg, country: Spain
incidence: 2/167, conc. range: ca. 10 µg / kg, country: Turkey
incidence: 11/29, conc. range: < 50-500 µg / kg, country: UK
incidence: 5/39, conc. range: ≤ 4.9-11.2 µg / kg, country: UK
incidence: 4/11, conc. range: ≤ 0.8 µg / kg, country: UK
incidence: 11/19, conc. range: ≤ 0.7 µg / kg, country: UK
incidence: 3/293, conc. range: 83-166 µg / kg, Ø conc.: 123 µg / kg, country: USA
incidence: 1/283, conc. range: 130 µg / kg, country: USA
incidence: 50/542, conc. range: 6-140 µg / kg, country: Yugoslavia
incidence: 50/191, conc. range: 45-5100 µg / kg, Ø conc.: 490 µg / kg, country: Yugoslavia
incidence: 2/48, conc. range: 14-90 µg / kg, Ø conc.: 40.6 µg / kg, country: Yugoslavia
→ ochratoxin B
incidence: 2/293, conc. range: traces, country: Canada
→ penicillic acid
incidence: 7/20*, conc. range: 5-231 µg / kg, Ø conc. 59 µg / kg, country: USA, *mold damaged
incidence: 48/48*, conc. range: 5-184 µg / kg, Ø conc.: 46 µg / kg, country: USA, *mold damaged
→ sterigmatocystin
incidence: 4/155, conc.: nc, country: South Africa
incidence: 10/167, conc. range: ca. 20 µg / kg, country: Turkey
incidence: 2/29, conc. range: > 10 µg / kg, country: UK
→ T-2 toxin
incidence: 1/52, conc.: 10 µg / kg, country: Germany
incidence: 4/56, conc. range: 100-200 µg / kg, country: Germany

incidence: 7/11, conc. range: 100-4400 µg / kg, country: Hungary
incidence: 5/150, conc. range: 500-5000 µg / kg, country: Hungary
incidence: 1*/nc, conc.: 4000 µg / kg, country: India, *moldy
incidence: 1/4, conc.: 0.8 µg / kg, country: Italy
incidence: 3/100, conc. range: nc, country: Italy
incidence: 3/162, conc. range: 47,000-992,000 µg / kg, ∅ conc.: 411,333 µg / kg, country: Poland
incidence: 9/118, conc. range: 78-650 µg / kg, country: Taiwan
incidence: 8/100, conc. range: nc, country: Yugoslavia
incidence: 1/1, conc.: 2000 µg / kg, country: USA
incidence: 15/100, conc. range: 900-2400 µg / kg, country: USA
incidence: 9/118, conc. range: 78-650 µg / kg, country: USA
incidence: 93/173, conc. range: 0.2-1 µg / kg, country: USA
incidence: 13/20, conc. range: ≤ 200 µg / kg, ∅ conc.: 74 µg / kg, country: New Zealand
T-2 tetraol
incidence: 1*/162, conc.: 36,200 µg / kg, country: Poland, *heavily damaged kernels
T-2 triol
incidence: 2/56, conc. range: 300 µg / kg, ∅ conc. 300 µg / kg, country: Germany
incidence: 2*/162, conc. range: 9700-14,500 µg / kg, ∅ conc.: 12,100 µg / kg, country: Poland, *heavily damaged kernels
→ zearalenols
incidence: nc/6, conc. range: 20-90 µg / kg, country: Italy
zearalenone
incidence: 15/20, ∅ conc.: 6 µg / kg, country: Argentina
incidence: 676/2271, conc. range: ≤ 2000 µg / kg, ∅ conc.: 165 µg / kg, country: Argentina
incidence: 16/55, conc. range: 200-750 µg / kg, country: Argentina
incidence: 9/150, conc. range: 40-350 µg / kg, ∅ conc.: 210 µg / kg, country: Argentina
incidence: 148/174, conc. range: ≤ 2070 µg / kg, ∅ conc.: 230 µg / kg, country: Australia
incidence: 3/3, conc. range: 1100-1300 µg / kg, ∅ conc.: 1200 µg / kg, country: Austria
incidence: 27/51, conc. range: 1-200 µg / kg, ∅ conc.: 70 µg / kg, country: Austria
incidence: 41/78, conc. range: ≤ 70 µg / kg, ∅ conc.: 9 µg / kg, country: Austria
incidence: 3/6, conc. range: 420-1000 µg / kg, ∅ conc.: 740 µg / kg, country: Austria
incidence: 1/1, conc.: 33 µg / kg, country: Canada
incidence: 23/81*, conc. range: 130-475 µg / kg, country: Canada, *domestic, maize and maize products
incidence: 1/61*, conc.: 200 µg / kg, country: Canada, *imported, maize and maize products
incidence: 62/75, conc. range: 10,000-175,000 µg / kg, country: France
incidence: 16/59, conc. range: 1-260 µg / kg, ∅ conc.: 50 µg / kg, country: Germany
incidence: 2/174, conc. range: 10-1200 µg / kg, country: Germany
incidence: 2/4*, conc. range: 49-92 µg / kg, ∅ conc.: 70.5 µg / kg, country: Germany, *organic produce
incidence: 8/11, conc. range: 700-7500 µg / kg, country: Hungary
incidence: 1*/nc, conc.: 16,000 µg / kg, country: India, *moldy
incidence: 2/16, conc. range: 11-12 µg / kg, ∅ conc.: 11.5 µg / kg, country: Indonesia
incidence: 1/3, ∅ conc.: 35 µg / kg, country: Italy
incidence: 31/111, conc. range: 51-670 µg / kg, country: Italy

incidence: nc/6, conc. range: 400-7400 µg / kg, country: Italy
incidence: 3/41, conc. range: 40-2000 µg / kg, country: Italy
incidence: 1/1, conc.: 40 µg / kg, country: Kenya
incidence: 1/15, conc.: 71 µg / kg, country: Korea
incidence: 32/36*, conc. range: 2-7300 µg / kg, Ø conc.: 600 µg / kg, country: Korea, *moldy
incidence: 7/35*, conc. range: 2-300 µg / kg, Ø conc.: 70 µg / kg, country: Korea, *healthy
incidence: 1/1, conc.: 400 µg / kg, country: Malawi
incidence: 6/139, conc. range: nc, country: Mexico
incidence: 5/9, Ø conc.: 819 µg / kg, country: Nepal
incidence: 69/91, conc. range: ≤ 500 µg / kg, country: New Zealand
incidence: 2/50, conc. range: 59-505 µg / kg, Ø conc.: 282 µg / kg, country: Philippines
incidence: 5/9*, conc. range: 50-2050 µg / kg, Ø conc.: 1048 µg / kg, country: Poland, *healthy and damaged kernels
incidence: 8/8, conc. range: 620-72,000 µg / kg, country: Portugal
incidence: 2/44, conc. range: 20-503 µg / kg, country: South Africa
incidence: 14/24, conc. range: 120-3280 µg / kg, Ø conc.: 1200 µg / kg*, 400 µg / kg**, country: South Africa, *low-prevalence EC area, **high-prevalence EC area
incidence: 2*/2, conc. range: 4000-8000 µg /kg, Ø conc.: 6000 µg / kg, country: South Africa, *moldy
incidence: 9/155, conc.: nc, country: South Africa
incidence: 2/9, conc. range: 40-80 µg / kg, Ø conc.: 60 µg / kg, country: Tanzania
incidence: 1/27, conc.: 923 µg / kg, country: Thailand
incidence: 6/76*, conc. range: 100-200 µg / kg (2 samples), > 200 µg / kg (4 sa), country: Uruguay, *and by-products
incidence: 4/nc, conc. range: 2310-35,600 µg / kg, country: Yugoslavia
incidence: 54/116, conc. range: 10-275,800 µg / kg, country: Yugoslavia
incidence: 5/191, conc. range: 43-10,000 µg / kg, country: Yugoslavia
incidence: 23/54, conc. range: 700-37,500 µg / kg, country: Yugoslavia
incidence: 4/29, conc. range: ca. 2000 µg / kg, country: UK
incidence: 7/73, conc. range: 49-303 µg / kg, country: USA
incidence: 6/576, conc. range: 450-800 µg / kg, Ø conc. 624 µg / kg, country: USA
incidence: 38/223, conc. range: 100-5000 µg / kg, Ø conc. 900 µg / kg, country: USA
incidence: 2/283, conc. range: 800- > 1250 µg / kg, country: USA
incidence: 5/293, conc. range: 450-750 µg / kg, country: USA
incidence: 17/20*, conc. range: 200-13,200 µg / kg, Ø conc.: 2700 µg / kg, country: USA, *moldy
incidence: 6/26, conc. range: 200-500 µg / kg, country: USA
incidence: 19/315, conc. range: < 100-210 µg / kg, country: USA
incidence: 18/315, conc.: 400 µg / kg, country: USA
incidence: 4/12, Ø conc.: 10 µg / kg, country: Yemen
→ cereals

Maize, boiled may contain the following → mycotoxins:
aflatoxin (→ aflatoxins)
incidence: 16/24, Ø conc.: 9 µg / kg, country: Philippines

Maize, brewers may contain the following → mycotoxins:
→ deoxynivalenol
incidence: 8/11, conc. range: 20-100 µg / kg (6 samples), 101-500 µg / kg (1 sa), country: UK

→ zearalenone
incidence: nc/17, ∅ conc.: 290 μg / kg, country: Zambia
incidence: nc/13, ∅ conc.: 680 μg / kg, country: Zambia

Maize, brewers flaked may contain the following → mycotoxins:
→ deoxynivalenol
incidence: 5/6, conc. range: 10-110 μg / kg, ∅ conc.: 50 μg / kg, country: UK

Maize, brewers grits may contain the following → mycotoxins:
→ deoxynivalenol
incidence: 2/3, conc. range: 40-140 μg / kg, ∅ conc.: 90 μg / kg, country: UK

Maize, canned may contain the following → mycotoxins:
aflatoxin (→ aflatoxins)
incidence: 3/4, conc. range: ≤ 25 μg / kg, ∅ conc.: 6 μg / kg, country: Philippines
→ fumonisin B_1
incidence: 1/1, conc.: 26 μg / kg, country: USA
hydrolyzed fumonisin B_1 (HBF_1)
incidence: 1/1, conc.: nc, country: USA

Maize, dried may contain the following → mycotoxins:
aflatoxin (→ aflatoxins)
incidence: 33/660, conc. range: ≤ 1152 μg / kg, ∅ conc.: 76 μg / kg, country: Philippines

Maize, fiber cereal may contain the following → mycotoxins:
→ fumonisin B_1
incidence: 1/1, conc.: 130 μg / kg, country: USA
incidence: 1/1, conc.: 60 μg / kg, country: Venezuela
→ fumonisin B_2
incidence: 1/1, conc.: 30 μg / kg, country: Venezuela

Maize, hominy may contain the following → mycotoxins:
→ fumonisin B_1
incidence: 1/1, conc.: 60 μg / kg, country: USA
→ fumonisin B_2
incidence: 1/1, conc.: 20 μg / kg, country: USA

Maize, infant cereal may contain the following → mycotoxins:
→ fumonisins (no specification)
incidence: 1/1, conc.: 200 μg / kg, country: USA

Maize, infant cream corn may contain the following → mycotoxins:
→ fumonisins (no specification)
incidence: 1/1, conc.: 200 μg / kg, country: USA

Maize, popped may contain the following → mycotoxins:
→ fumonisin B_1
incidence: 3/5, conc. range: ≤ 300 μg / kg, country: The Netherlands

Maize, preharvest may contain the following → mycotoxins:
→ zearalenone
incidence: 1/116, conc.: < 5000 μg / kg, country: Spain

Maize, puffed may contain the following → mycotoxins:
→ fumonisin B_1
incidence: 6/6, conc. range: 790-6100 μg / kg, ∅ conc.: 3145 μg / kg, country: Italy
→ fumonisin B_2
incidence: 6/6, conc. range: 110-740 μg / kg, ∅ conc. 397 μg / kg, country: Italy

Maize, quality-protein may contain the following → mycotoxins:

→ fumonisin B_1
incidence: nc/12, conc. range: ≤ 2040 µg / kg, ∅ conc.: 410 µg / kg (all samples), country: South Africa
incidence: nc/59, conc. range: ≤ 4400 µg / kg, ∅ conc.: 340 µg / kg (all samples), country: South Africa
→ fumonisin B_2
incidence: nc/12, conc. range: ≤ 1090 µg / kg, ∅ conc.: 120 µg / kg (all samples), country: South Africa
incidence: nc/59, conc. range: ≤ 1290 µg / kg, ∅ conc.: 110 µg / kg (all samples), country: South Africa
→ fumonisin B_3
incidence: nc/12, conc. range: ≤ 60 µg / kg, ∅ conc.: < 10 µg / kg (all samples), country: South Africa
incidence: nc/59, conc. range: ≤ 800 µg / kg, ∅ conc.: 40 µg / kg (all samples), country: South Africa

Maize, shelled may contain the following → mycotoxins:
→ aflatoxins
incidence: 36/1594, conc. range: ≤ 37 µg / kg, ∅ conc.: 9 µg / kg, country: USA
incidence: 21/60, conc. range: ≤ 348 µg / kg, ∅ conc.: 66 µg / kg, country: USA
incidence: 394/1283, conc. range: ≤ 306 µg / kg, ∅ conc.: 35 µg / kg, country: USA
incidence: 152/297, conc. range: ≤ 3190 µg / kg, ∅ conc.: 50 µg / kg, country: USA
incidence: 35/81, conc. range: ≤ 710 µg / kg, ∅ conc.: 49 µg / kg, country: USA
incidence: 16/34, conc. range: ≤ 145 µg / kg, ∅ conc.: 17 µg / kg, country: USA
incidence: 4/169, conc. range: ≤ 5 µg / kg, ∅ conc.: 2 µg / kg, country: USA
incidence: 432/1385, conc. range: ≤ 3300 µg / kg, ∅ conc.: 242 µg / kg, country: USA
incidence: 44/62, conc. range: ≤ 1524 µg / kg, ∅ conc.: 142 µg / kg, country: USA
incidence: 18/31, conc. range: ≤ 631 µg / kg, ∅ conc.: 82 µg / kg, country: USA
incidence: 26/11, conc. range: ≤ 16 µg / kg, ∅ conc.: 8 µg / kg, country: USA
incidence: 116/1395, conc. range: ≤ 1290 µg / kg, ∅ conc.: 47 µg / kg, country: USA
incidence: 24/148, conc. range: ≤ 364 µg / kg, ∅ conc.: 57.8 µg / kg, country: USA
→ deoxynivalenol
incidence: 64/96, conc. range: tr-500 µg / kg (53 samples), 500-1000 µg / kg (10 sa), 1000-2000 µg / kg (1 sa), country: USA

Maize, steeped may contain the following → mycotoxins:
→ fumonisins
Experimental studies showed an increased fumonisin concentration in the steeping water with a similar decrease in the → maize kernels. The different rates of interchange of fumonisin B_1 and FB_2 between the solid matrix and the water solution may be explained by the different polarities of the fumonisins. Although FB_1 and FB_2 have been detected in both the germ and the remaining kernel, the germ fraction contained lower fumonisin levels.

Maize, sweet may contain the following → mycotoxins:
→ fumonisin B_1
incidence: 11/40, conc. range: < 10-190 µg / kg, country: Germany
incidence: 5/5, conc. range: 60-790 µg / kg, ∅ conc.: 298 µg / kg, country: Italy
incidence: 1/7, conc.: 70 µg / kg, country: Switzerland
incidence: 12/24*, conc. range: ≤ 1089 µg / kg, ∅ conc.: 400 µg / kg, country: Thailand, *canned
→ fumonisin B_2
incidence: 6/24*, conc. range: ≤ 658 µg / kg, ∅ conc.: 64.5 µg / kg, country: Thailand, *canned
→ fumonisins (FB_1, FB_2, FB_3)

incidence: 1/22, conc.: 11 µg / kg, country: UK

Maize-based thickeners may contain the following → mycotoxins:
→ fumonisins (FB_1, FB_2, FB_3)
incidence: 4/21, conc. range: 14-110 µg / kg, ∅ conc.: 23 µg / kg, country: UK

Maize bran may contain the following → mycotoxins:
→ aflatoxin
incidence: 2/2, conc. range: 37-71 µg / kg, ∅ conc.: 54 µg / kg, country: Philippines
→ fumonisin B_1
incidence: 3/4, conc. range: 60-330 µg / kg, ∅ conc.: 168 µg / kg, country: USA
incidence: 1/1, conc.: 290 µg / kg, country: USA
→ fumonisin B_2
incidence: 3/3, conc. range: 10-40 µg / kg, ∅ conc.: 23.3 µg / kg, country: USA
incidence: 1/1, conc.: 70 µg / kg, country: USA
→ bran

Maize chips may contain the following → mycotoxins:
→ fumonisin B_1
incidence: 2/2, conc.: tr-37 µg / kg, country: Canada
incidence: 3/9, conc. range: ≤ 160 µg / kg, country: The Netherlands

Maize flakes may contain the following → mycotoxins:
→ fumonisin B_1
incidence: 2/5, conc. range: 10 µg / kg, ∅ conc.: 10 µg / kg, country: USA
→ cereal flakes, → corn flakes, → oat flakes

Maize flour Pattern of → zearalenone distribution in → maize kernels is the same as in the case of aflatoxin (→ aflatoxins). After dry → milling the largest amounts were found in the high fat fractions (oil and feed usage).
Processing of zearalenone-contaminated corn (120 µg / kg) led to 15 µg zearalenone / kg in the starch. The starch of wet-milled maize did not show any zearalenone contamination but gluten and solubes contained about 50% and ca. 20% zearalenone from the whole maize.
Wet milling of maize contaminated with → nivalenol, → deoxynivalenol and zearalenone caused a significant accumulation of the highly water-soluble NIV and DON in the concentrated steep liquor fractions (≤ 8800 µg / kg). Low levels were observed in the solid (germ, fibre and gluten) fractions (< 300 µg / kg). In contrast, the relatively water insoluble zearalenone accumulated in the solids (2200-4800 µg / kg), while only 600 µg / kg were found in the concentrated steep liquor. The starch fractions were almost free of → mycotoxins.
Dry milling of deoxynivalenol contaminated maize resulted in the accumulation of the mycotoxin in the maize germ meal (animal feed).
Most of → T-2 toxin (> 60%) was removed by wet milling with the steep and process water. The starch contained less than 5% while the remainder is found in the germ, gluten und fibre.
In a laboratory simulated scale aflatoxin, → fumonisins, T-2 toxin, and zearalenone accumulated in the solubes fractions during milling. While the fumonisins and zearalenol also concentrated in the gluten, aflatoxin and fumonisins were found in the fibre fraction, too.
Maize flour may contain the following mycotoxins:
→ aflatoxin B_1
incidence: 1/5, conc.: 15 µg / kg, country: Japan
incidence: 1/4, conc.: 31 µg / kg, country: Japan

incidence: 11/11, conc. range: 3.7-37 µg / kg, ∅ conc.: 18.95 µg / kg, country: Thailand
→ aflatoxin B_2
incidence: 1/5, conc.: 5.2 µg / kg, country: Japan
incidence: 1/4, conc.: 5.3 µg / kg, country: Japan
incidence: 11/11, conc. range: 2.3-9.9 µg / kg, ∅ conc.: 5.96 µg / kg, country: Thailand
→ citrinin
incidence: 1/5 conc.: 27 µg / kg, country: Japan
incidence: 1/4, conc.: 73 µg / kg, country: Japan
incidence: 14/23, conc. range: ≤ 1390 µg / kg, country: Japan
incidence: 11/11, conc. range: 10-98 µg / kg, ∅ conc.: 58.9 µg / kg, country: Thailand
deoxynivalenol
incidence: nc, ∅ conc.: 180 µg / kg, country: Canada
incidence: 1/2, conc.: 240 µg / kg, country: UK
incidence: 5/5, conc. range: 20-50 µg / kg, country: UK
incidence: nc/4, conc. range: 17-67 µg / kg, country: UK
→ fumonisin B_1
incidence: 4/4, conc. range: 35-255 µg / kg, ∅ conc.: 185 µg / kg, country: Botswana
incidence: 11/39, conc. range: < 100-1600 µg / kg, ∅ conc.: 550 µg / kg, country: Canda
incidence: 3/4, conc. range: 60-200 µg / kg, ∅ conc.: 100 µg / kg, country: China
incidence: 1/1, conc.: 3540 µg / kg, country: Italy
incidence: 2/2, conc. range: 60-70 µg / kg, ∅ conc.: 65 µg / kg, country: South Africa
incidence: nc/3, conc. range: 0-310 µg / kg, ∅ conc.: 100 µg / kg (all samples), country: South Africa
incidence: nc/13, conc. range: 40-3910 µg / kg, ∅ conc.: 550 (all samples), country: South Africa
incidence: 1/3, conc. range: 50-70 µg / kg, country: Spain
incidence: 5/25, conc. range: < 30-330 µg / kg, ∅ conc.: 130 µg / kg, country: Spain
incidence: 1/2, conc.: 608 µg / kg, country: Thailand
incidence: 6/6, conc. range: 480-880 µg / kg, ∅ conc.: 660 µg / kg, country: Thailand
incidence: 5/7*, conc. range: 40-90 µg / kg, ∅ conc.: 58 µg / kg, country: The Netherlands, *mixes
incidence: 2/6*, conc. range: 8-25 µg / kg, country: The Netherlands, *mixes
incidence: 1/1, conc.: 740 µg / kg, country: Zambia
incidence: 4/4, conc. range: 55-1910 µg / kg, ∅ conc.: 625 µg / kg, country: Zimbabwe
→ fumonisin B_2
incidence: 2/4, conc. range: 75-85 µg / kg, ∅ conc.: 80 µg / kg, country: Botswana
incidence: 1/1, conc.: 840 µg / kg, country: Italy
incidence: nc/13, conc. range: 0-810 µg / kg, ∅ conc.: 90 (all samples), country: South Africa
incidence: nc/25, conc. range: 50-60 µg / kg, country: Spain
incidence: nc/6, conc. range: 120-240 µg / kg, ∅ conc.: 160 µg / kg, country: Thailand
incidence: 1/1, conc.: 380 µg / kg, country: Zambia
incidence: 2/4, conc. range: 150-620 µg / kg, ∅ conc.: 385 µg / kg, country: Zimbabwe
→ fumonisin B_3
incidence: 1/4, conc.: 30 µg / kg, country: Botswana
incidence: nc/13, conc. range: 0-470 µg / kg, ∅ conc.: 40 (all samples), country: South Africa
incidence: 1/1, conc.: 85 µg / kg, Zambia

incidence: 2/4, conc. range: 55-205 µg / kg, ∅ conc.: 130 µg / kg, country: Zimbabwe
→ fumonisins (FB_1, FB_2)
incidence: 1/4, conc.: 218 µg / kg, country: UK
→ moniliformin
incidence: 6/6, conc. range: < 50-250 µg / kg, country: UK, USA
→ ochratoxin A
incidence: 4/13, conc. range: 50-200 µg / kg, country: UK
incidence: 1/4, conc.: 0.6 µg / kg, country: UK
zearalenone
incidence: 1/4, conc.: 100 µg / kg, country: Botswana
incidence: nc/4, conc. range: 6.5-40.8 µg / kg, country: UK
→ flour

Maize grits (Syn.: polenta, semolina) may contain the following → mycotoxins:
→ aflatoxin B_1
incidence: 14/35, conc. range: 0.5-1 µg / kg (8 samples), 1-3 µg / kg, (6 sa), country: Switzerland
→ deoxynivalenol
incidence: 3/3, conc. range: 130-910 µg / kg, ∅ conc.: 640 µg / kg, country: Germany
incidence: 1/1*, conc.: 170 µg / kg, country: Germany, *organic produce
→ fumonisin B_1
incidence: 1/3, conc.: 800 µg / kg, country: Canada
incidence: 1/1, conc.: 3760 µg / kg, country: Italy
incidence: 6/6, conc. range: 420-3730 µg / kg, ∅ conc.: 2152 µg / kg, country: Italy
incidence: 6/6, conc. range: 45.6-1230 µg / kg, country: Italy
incidence: 20/20, conc. range: 150-3760 µg / kg, ∅ conc.: 1380 µg / kg, country: Italy
incidence: 2/2, conc. range: < 10-20.8 µg / kg, country: Germany
incidence: 5/5, conc. range: < 10-33.1 µg / kg, country: Germany
incidence: 14/17, conc. range: 200-2600 µg / kg, ∅ conc.: 500 µg / kg, country: Japan
incidence: 10/18, conc. range: 0-190 µg / kg, ∅ conc.: 125 µg / kg, country: South Africa
incidence: nc/8, conc. range: 0-740 µg / kg, ∅ conc.: 130 µg / kg (all samples), country: South Africa
incidence: nc/73, conc. range: 0-1380 µg / kg, ∅ conc.: 140 µg / kg (all samples), country: South Africa
incidence: 3/15, conc. range: 50-90 µg / kg, country: Spain
incidence: 34/55, conc. range: 0-790 µg / kg, ∅ conc.: 260 µg / kg, country: Switzerland
incidence: 5/5, conc. range: 250-1820 µg / kg, ∅ conc.: 830 µg / kg, country: Thailand
incidence: 2/3, conc. range: ≤ 40 µg / kg, country: The Netherlands
incidence: 10/10, conc. range: 105-2545 µg / kg, ∅ conc.: 601 µg / kg, country: USA
incidence: 4/4, conc. range: 140-270 µg / kg, ∅ conc.: 198 µg / kg, country: USA
incidence: 1/3, conc.: 80 µg / kg, country: USA
incidence: 5/5, conc. range: 140-270 µg / kg, ∅ conc.: 200 µg / kg, country: USA
→ fumonisin B_2
incidence: 1/1, conc.: 910 µg / kg, country: Italy
incidence: 6/6, conc. range: 80-840 µg / kg, ∅ conc.: 477 µg / kg, country: Italy
incidence: nc/20, conc. range: 60-910 µg / kg, ∅ conc.: 370 µg / kg, country: Italy

incidence: 5/17, conc. range: 300-2800 µg / kg, ∅ conc.: 1000 µg / kg, country: Japan
incidence: 4/18, conc. range: 0-120 µg / kg, ∅ conc.: 85 µg / kg, country: South Africa
incidence: nc/8, conc. range: 0-70 µg / kg, country: South Africa
incidence: nc/73, conc. range: 0-420 µg / kg, ∅ conc.: 20 µg / kg (all samples), country: South Africa
incidence: 13/55, conc. range: 0-160 µg / kg, ∅ conc.: 100 µg / kg, country: Switzerland
incidence: nc/5, conc. range: 70-400 µg / kg, ∅ conc.: 190 µg / kg, country: Thailand
incidence: 5/10, conc. range: 0-1065 µg / kg, ∅ conc.: 375 µg / kg, country: USA
incidence: 3/4, conc. range: 60-110 µg / kg, ∅ conc.: 86.6 µg / kg, country: USA
incidence: nc/5, conc. range: 10-111 µg / kg, ∅ conc.: 80 µg / kg, country: USA
→ fumonisin B_3
incidence: nc/73, conc. range: 0-160 µg / kg, country: South Africa
→ fumonisins (FB_1, FB_2, FB_3)
incidence: 4/4, conc. range: 20-1200 µg / kg (HPLC), ∅ conc.: 400 µg / kg, country: Germany
incidence: 16/20, conc. range: 16-2124 µg / kg, ∅ conc.: 531 µg / kg, country: UK
fumonisins (no specification)
incidence: 3/3, conc. range: 3.6-2600 µg / kg, ∅ conc.: 869 µg / kg, country: Germany
→ ochratoxin A
incidence: 1/4, conc.: < 5 µg / kg, country: UK
→ barley grits, → rye grits, → wheat grits

Maize malt may contain the following → mycotoxins:
aflatoxin (→ aflatoxins)
incidence: 1/13, conc.: 1.71 µg / kg, country: Zambia
→ zearalenone
incidence: nc/13, conc. range: 800-4000 µg / kg, ∅ conc.: 680 µg / kg, country: Zambia
→ barley malt

Maize meal For the US-market it could be shown that maize meal may contain mean levels up to and above 1 mg / kg → fumonisin B_1 while other maize products e.g. → maize grits usually show a lower contamination.
Maize meal spiked with → fumonisins was completely free of fumonisins after heating to 220 °C for 25 min.
Maize meal may contain the following → mycotoxins:
→ aflatoxins (AFB_1, AFB_2, AFG_1, AFG_2)
incidence: nc/4, conc. range: 0.2-0.7 µg / kg, country: UK
aflatoxins
incidence: 1/2*, conc.: 129 µg / kg, country: USA, *imported
→ deoxynivalenol
incidence: 45/50, conc. range: 0-250 µg / kg, country: USA
incidence: nc, ∅ conc.: 110 µg / kg, country: Canada
incidence: 2/2*, conc. range: 500-870 µg / kg, ∅ conc.: 685 µg / kg, country: Germany, *organic product
fumonisin B_1
incidence: nc/3, conc. range: < 50-1150 µg / kg, country: Austria
incidence: 5/5, conc. range: 180-450 µg / kg, country: Botswana
incidence: nc/15, conc. range: < 50-210 µg / kg, country: Bulgaria
incidence: 1/2, conc.: 50 µg / kg, country: Canada
incidence: 18/53*, conc. range: < 100-3500 µg / kg, ∅ conc.: 530 µg / kg, country: Canada *and semolina

incidence: 2/2, conc. range: 1780-2980 µg / kg, ∅ conc.: 2380 µg / kg, country: Egypt
incidence: 1/1, conc.: 1240 µg / kg, country: France
incidence: nc/3, conc. range: < 50-110 µg / kg, country: Kenya
incidence: 1/2, conc.: 660 µg / kg, country: Peru
incidence: 46/52, conc. range: < 50-475 µg / kg, ∅ conc.: 138 µg / kg, country: South Africa
incidence: nc/81, conc. range: 0-3900 µg / kg, ∅ conc.: 200 µg / kg (all samples), country: South Africa
incidence: nc/127, conc. range: 0-2850 µg / kg, ∅ conc.: 290 µg / kg (all samples), country: South Africa
incidence: 2/7, conc. range: 0-110 µg / kg, ∅ conc.: 85 µg / kg, country: Switzerland
incidence: 27/27*, conc. range: < 10-2200 µg / kg, ∅ conc.: 260 µg / kg, country: Switzerland, *and grits
incidence: 15/16, conc. range: < 50-2790 µg / kg, ∅ conc.: 1048 µg / kg, country: USA
incidence: 2/2*, conc. range: ca. 210-360 µg / kg, ∅ conc.: 290 µg / kg, country: USA
incidence: 3/3*, conc. range: 600-1200 µg / kg, ∅ conc.: 800 µg / kg, country: USA
incidence: 10/12*, conc. range: < 100-1200 µg / kg, ∅ conc.: 550 µg / kg, country: USA, *maize meal, white
incidence: 4/4*, conc. range: ca. 560-840 µg / kg, country: USA
incidence: 3/3*, conc. range: 500-1000 µg / kg, ∅ conc.: 700 µg / kg, country: USA
incidence: 5/6*, conc. range: < 100-1710 µg / kg, ∅ conc.: 980 µg / kg, country: USA, *maize meal yellow
incidence: 10/13, conc. range: 430-2050 µg / kg, ∅ conc.: 848 µg / kg, country: USA
incidence: 11/13, conc. range: < 80-2800 µg / kg, ∅ conc.: 970 µg / kg (all samples), country: USA
incidence: 16/16, conc. range: 280-2050 µg / kg, ∅ conc.: 860 µg / kg, country: USA
incidence: 1/1*, conc.: 2850 µg / kg, country: USA, *maize meal white, self rising
incidence: 5/5*, conc. range: 400-1300 µg / kg, ∅ conc.: 890 µg / kg, country: USA, *maize meal yellow
incidence: 1/1*, conc.: 6320 µg / kg, country: USA, *maize meal blue
incidence: 1/1*, conc.: 40 µg / kg, country: Venezuela, *maize meal white
incidence: 4/4, conc. range: 70-1880 µg / kg, ∅ conc.: 718 µg / kg, country: Venezuela
incidence: 3/3, conc. range: 1060-3630 µg / kg, country: Zimbabwe

→ fumonisin B_2

incidence: 5/5, conc. range: < 50-120 µg / kg, country: Botswana
incidence: nc/15, conc. range: 50-150 µg / kg, country: Bulgaria
incidence: 2/2, conc. range: 470-780 µg / kg, ∅ conc.: 625 µg / kg, country: Egypt
incidence: 1/1, conc.: 390 µg / kg, country: France
incidence: nc/3, conc. range: < 50-140 µg / kg, country: Kenya
incidence: 1/2, conc.: 135 µg / kg, country: Peru
incidence: 11/52, conc. range: < 50-131 µg / kg, ∅ conc.: 83 µg / kg, country: South Africa
incidence: nc/81, conc. range: 0-760 µg / kg, ∅ conc.: 100 µg / kg (all samples), country: South Africa
incidence: nc/127, conc. range: 0-910 µg / kg, ∅ conc.: 70 µg / kg (all samples), country: South Africa
incidence: nc/27*, conc. range: < 10-590 µg / kg, ∅ conc.: 60 µg / kg, country: Switzerland, *and grits

incidence: 13/16, con. range: 0-920 µg / kg, ∅ conc.: 298 µg / kg, country: USA
incidence: 2/2*, conc. range: ca. 33-58 µg / kg, ∅ conc.: 40.5 µg / kg, country: USA
incidence: nc/12*, conc. range: < 100-520 µg / kg, ∅ conc.: 210 µg / kg, country: USA, *maize meal white
incidence: 4/4*, conc. range: ca. 120-414 µg / kg, country: USA
incidence: nc/6*, conc. range: < 100-470 µg / kg, ∅ conc.: 380 µg / kg, country: USA, *maize meal yellow
incidence: 10/13, conc. range: 50-360 µg / kg, ∅ conc.: 192 µg / kg, country: USA
incidence: nc/13, conc. range: < 100-1000 µg / kg, ∅ conc.: 300 µg / kg (all samples), country: USA
incidence: nc/16, conc. range: 50-530 µg / kg, ∅ conc.: 200 µg / kg, country: USA
incidence: 1/1*, conc.: 10 µg / kg, country: Venezuela, *maize meal white
incidence: 4/4, conc. range: 20-530 µg / kg, ∅ conc.: 198 µg / kg, country: Venezuela
incidence: nc/3, conc. range: 240-910 µg / kg, country: Zimbabwe
→ fumonisin B_3
incidence: 5/5, conc. range: < 50-120 µg / kg, country: Botswana
incidence: 1/1, conc.: 180 µg / kg, country: France
incidence: nc/81, conc. range: 0-150 µg / kg, ∅ conc.: < 100 µg / kg (all samples), country: South Africa
incidence: nc/127, conc. range: 0-460 µg / kg, ∅ conc.: 30 µg / kg (all samples), country: South Africa
incidence: 2/2*, conc. range: "present", country: USA, *maize meal white
incidence: 4/4*, conc. range: "present", country: USA, *maize meal yellow
incidence: nc/3, conc. range: 130-230 µg / kg, country: Zimbabwe
fumonisins (FB_1, FB_2)
incidence: 1/12, conc.: < 1000 µg / kg, country: UK
fumonisins
incidence: 1/2, conc.: 7.1 µg / kg, country: Germany
incidence: 1/1, conc.: 1300 µg / kg (HPLC), country: Germany
incidence: 3/3, conc. range: 1500-4700 µg / kg, ∅ conc.: 2933 µg / kg, country: USA
incidence: 2/2*, conc. range: no exact data, country: USA, *maize meal, blue
incidence: 7/12*, conc. range: no exact data, country: USA, *maize meal yellow
incidence: 6/6*, conc. range: 450-4750 µg / kg, ∅ conc.: 1558 µg / kg, country: USA *maize meal yellow
incidence: 4/4*, conc. range: 650-7450 µg / kg, ∅ conc.: 3075 µg / kg, country: USA, *maize meal white
→ moniliformin
incidence: 27/27, conc. range: 50-180 µg / kg, ∅ conc.: 85.6 µg / kg, country: France, UK, USA
→ zearalenone
incidence: 2/2*, conc. range: 38-65 µg / kg, ∅ conc.: 56.5 µg / kg, country: Germany, *organic product
incidence: 12/50, conc. range: nc, country: Mexico
incidence: 9/11, conc. range: 11-69 µg / kg, ∅ conc.: 33.1 µg / kg, country: USA
incidence: 7/9, conc. range: 3.2-120 µg / kg, ∅ conc.: 23 µg / kg, country: USA
→ sorghum meal

Maize muffin may contain the following → mycotoxins:
→ fumonisins
incidence: 1/1, conc.: 300 µg / kg, country: USA

Maize pop cereal may contain the following → mycotoxins:

→ fumonisins (no specification)
incidence: 1/1, conc.: 200 µg / kg, country: USA

Maize products (no specification)
Fumonisin concentration in refined → maize products may be lowered during the process of → milling.
Maize products may contain the following → mycotoxins:
→ aflatoxins
incidence: 1/23, conc.: 37 µg / kg, country: Japan
incidence: 15/41, conc.: > 30- ≤ 400 µg / kg, country: Philippines
incidence: 19/139, conc. range: ≤ 53 µg / kg, ∅ conc.: 19.6 µg / kg, country: USA
→ fumonisin B_1
incidence: 5/71, conc. range: < 100-1200 µg / kg, ∅ conc.: 330 µg / kg, country: Canada
incidence: 2/2, conc. range: 1780-2980 µg / kg, ∅ conc.: 2380 µg / kg, country: Egypt
incidence: 5/8, conc. range: < 10-60 µg / kg, ∅ conc.: 50 µg / kg, country: Italy
incidence: nc/4, conc. range: 0-660 µg / kg, ∅ conc.: 165 µg / kg, country: Peru
incidence: 2/6, conc. range: 41.4-73 µg / kg, ∅ conc.: 57.2 µg / kg, country: Thailand
incidence: 2/8, conc. range: 0-91 µg / kg, ∅ conc.: 84 µg / kg, country: South Africa
incidence: nc/68, conc. range: 0-475 µg / kg, ∅ conc: 105 µg / kg, country: South Africa
incidence: 2/20, conc. range: 60-200 µg / kg, ∅ conc.: 130 µg / kg, country: Spain
incidence: 4/4, conc. range: 85-700 µg / kg, ∅ conc.: 409 µg / kg, country: USA
incidence: 4/4, conc. range: 20-320 µg / kg, ∅ conc.: 170 µg / kg, country: USA
incidence: 4/9, conc. range: < 10-120 µg / kg, ∅ conc.: 70 µg / kg, country: USA
incidence: nc/29, conc. range: 0-2790 µg / kg, ∅ conc.: 711 µg / kg, country: USA
incidence: 3/5, conc. range: < 50-1210 µg / kg, ∅ conc.: 540 µg / kg, country: USA
→ fumonisin B_2
incidence: 2/2, conc. range: 410-780 µg / kg, ∅ conc.: 595 µg / kg, country: Egypt
incidence: nc/8, conc. range: < 10-20 µg / kg, country: Italy
incidence: nc/4, conc. range: 0-135 µg / kg, ∅ conc.: 34 µg / kg, country: Peru
incidence: nc/68, conc. range: 0-120 µg / kg, ∅ conc: 21 µg / kg, country: South Africa
incidence: 3/4, conc. range: 0-240 µg / kg, ∅ conc.: 148 µg / kg, country: USA
incidence: nc/9, conc. range: 10-30 µg / kg, ∅ conc.: 20 µg / kg, country: USA
incidence: nc/29, conc. range: 0-2790 µg / kg, ∅ conc.: 711 µg / kg, country: USA
→ ochratoxin A
incidence: 1/23, conc.: 73 µg / kg, country: Japan

Maize screenings (Syn.: corn screenings)
Compared to the intact corn fumonisin levels in → maize screenings can be about 10 times higher (Iowa corn). It seems that there is no size-related segregation of fumonisin contents in corn screenings. The accumulation of → fumonisins in corn screenings may be a source of concern since they are used in feed formulas for livestock.

Maize snacks may contain the following → mycotoxins:
→ fumonisin B_1
incidence: 2/11, conc. range: 50-200 µg / kg, country: Spain

incidence: 26/78, conc. range: $\leq$ 2395 µg / kg, ∅ conc.: 456 µg / kg, country: Thailand
→ fumonisin B_2
incidence: 16/78, conc. range: $\leq$ 715 µg / kg, ∅ conc.: 145 µg / kg, country: Thailand
→ fumonisins (FB_1, FB_2, FB_3)
incidence: 31/40, conc. range: 11-220 µg / kg, ∅ conc.: 46 µg / kg, country: UK

Maize starch may contain the following → mycotoxins:
→ aflatoxin
incidence: 6/9, conc. range: $\leq$25 µg / kg, ∅ conc.: 12 µg / kg, country: Philippines
→ fumonisins (no specification)
incidence: 1/1, conc.: 500 µg / kg, country: USA

Majoran may contain the following → mycotoxins:
→ ochratoxin A
incidence: 1/1, conc.: 28 µg / kg, country: Austria
→ spices

Malaga → Wine

Malt (malting)
The mycotoxin contamination of malt with e.g. → deoxynivalenol, → nivalenol and / or → zearalenone is due to the use of natural contaminated → grains and / or growth of certain fungi during various stages of the malting production.
→ Mycotoxins may impair malt processing. → T-2 toxin, added before malting, inhibited coleoptile and rootlet elongation in germinating acid-dehusked → barley depending on the concentration used. To some extent this mycotoxin also retarded de novo synthesis of α-amylase. → Diacetoxyscirpenol and deoxynivalenol act in the same way.
The apparent loss of zearalenone ($\approx$ 75%) and T-2 toxin ($\approx$ 54%) during malting might be due to the binding of the mycotoxins to substances extracted from barley into steep liquor and present in kilned malt or might be caused by binding to microoganisms or degradation products of malt sugars. DON losses amounted up to almost 80%.
Since substantial to total losses of → ochratoxin A and → citrinin during malting have been reported, it is concluded that neither OTA nor citrinin are likely to enter the brewing process from malted barley but from brewing adjuncts (OTA).
Malt may contain the following mycotoxins:
ochratoxin A
incidence: 3/11*, conc. range: 0.1-0.92 µg / kg, country: Germany, *partly imported
incidence: 1/2, conc. range: 1.5-9.99 µg / kg, country: Germany
→ beer

Mandarin fruits may contain the following → mycotoxins:
→ alternariol
incidence: 2/3*, conc. range: 1000-5200 µg / kg, ∅ conc: 3100 µg / kg, country: Italy
→ alternariol methyl ether
incidence: 2/3*, conc. range: 550-1400 µg / kg, ∅ conc.: 975 µg / kg, country: Italy
→ tenuazonic acid
incidence: 3/3*, conc. range: 21,000-173,900 µg / kg, ∅ conc.: 94,033 µg / kg, country: Italy
*samples visibly affected by → Alternaria rot
→ fruits

Mango (pickled in salt)
may contain the following → mycotoxins:
→ aflatoxin B_1
incidence: 4/8*, ∅ conc.: 52 µg / kg, country: India

incidence: 26/40**, Ø conc.: 210 μg / kg, country: India
→ aflatoxin B_2
incidence: 4/8*, Ø conc.: 5 μg / kg, country: India
incidence: 26/40**, Ø conc.: 32 μg / kg, country: India
→ aflatoxin G_1
incidence: 4/8*, Ø conc.: 24 μg / kg, country: India
incidence: 26/40**, Ø conc.: 184 μg / kg, country: India
→ aflatoxin G_2
incidence: 4/8*, Ø conc.: traces, country: India
incidence: 26/40**, Ø conc.: 15 μg / kg, country: India
* stored in bottles, **stored in polythene bags
→ fruits

Manioc may contain the following → mycotoxins:
→ aflatoxins (no specification)
incidence: 1/8, conc.: nc, country: Mocambique

Marchpane → marzipan

Marzipan (almond paste)
Blanched → almonds for marzipan manufacture should be processed immediately after blanching. If the period of storage prior to blending with sugar and drying is too long, fungal infection may occur with subsequent aflatoxin contamination. 3 days of storage at 28 °C are almost critical.
Marzipan may contain the following → mycotoxins:
→ aflatoxin B_1
incidence: 1/168, conc.: 39 μg / kg, country: Finland
incidence: 3/12, conc. range: tr-2 μg / kg, country: Germany
→ aflatoxin B_2
incidence: 1/168, conc.: 7 μg / kg, country: Finland
incidence: 1/16, conc.: < 1 μg / kg, country: Germany
→ aflatoxin G_1
incidence: 1/12, conc.: traces, country: Germany
→ almonds, → persipan

Masa is tortilla → flour which has traditionally been treated with $Ca(OH)_2$ and heat (nixtamalization). This processing, which improves the nutritive value of → maize, may decontaminate fumonisin-contaminated maize because the fumonisin levels in → maize products made with masa usually are low.
Masa may contain the following → mycotoxins:
→ fumonisin B_1
incidence: 2/3, conc. range: 590-1800 μg / kg, Ø conc.: 1195 μg / kg, country: Mexico
incidence: 3/3, conc. range: 40-380 μg / kg, Ø conc.: 163 μg / kg, country: USA
incidence: 8/8, conc. range: 63-689 μg / kg, Ø conc.: 262 μg / kg, country: USA / Mexico
→ fumonisin B_2
incidence: 2/3, conc. range: 110-1380 μg / kg, Ø conc.: 553 μg / kg, country: Mexico
incidence: 1/3, conc.: 60 μg / kg, country: USA
hydrolyzed fumonisin B_1 (HBF_1)
incidence: 1/2, conc.: 100 μg / kg, country: Mexico
incidence: 2/3, conc. range: 20-100 μg / kg, Ø conc.: 60 μg / kg, country: USA
incidence: 6/8, conc. range: 21-178 μg / kg, Ø conc.: 64 μg / kg, country: USA/Mexico
→ maize

Meat and meat products
Probably there is little or no danger in the transmission of → mycotoxins into the muscle tissues of most animals consuming feed contaminated with mycotoxins. This is due to the fact that the transfer ratios are obviously high; the transfer ratio for → aflatoxin B_1 (μg / kg mycotoxin in feed: μg / kg mycotoxin in tissue) is in the range of 1000-14,000. In general, it is expected that animals exposed to such high mycotoxin concentrations suffer from obvious disease symptoms or even die. It is most unlikely that such animals enter the food chain and therefore animal tissues do not contribute substantially to mycotoxin intake of humans. However, → ochratoxin A in kidneys, → sausages, and black pudding prepared from pigs may represent an exception.
Feeding experiments with aflatoxin B_1, ochratoxin A, → patulin, → penicillic acid, → sterigmatocystin, → T-2 toxin, and → zearalenone labeled with radioactive elements show a major excretion rate for the mycotoxins and / or their metabolites within 24 h. Only minor levels could be detected in the edible tissue. Extensive breakdown of the mycotoxins is not expected since a negligible amount of radioactivity could be detected in the expired air. Elaborated data suggest that the structure modifications are due to mixed-function oxidases and the high specific activity implies that the liver and biliary system is mainly responsible for the elimination process.
The only two mycotoxins that might be important in domestic animals are aflatoxin B_1 (→ milk / → aflatoxin M_1) in cows and ochratoxin A in liver, kidneys and meat products, e.g. certain sausages of pigs. There are three possible ways of mycotoxin contamination of meat / meat products:
i) Mold growth on the product surface leading to direct mycotoxin contamination which is of minor importance. Contamination rate of fermented meat products such as salamis or country cured → ham with the most dangerous → Aspergillus flavus Link and → Aspergillus parasiticus Speare is rare. If present these molds must successfully compete against starter cultures (e.g. *Penicillium nalgiovense*) or the house flora. In addition, relative humidities < 80%, temperatures between 10 and 15 °C, as well as smoking and curing ingredients prevent or reduce aflatoxin production. It is not expected that human exposure to → aflatoxins by this route is of great importance.
ii) Meat products may contain mycotoxins due to the use of naturally-contaminated → spices and spice mixtures (see → meat, luncheon)
iii) However, transmission (→ carry-over) of mycotoxins by the intake of contaminated feedstuff, especially in the case of OTA, is more serious. A period of approximately four weeks is necessary to reduce OTA to nondetectable levels (< 2 μg / kg) in swine kidneys.
Among the domestic animals ruminants, e.g. cattle, are not very susceptible to ochratoxin A. Since OTA is a phenylalanine amide of isocoumarin the rumen flora decomposes the mycotoxin by proteolytic enzymes. However, pigs are extremely sensitive to OTA which posseses a relatively high serum half-live of 72-120 h. Blood / plasma contains the highest OTA concentrations followed by kidneys, liver, muscles, fat (in decreasing order). On average, OTA levels in lean muscle is half of that in the kidney. OTA levels in e.g. raw sausages, liver sausage-type, Frankfurter type sausage, ham, bacon depend on the particular recipe of the meat product (proportion of pork, beef, etc.). If liver tissue or blood is used, OTA concentrations greatly increase in the corresponding meat products, such as Bologna-type sausages, blood sausages or black pudding. In contrast, pork as well as poultry meat contain, if at all, low

levels of OTA and therefore do not constitute a significant health problem for humans (see Figure Meat).
The stability of OTA prevents its reduction during the stages of meat-product manufacturing, such as heating and ripening as well as storage. Only frying or boiling lowered the OTA concentration by as much as 40% (blood-pudding, kidneys, muscular tissues) depending on the water content and the inner temperature of the treated product. No OTA losses occurred in adipose tissues. Compared to food of plant origin, meat and meat products usually show lower levels of OTA contamination.
Contamination problems with the milk aflatoxin M_1 arise because feed consumption and lactation are concurrent events without any withdrawal period. However, the four "primary" aflatoxins B_1, B_2, G_1, and G_2 are rapidly metabolized. In consequence, none of them or only low levels are found in animal tissues or milk. In addition, despite the toxicity of AFM_1, the macromolecule-bound AFB_1 derivatives in meat are at least 4000 times less active than AFB_1. The water-soluble conjugates are at least 100 times less potent compared to AFB_1. From the present data it is concluded that there is a negligible carcinogenic risk for humans who consume aflatoxin contaminated liver or meat compared with certain foodstuffs of plant origin, e.g. → nuts.
According to Frisvad (1988) the following mycotoxins may be found in meat (and eggs): aflatoxins, → citrinin, → cyclopiazonic acid, ochratoxin A, patulin, → penicillic acid, penitrem A (→ penitrems), → rugulosin, → sterigmatocystin, → viomellein, → xanthomegnin.
Meat and meat products may contain the following mycotoxins:
→ aflatoxins (no specification)
incidence: 2/19, conc. range: < 1 µg / kg, country: UK
→ citrinin
incidence: 9/23, conc. range: < 100 µg / kg, country: UK
→ ochratoxin A
incidence: 7/33, conc. range: 0-4 µg / kg, country: UK
incidence: 6/6*, conc. range: 0.1-2.2 µg / kg, country: Tunesia, *and fish (mackerel)
→ patulin
incidence: 7/24, conc. range: 0-200 µg / kg, country: UK

Meat, luncheon Detection of → aflatoxins in luncheon → meat results from the use of mycotoxin contaminated → spices and / or the incorporation of aflatoxin producers.
Luncheon meat may contain the following → mycotoxins:
→ aflatoxin B_1
incidence: 1/25, conc.: 4 µg / kg, country: Egypt
→ aflatoxin B_2
incidence: 1/25, conc.: 2 µg / kg, country: Egypt

Mechanical damage of the seed or fruit coat favors the penetration of molds into → grains / → fruits as a prerequisite for

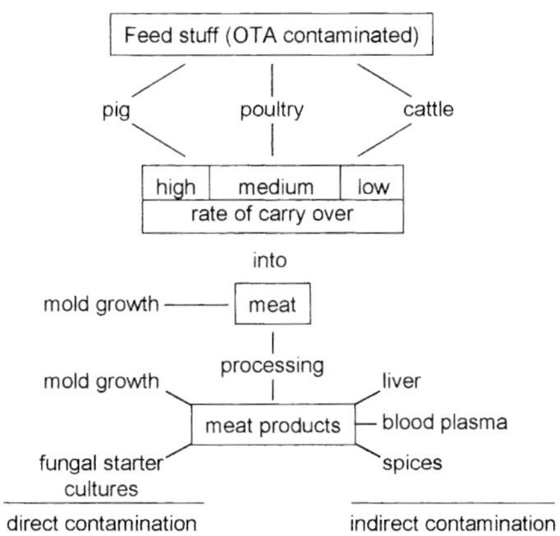

Meat. Direct and indirect OTA contamination of meat and meat products

mycotoxin (→ mycotoxins) contamination.

Melon may contain the following → mycotoxins:
→ alternariol methyl ether
incidence: 1/1*, conc.: 51 µg / kg, country: Italy
*sample visibly affected by → Alternaria rot
→ fruits

Melon balls snacks may contain the following → mycotoxins:
→ aflatoxins
incidence: 4*/40, conc. range: 10-40 µg / kg, country: Nigeria
*all samples contained → aflatoxin B_1; 2 sa additionally → aflatoxin B_2 and / or → aflatoxin G_1

Melon seeds may contain the following → mycotoxins:
→ aflatoxins (no specification)
incidence: 2/4*, conc. range: $\leq$ 29 µg / kg, Ø conc.: 26 µg / kg, country: USA, *imported

Microbial interactions The presence of competing fungi / bacteria and their effects on toxin production are not predictable. In general, development and mycotoxin formation of → Aspergillus spp. and → Penicillium spp. is considerably reduced if other competing microorganisms are present.

Milk, camel may contain the following → mycotoxins:
→ aflatoxin M_1
incidence: 6/20, conc. range: 0.25-0.8 µg / l, country: UAE

Milk, cow Of all animal products milk, one of the best natural foods and the primary nutrient for children, including infants, is most frequently contaminated with aflatoxin residues. Therefore, a theoretical hazard associated with → aflatoxin M_1 in commercially available milk, → human breast milk, and milk products does exist mainly because the growing young are very susceptible to the adverse effects of → mycotoxins. They usually have a relatively low body weight, showing a high cell activity whereas the immune system is only partially developed. Due to the slower rate of biotransformation of carcinogens in infants a longer circulation time of the chemicals may be the consequence.
In the early 1960s a toxic factor in milk was discovered after feeding lactating cows with aflatoxin-contaminated diet. The toxic factor named aflatoxin M occurred in milk 4-5 h after ingestion of the contaminated feed. Structure elucidation of the milk metabolite AFM_1 succeeded in 1966. Subsequent studies revealed that AFM_1 is the major aflatoxin in milk although other hydroxylated → aflatoxin B_1 metabolites such as → aflatoxin M_2, → aflatoxin M_4, → aflatoxin Q_1, and → aflatoxicol have been detected. However, these aflatoxin derivatives occur in very low concentrations (two to three orders of magnitude lower compared to AFM_1).
Transmission of other → mycotoxins such as → deoxynivalenol, → fumonisins, → ochratoxin A, → sterigmatocystin, → T-2 toxin, and → zearalenone in milk and → milk products has been studied / detected. In the case of → trichothecenes relatively high concentrations have to be ingested to produce detectable toxin residues in milk. Compared to DON marginally more T-2 toxin appears to be transmitted into milk probably due to its lipophilic nature. However, in cattle this mycotoxin is extensively and very rapidly metabolized. Instead of relatively high oral doses (0.5-3.6 mg / kg) no substantial accumulation of any residues in milk, organs or tissues could be observed.

Their low toxicity and / or limited presence in milk probably makes these mycotoxins of little significance for human health.
Contaminated concentrate feeds are mainly responsible for aflatoxin contamination in milk and dairy products. The increased AFM_1 contamination rate in milk during the winter months was due to the major usage of compound feeds in the cold season imported from tropical and subtropical countries. In contrast, in spring and summer time home grown non contaminated roughage, summer forage, and pasture are available.
Strict control measures for locally manufactured and imported feedstuffs are a prerequisite of low aflatoxin levels in milk and → dairy products. In this connection the Commission of the European Communities further tightened the acceptable level for AFB_1 in feedstuff in dairy cattle from 20 to 10 μg / kg in 1984 to 5 μg / kg in 1991. To prevent AFM_1 contamination in milk feeding of → peanuts which are frequently contaminated by AFB_1 to lactating cows has been forbidden by the Swiss legislation.
Transmission rate (→ carry over) of aflatoxin B_1 that is ingested in the feed and excreted as the 4-hydroxylated derivative AFM_1 in milk varies among individual animals but is linearly correlated with milk yield and roughly amounted 1.5% (0.35-3%), e.g. 300 μg aflatoxin B_1 / kg feed will result in ≈ 4.5 μg aflatoxin M_1 / l milk. A rapid increase in aflatoxin concentration was observed when a high intake of the mycotoxin reduced the milk yield.
Metabolization of AFB_1 is due to the hepatic microsomal mixed-function oxidase system, but, depending on species, several other metabolic conversions are possible (e.g. metabolism rate in the liver, excretion rate by other routes). AFM_1 is found in cow's milk as early as 4 h after ingestion of the contaminated feed. A significant decrease in aflatoxin concentration occurs 24-48 h after exposition. 3 to 5 days after aflatoxin-free rations are given aflatoxin values decreased to zero in the milk.
Heating, e.g. pasteurization (→ milk, pasteurized) or sterilization (→ milk, sterilized), does not cause any destruction of the thermoresistant AFM_1 molecule in milk although different results have been reported (63% pasteurization, 80% sterilization). Data about the reduction of AFM_1 concentration in cold treated or frozen milk are contradictory and not conclusive. AFM_1 contamination of (processed) milk indicates the level of AFB_1 in animal feed.
There is no homogeneous distribution of AFM_1 in milk. Since the semipolar AFM_1 is primarily bound to casein it is estimated that about 30% of AFM_1 are associated with the nonfat milk solids. The enrichment of AFM_1 in the nonfat fraction resulted from processes which involve fat (→ cream) separation. When butter is made from naturally contaminated cream, the AFM_1 concentration in the butter amounted to a little more than 20%, while the major portion of AFM_1 is found in buttermilk (→ milk-, butter).
Skim-milk manufacturing may lead to the accumulation of about 80% of AFM_1 in that portion. Lower levels of AFM_1 (60-75%) may be found in concentrated milk. No AFM_1 reduction was observed during the manufaturing of cheese and yogurt.
Although aflatoxin B_1 is also a contaminant of milk very much lower levels compared to AFM_1 have been found.
Compared to raw farm milk, the rate of AFM_1 contamination in commercial milk is often higher, resulting from the addition of a few contaminated samples to uncontaminated bulk milk. However, for the same reason commercial milk in general shows low levels of AFM_1 contamination.

Detectable levels of OTA in cow's milk only result from a daily dose of OTA higher than 1.66 mg / kg bw. Lower amounts did not lead to the detection of OTA in milk mainly due to the hydrolysis of this mycotoxin (ochratoxin α) by the microflora in the rumen of the cow. It seems that milk is not an important contributor to OTA intake but taking into account that e.g. children consume large quantities of milk, even low OTA levels (0.01-0.04 μg / l milk) might significantly increase the daily OTA intake.
Milk may contain the following mycotoxins:
aflatoxin B_1
incidence: 2/1150, conc. range: 0.28-0.36 μg / l, country: Spain
incidence: 5/105, conc. range: ≤ 2500 μg / l, country: Yugoslavia
aflatoxin M_1
incidence: 9/12*, conc. range: 0.002-0.05 μg / l, country: Austria, *raw
incidence: 32/88*, conc. range: 0.001-0.01 μg / l, country: Austria, *commercial
incidence: 42/68, conc. range: 0.02-0.2 μg / l, country: Belgium
incidence: 46/145, conc. range: < 0.02-0.5 μg / l, country: Belgium
incidence: 3/6, conc. range: 0.025-0.5 μg / l, country: Brazil
incidence: 4/224, conc. range: tr-0.002 μg / l, country: Brazil
incidence: 1/100*, conc.: 0.2 μg / l, country: Brazil, *commercial
incidence: 9/50*, conc. range: 0.1-1.68 μg / l, country: Brazil, *farm
incidence: 22/85, conc. range: > 0.5 μg / l, country: Cuba
incidence: 5/77*, conc. range: tr-0.38 μg / l, country: Czechoslovakia, *raw
incidence: 27/89*, conc. range: < 0.5 μg / l, country: Czechoslovakia, *raw
incidence: 25/191, conc. range: 0.05-0.1 μg / l, country: Czechoslovakia
incidence: 9/67, conc. range: 0.05-0.1 μg / l, country: Czechoslovakia
incidence: 43/403, conc. range: 0.025-0.1 μg / l (37 samples), 0.1-0.5 μg / l (6 sa), country: Czechoslovakia
inicdence: 46/376, conc. range: 0.025-0.1 μg / l (44 samples), > 0.1 μg / l (2 sa), country: Czechoslovakia
incidence: 9/117*, conc. range: 0.05-0.1 μg / l, country: Czechoslovakia, *commercial
incidence: 11/88, conc. range: < 0.001-0.023 μg / l, country: France
incidence: 168/380, conc. range: 0.05-1.15 μg / l, country: France
incidence: 32/102, conc. range: 0.5-5 μg / l, country: France
incidence: 5489/5489, conc. range: 0-0.05 μg / l (5.284 samples), 0.05-0.5 μg / l (200 sa), > 0.5 μg / l (5 sa), country: France
incidence: 757/757, conc. range: 0-0.05 μg / l (659 samples), 0.05-0.5 μg / l (84 sa), > 0.5 μg / l (14 sa), country: France
incidence: 70/112, conc. range: < 0.01-16.1 μg / l, country: France
incidence: 31/225, conc. range: < 0.001-0.01 μg / l, country: Germany
incidence: 16/25, conc. range: 0.04-0.13 μg / l, country: Germany
incidence: 21/48, conc. range: 0.04-0.25 μg / l, country: Germany
incidence: 7/13, conc. range: 0.05-0.13 μg / l, country: Germany
incidence: 79/419, conc. range: 0.05-0.54 μg / l, ∅ conc.: 0.12 μg / l, country: Germany
incidence: 118/260, conc. range: 0.05-0.33 μg / l, country: Germany
incidence: 4/60, conc. range: 1.7-6.5 μg / l, ∅ conc.: 3.6 μg / l, country: Germany
incidence: 265/279, conc. range: 0.0003-0.68 μg / l, country: Germany
incidence: 624/6445, conc. range: 0.01->0.05 μg / l, country: Germany
incidence: 1507/1507, conc. range: 0-0.05 μg / l (1504 samples), > 0.05 μg / l (3 sa), country: Germany

incidence: 388/388, conc. range: 0-0.01 µg / l (387 samples), > 0.01 µg / l (1 sa) country: Germany
incidence: 28/61, conc. range: 0.04-0.25 µg / l, country: Germany
incidence: 4/36, conc. range: 1.7-6.5 µg / l, Ø conc.: 3.6 µg / l, country: Germany
incidence: 4/99*, conc. range: 0.1-0.13 µg / l, country: Greece, *raw
incidence: 72/81, conc. range: 0.0005-0.001 µg / l (31 samples), 0.0025-0.005 µg / l (32 sa), > 0.005-0.177 µg / l (9 sa)., country: Greece
incidence: 89/504, conc. range: 0.1-3.5 µg / l, country: India
incidence: 3/21, conc. range: ≤ 13.3 µg / l, Ø conc.: 1159 µg / l, country: India
incidence: 48/52, conc. range: ≤ 23 µg / l, country: Iran
incidence: 38*/95, conc. range: 8-500 µg / l, country: Iran, *mainly AFM_1 and to a minor degree AFM_2
incidence: 12/18, conc. range: 0.005-0.03 µg / l, country: Italy
incidence: 76/106, conc. range: 0.004-0.28 µg / l, country: Italy
incidence: 34/82, conc. range: tr-0.569 µg / l, country: Italy
incidence: 46/59, conc. range: tr-0.378 µg / l, country: Italy
incidence: 24/27, conc. range: 0.005-0.065 µg / l, country: Italy
incidence: 136/159, conc. range: < 0.001-0.1 µg / l, Ø conc.: 0.01 µg / l, country: Italy
incidence: 5/31, conc. range: 0.03-0.07 µg / l, country: Italy
incidence: 9/52*, conc. range: 0.005-0.146 µg / l, country: Italy, *raw
incidence: 8/31*, conc. range: 0.005-0.091 µg / l, country: Italy, *raw
incidence: 24/57*, conc. range: 0.3-0.93 µg / l, country: Italy, *raw
incidence: 3/60*, conc. range: 0.1-0.28 µg / l, country: Italy, *raw
incidence: 5/107, conc. range: 0.024-0.094 µg / l, country: Italy
incidence: 66/107, conc. range: 0.006-0.101 µg / l, country: Italy
incidence: 56/107, conc. range: 0.003-0.06 µg / l, country: Italy
incidence: 1/50, conc.: 0.4 µg / l, country: Italy
incidence: 19/22*, conc. range: 0.18-0.434 µg / l, country: Italy, *commercial
incidence: 30/276*, conc. range: 0.01-0.2 µg / l, country: Italy, *raw
incidence: 4/4, conc. range: 1.3-6.8 µg / l, country: Norway
incidence: 11/22, conc. range: 0.01-0.25 µg / l, country: Poland
incidence: 5/21, conc. range: 0.02-0.2 µg / l, country: South Africa
incidence: 14/47, conc. range: 0.02-0.1 µg / l, country: Spain
incidence: 61/61*, conc. range: < 0.01 µg / l (49 samples), 0.01-0.02 µg / l (10 sa), 0.02-0.04 µg / l (2 sa), country: Spain, *raw
incidence: 1/84, conc. range: 0.05-0.1 µg / l, country: Sweden
incidence: 13/13, conc. range: 0.005-0.36 µg / l, country: Sweden
incidence: 16/163, conc. range: 0.05-2 µg / l, country: Switzerland
incidence: 40/230, conc. range: 0.05-3 µg / l, country: Switzerland
incidence: 8/91, conc. range: 0.001-0.609 µg / l, country: Switzerland
incidence: 2/38, conc. range: 0.01-0.05 µg / l, country: Switzerland
incidence: 84/105*, conc. range: 0.015-0.09 µg / l, country: The Netherlands, *and UHT
incidence: 74/95, conc. range: < 0.09-0.5 µg / l, country: The Netherlands
incidence: 85/278, conc. range: 0.03-0.52 µg / l, country: UK
incidence: 24/409, conc. range: 0.02-0.05 µg / l (10 samples), 0.05-0.1 µg / l (6 sa), > 0.1 µg / l (8 sa), country: UK
incidence: 7/22, conc. range: 0.2-0.5 µg / l (6 samples), > 0.5 µg / l (1 sa), country: Uruguay

incidence: 192/302, conc. range: < 0.1 µg / l (15 samples), 0.1-0.4 µg / l (158 sa), 0.5-3.9 µg / l (19 sa), country: USA
incidence: 554/816, conc. range: 0.1-2 µg / l, country: USA
incidence: 116/912, conc. range: 0.1-2 µg / l, country: USA
incidence: 144/624, conc. range: 0.1-2 µg / l, country: USA
incidence: 107/847, conc. range: 0.1-2 µg / l, country: USA
incidence: 235/786, conc. range: 0.1-2 µg / l, country: USA
incidence: 99/168, conc. range: 0.1-2 µg / l, country: USA
→ fumonisin B_1
incidence: 1/165, conc. 1.3 µg / l, country: USA
ochratoxin A
incidence: 4/36, conc. range: 0.007-0.030 µg / l, country: Germany
incidence: 9/50, conc. range: 1.7-6.6 µg / l, country: Italy
incidence: 6/40*, conc. range: 0.011-0.058 µg / l, country: Norway, *conventional
incidence: 5/47*, conc. range: 0.015-0.028 µg / l, country: Norway, *organic
incidence: 5/36, conc. range: 0.01-0.04 µg / l, country: Sweden
→ cheese, → human breast milk

Milk (raw or dried, for infant formulae) may contain the following → mycotoxins:
→ aflatoxin M_1
incidence: 46/376, conc. range: < 0.5 µg / l, country: Czechoslovakia
incidence: 2/376, conc. range: > 0.1 µg / l, country: Czechoslovakia
incidence: 1/56, conc.: ca. 0.67 µg / kg, country: Germany
incidence: 7/7, conc. range: 0.679-1.96 µg / l, country: Italy
incidence: 58/233, conc. range: 0.2-0.8 µg / l, country: Italy

Milk (skim milk, dried) may contain the following → mycotoxins:
→ aflatoxin M_1
incidence: 16/28, conc. range: 0.08-1.5 µg / l, country: Germany
incidence: 8/93, conc. range: 0.05-0.4 µg / l, country: USA
incidence: 17/27, conc. range: tr- > 2 µg / l, country: USA

Milk, pasteurized Pasteurization seems to have only a minor effect on inactivating → aflatoxin M_1 in milk whereas sterilization will cause some losses of AFM_1.
Pasteurized milk may contain the following → mycotoxins:
aflatoxin M_1
incidence: 4/204*, conc. range: 0.073-0.37 µg / l, ∅ conc.: 0.155 µg / l, country: Brazil
*includes pasteurized → milk, → milk powder and → milk products
incidence: 16/314, conc. range: < 0.5 µg / l, country: Czechoslovakia
incidence: 9/9, conc. range: ≤ 20.1 µg / l, country: Iran
incidence: 59/66, conc. range: 0.004-0.15 µg / l, country: Italy
incidence: 61/68, conc. range: 0.005-0.05 µg / l, country: Italy
incidence: 27/30, con. range: 0.003-0.022 µg / l, country: Italy
incidence: 7/143, conc. range: 0.1-0.4 µg / l, country: Portugal
incidence: 2/24, conc. range: 0.02-0.04 µg / l, country: Spain

Milk, sterilized Sterilization of milk will cause some losses in → aflatoxin M_1 levels.
Sterilized milk may contain the following → mycotoxins:
aflatoxin M_1
incidence: 5/33, conc. range: 0.01-0.04 µg / l, country: Spain

Milk, UHT may contain the following → mycotoxins:

→ aflatoxin M_1
incidence: 10/60, conc. range: 0.1-0.5 μg / l, country: Italy
incidence: 28/32, conc. range: 0.007-0.050 μg / l, country: Italy
incidence: 14/47, conc. range: 0.02-0.1 μg / l, country: Italy
incidence: 33/33, conc. range: < 0.010 μg / l (28 samples), 0.01-0.025 μg / l (5 sa), country: Spain
incidence: 12/76*, conc. range: 0.02-0.04 μg / kg, country: Spain, *includes 24 semiskimmed samples, all not contaminated

Milk powder The production of dry milk may lead to a decrease in → aflatoxin M_1 concentration of about 85% compared to the raw milk.
Milk powder may contain the following → mycotoxins:
→ aflatoxin B_1
incidence: 4/4*, conc. range: 320-5400 μg** / kg, ∅ conc.: 3193 μg / kg, country: Czechoslovakia, *leftover, ** in surface layers
incidence: 5/5*, conc. range: 42-550 μg / kg, country: Czechoslovakia, *leftover
incidence: 1/18, conc.: 6.4 μg / kg, country: Germany
aflatoxin M_1
incidence: 8/210, conc. range: 0.2-nc μg / kg, country: Austria
incidence: 468/837, conc. range: 0.03-0.69 μg / kg, country: Austria
incidence: 33/300. conc. range: 0.1-1 μg / kg, ∅ conc.: 0.27 μg / kg, country: Brazil
incidence: 21/28, conc. range: 0.015-0.464 μg / kg, ∅ conc.: 0.1 μg / kg, country: China
incidence: 1/15, conc.: 15 μg / kg, country: Denmark
incidence: 129/222, conc. range: 0.050-5.2 μg / kg, country: France
incidence: nc/183, conc. range: ≤ 15.4 μg / kg, ∅ conc.: 1.79 μg / kg, country: France
incidence: nc/55, conc. range: ≤ 1.36 μg / kg, ∅ conc.: 0.225 μg / kg, country: France
incidence: 47/95, conc. range: 0.100-2.55 μg / kg, country: Germany
incidence: 7/80, conc. range: 0.67-2 μg / kg, country: Germany
incidence: 30/41, conc. range: 0.2-2 μg / kg, ∅ conc.: 0.5 μg / kg, country: Germany
incidence: 74/120, conc. range: 0.02-0.4 μg / kg, country: Germany
incidence: 8/166, conc. range: 0.67-2.0 μg / kg, country: Germany
incidence: 7/120, conc. range: 0.05-0.13 μg / kg, country: Germany
incidence: 36/55*, conc. range: tr-4 μg / kg, country: Germany, *27 samples of skim milk and 28 samples of whole milk powder
incidence: 58/233, conc. range: 0.002-0.008 μg / kg, country: Italy
incidence: 81/97, conc. range: < 0.001-0.1013 μg / kg, ∅ conc.: 0.0218 μg / kg, country: Italy
incidence: 4/21, conc. range: 0.030-0.25 μg / kg, country: Italy
incidence: 6/13, conc. range: 0.050-0.1 μg / kg, country: Italy
incidence: 9/9, conc. range: 0.01-0.28 μg / kg, country: Italy
incidence: 10/10, conc. range: 0.015-0.1 μg / kg, country: Italy
incidence: 2/6, conc. range: 0.015-0.035 μg / kg, country: Italy
incidence: 3/18, conc. range: 0.040-0.090 μg / kg, ∅ conc.: 0.066 μg / kg, country: Italy
incidence: 3/12, conc. range: traces, country: Italy
incidence: 3/3, conc. range: 0.015-0.085 μg / kg, country: Poland
incidence: 35/277, conc. range: < 0.03 μg / kg (24 samples), 0.01-0.02 μg / kg (6 sa), 0.02-0.04 μg / kg (5 sa), country: UK

incidence: 213/213, conc. range: < 0.1-0.8 µg / kg, country: UK
incidence: 5/10, conc. range: 0.015-0.243 µg / kg, country: USA
incidence: ?/5, conc. range: 3.83-5.74 µg / kg, ∅ conc.: 4.91 µg / kg, country: USA
incidence: 24/320, conc. range: 0.1-0.4 µg / kg, country: USA
incidence: 192/302, conc. range: tr-3.9 µg / kg, country: USA
→ milk

Milk products → Dairy products

Millet may contain the following → mycotoxins:
→ 15-acetylscirpentriol
incidence: nc, ∅ conc.: 400 µg / kg, country: USA
→ aflatoxins (AFB_1, AFB_2)
incidence: 9/9*, conc. range: 1-27 µg / kg, ∅ conc.: 9.8 µg / kg, country: Gambia, *millet, Sanyo (*Pennisetum typhoideum*)
aflatoxins (no specification)
incidence: 9*/55, conc. range: 1-100 µg / kg, country: Uganda, *6 samples contained AFB_1, 4 AFB_2, 2 AFG_1, 1 AFG_2
incidence: nc, ∅ conc.: 0.3 µg / kg, country: USA
→ cyclopiazonic acid
incidence: 2/2*, conc. range: nc, country: India, *kodo millet
→ deoxynivalenol
incidence: 1/4*, conc.: 229 µg / kg, country: Korea, *Indian millet
incidence: nc, ∅ conc.: 300 µg / kg, country: USA
→ nivalenol
incidence: 1/4*, conc.: 340 µg / kg, country: Korea, *Indian millet
incidence: nc, ∅ conc.: 1200 µg / kg, country: USA
→ ochratoxin A
incidence: 1/2, conc.: ≤ 0.3 µg / kg, country: The Netherlands
→ zearalenone
incidence: nc, ∅ conc.: 300 µg / kg, country: USA
→ cereals, → sorghum

Millet meal may contain the following → mycotoxins:
→ deoxynivalenol
incidence: 1/1*, conc.: 720 µg / kg, country: Papua New Guinea, *imported
→ nivalenol
incidence: 1/1*, conc.: 1540 µg / kg, country: Papua New Guinea, *imported
→ zearalenone
incidence: 1/1*, conc.: 440 µg / kg, country: Papua New Guinea, *imported

Milling The influence of milling on the mycotoxin contamination of the → flour fractions compared to the whole → grains differs mainly due to the relative distribution of the → mycotoxins throughout the kernel. In some cases the degree of kernel contamination is also decisive. E.g. grains showing an overall high → deoxynivalenol contamination but predominantly in the surface layer will yield a flour low in DON concentration compared to grains having a lower overall DON contamination but predominantly colonized / contaminated by the mycotoxigenic fungus / mycotoxins in the endosperm.
The behavior of → trichothecenes during milling is shown in the table Influence of baking (see Table Milling). During wet-milling of maize the pattern of distribution for deoxynivalenol, → nivalenol, and → zearalenone follows the physical solubility. As they are highly water-soluble, DON and NIV accumulated in the steep liquor whereas low levels were found in the solid fractions (germ, fibre and gluten). The distribution of the relatively insoluble zearalenone was quite the opposite. Compared to the original concentration in the whole grains levels of three important → Fusarium mycotoxins,

→ nivalenol, deoxynivalenol and → zearalenone, in the flour fraction is lowered between 15-100% during milling.
If fumonisin (→ fumonisins) contaminated → maize is milled the mycotoxins persist in the wet-milled products. Between 10-40% of the original fumonisin concentration is found in the fiber, gluten and germ fraction.
In a milling study a major portion (60-80%) of → aflatoxins (B_1, B_2, G_1, G_2) occurred in the combined bran and polish fraction. These fractions contained 10 times more of the aflatoxins than the milled kernels.
No essential reduction of → ochratoxin A concentration was observed in the wholemeal compared to the cleaned → wheat kernels. Similar observations were made when white or wholemeal flour were baked into → bread. White flour from hard and soft wheat contained only 30 and 60% respectively of the ochratoxin of the uncleaned wheat which was mainly found in the → bran and offal fractions. Due to scouring which removes a proportion of the pericarp (bran coat) prior to milling, OTA levels were significantly removed (three-fold) for both hard and soft wheat.
Compared to milled → rice an increase of the → citrinin concentration in the bran and polish fraction has been observed, more pronounced in highly contaminated samples. However, it seems that citrinin will survive the milling process at least to some extent because this mycotoxin has been found in → maize flour from e.g. Thailand.
→ cereals

Miso may contain the following → mycotoxins:
→ aflatoxin B_1
incidence: 3/20, conc. range: 1400 µg/kg, country: Brazil
→ fermented products, → oriental fermentations

Mitosporic fungi (Syn.: Deuteromycetes, Deuteromycotina, Fungi Imperfecti, conidial fungi, asexual fungi) Artificial assemblage of mitosporic fungi with unknown meiotic states: e.g. → Alternaria, → Arthrinium, → Aspergillus, → Cladosporium, → Fusarium, → Penicillium

Modified atmosphere Compared to fungal growth mycotoxin production (→ mycotoxins) is more susceptible to low O_2 and high CO_2 atmospheres (→ atmosphere). A significant reduction in mycotoxin formation of some → Fusarium spp., → Aspergillus spp. and → Penicillium spp. could be achieved attaining CO_2 concentrations between 20 and 60%. High CO_2 levels are more effective in preventing mycotoxin formation than reduction in O_2 content.

Milling. Influence of baking and other heat processing on stability of trichothecenes (Scott 1990, modified)

Product	Mycotoxin	Mycotoxin level
Bread	Deoxynivalenol	No losses, except some iso-DON formed
Cookies	Deoxynivalenol	No losses
Doughnuts (yeast)	Deoxynivalenol	Increase
Popcorn	Deoxynivalenol	Minor losses
Bread	Nivalenol	Minor losses
Bread	Other trichothecenes	Comparable to DON

Mold ripened cheese manufactured with → Penicillium roquefortii Thom and / or → Penicillium camembertii Thom may contain → cyclopiazonic acid, → mycophenolic acid, and roquefortines (→ roquefortine A & B, → roquefortine C). However, adequate ripening and storage conditions (6-10 °C Roquefort, 14-19 °C Camembert) and those prevailing in the cheese during ripening greatly inhibit the enrichment of these → mycotoxins. Contamination with → mycophenolic acid is prevented by using strains unable to produce this mycotoxin. Only small amounts of these only weakly toxic and non carcinogenic mycotoxins can be found in → cheese. It is most unlikely that the consumption of mold ripened cheese is associated with hazards to human health.
→ cheese, Blue, → cheese, Camembert, → cheese, Roquefort

Moldy corn poisoning may be due to → maize infected with → Aspergillus flavus Link and → Penicillium *rubrum* and contaminated with → aflatoxin B_1 fed to pigs and → cattle. In these domestic animals hepatic lesions occurred.
→ Turkey "X" disease

Moldy corn toxicosis (Syn.: → Hemorrhagic syndrome) The problem of moldy corn toxicosis occurred in the early 1960s in the midwestern states of the U.S. However, other countries also reported from this irregularly occurring, long-standing disease.
Symptoms in farm animals (→ cattle, pigs, → poultry) mainly occurred in the digestive tract and included a general loss of appetite, dysentery often accompanied by bloody feces, reduction in milk yield, unthriftiness, and loss of weight. Massive hemorrhages (→ hemorrhage) throughout the body (e.g. bladder, heart, intestines, kidneys, lungs) were noted. Death only occurred in some cases. → Trichothecenes such as → diacetoxyscirpenol and → T-2 toxin produced at low temperatures by a highly toxic strain of → Fusarium *tricinctum* isolated from moldy sweet corn (→ maize) were suspected as causatives of hemorrhagic symptoms of farm animals in the USA. Moldy grain which induced moldy corn toxicosis in pigs was fed to dogs. The symptoms were almost the same as to those observed in pigs and resembled a disease called "hepatitis X". Based on the closely related if not identical syndromes, it was concluded that → ATA and moldy corn toxicosis have the same origin, viz. T-2 toxin and diacetoxyscirpenol primarily produced by → Fusarium sporotrichioides Sherb.

Moldy sweet potato toxicosis is due to a host parasite interaction (sweet potato / → Fusarium *solani*) leading to the production of phytoalexins such as 4- and 1-ipomeanol, ipomeanine, 1,4-ipomeadiol. They are catabolized by the fungus to lung-toxic metabolites wich interfere with the respiration of → cattle. Cases of death occured.
A chronic respiratory disease has also been reported from New Guinea where humans consume large quantities of sweet → potatoes. Since 4-ipomeanol (as well as ipomeamarone) occurred in slightly blemished sweet potatoes destined for sale in US supermarkets, it is possible that these phenolic compounds are also responsible for the etiology of this human disease.

Monascidin A (Syn.: → citrinin)

Moniliformin is a naturally occurring sodium or potassium salt of 1-hydroxycyclobut-1-ene-3,4-dione (see Figure Moniliformin). This mycotoxin (→ mycotoxins) was first isolated from → maize in

1973 contaminated by → Fusarium moniliforme Sheldon. During a study to determine the molecular structure of the toxin the corresponding strain losts its ability to produce the metabolite in culture. Isolation and structure elucidation eventually succeeded from a high-producing strain of *F. moniliforme* as a contaminant of → millet in Nigeria. Since this strain produced chlamydospores it was recently identified as *F. nygamai*. In contrast to other → Fusarium mycotoxins moniliformin occurs only in a very few crops.

CHEMICAL DATA
Empirical formula: C_4HO_3 Na / K, molecular weight: 120 / 136

FUNGAL SOURCES
At least 15 *Fusarium* species including *Fusarium anthophilum*, → Fusarium avenaceum (Fr.) Sacc., *F. chlamydosporum*, → Fusarium culmorum (Wm. G. Smith) Sacc., → Fusarium moniliforme Sheldon (most of the strains either produce only small amounts or none moniliformin), *F. nyagamai*, → Fusarium oxysporum Schlecht. emend. Snyd. & Hansen, → Fusarium proliferatum (Matsushima) Nirenberg, *F. sporotrichioides*, *F. subglutinans* are moniliformin producers.

NATURAL OCCURRENCE
→ maize, → maize flour, → maize meal, → oats, → rye, → triticale, → wheat
There are not many data about the occurrence of moniliformin in → food.

TOXICITY
rapid death (ducklings 1 h, rats 3 h) of experimental animals occurred (mycocardial degeneration / → edema, respiratory distress, and necrosis (liver, kidney)). Action similar to that of arsenite.
LD_{50} (po): 41.57 mg and 50.00 mg / kg bw female and male rats, respectively.

DETECTION
GC, HPLC, spectroscopy, TLC

POSSIBLE MYCOTOXICOSIS
Keshan disease / China

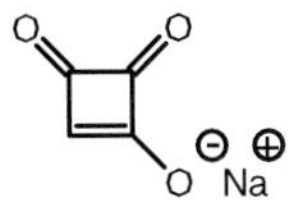

Moniliformin

FURTHER COMMENTS
Although moniliformin occurred ten times more abundantly in foodstuff from areas with a high incidence of human → esophageal cancer in the Transkei compared to low-incidence areas, the → fumonisins are most probably involved in the etiology of this disease.
Stability: A moderate stability of moniliformin has been found at room temperature, with 68-77% remaining after 6 days. However, heating at 100 °C for 0,5 h caused a 45% destruction in maize.

Monoacetoxyscirpenol is a 15-acetoxy-3α,4β-dihydroxy-12,13-epoxytrichothec-9-ene which belongs to the → trichothecenes (→ mycotoxins) (see Figure Monoacetoxyscirpenol).

CHEMICAL DATA
Empirical formula: $C_{17}H_{24}O_6$, molecular weight: 324

FUNGAL SOURCES
→ Fusarium sambucinum Fuckel, *F. semitectum*

NATURAL OCCURRENCE
→ oats

TOXICITY
bilateral inflammation of the beak area, gastrointestinal hemorrhaging (birds) (→ hemorrhage), dermatotoxic (rat)
LD_{50} (sc): 0.752 mg / kg bw rat (20-day-old, white, female, weanling)

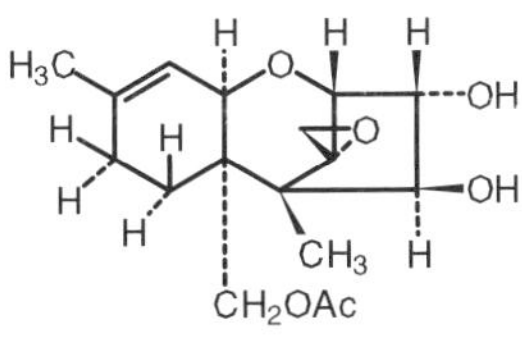

Monoacetoxyscirpenol

Detection
GC-MS

Monodeacetylanguidin → 15-acetylscirpentriol

Mselenin joint disease This progressive, crippling osteoarthropathic disease is common among blacks in the Mseleni area in Kwazulu (southern Africa). In general, the disease progresses further in women than in men. Life expectancy is not affected but the mobility of the individuals is limited to various degrees. It is estimated that in the endemic region about three thousand people from the Zulu and Toga tribes are affecetd (38.9% women, 11.1% men).
Since the diet in the endemic region mainly consists of products of plant origin such as → cowpeas, → peanuts, → sweet potatoes, → maize, → melons, pumpkins, and → vegetables and wild → fruits lack of calcium, magnesium, and manganese have been proposed as possible causal agents in the etiology of the disease.
However, fungi and their → mycotoxins may also be involved. → Fusarium moniliforme Sheldon was predominant on maize from the endemic region, 96.3% of the samples were infected. Several other Fusaria, e.g. → Fusarium equiseti (Corda) Sacc. sensu Gordon, → Fusarium oxysporum Schlecht. emend. Snyd. & Hansen, and → Fusarium poae (Wollenw.) Peck which might be implicated in the etiology of bone growth diseases in man and animals could be isolated. In addition, different species of the genera *Acremonium*, *Lasiodiplodia*, *Macrophomina*, *Nigrospora*, and → Penicillium frequently occurred on maize and peanuts. Foodstuffs (maize, groundnuts) of affected households showed a higher contamination with these fungi than nonaffected ones.

Muesli may contain the following → mycotoxins:
→ aflatoxin B_1
incidence: 4*/14, conc. range: 20-120 µg / kg, country: Germany, *moldy
→ aflatoxins**
incidence: 3/7, conc. range : nc, country: UK
→ ochratoxin A***
incidence: 2/26, conc. range: 0.4-0.5 µg / kg, ∅ conc.: 0.45 µg / kg, country: Germany
incidence: 6/50, conc. range: ≤ 3.9 µg / kg, country: UK
incidence: 3/7, conc. range: nc, country: UK
→ trichothecenes****
incidence: 1/1, conc.: nc, country: UK
** max. level: ≤ 25 µg / kg, *** max. level: ≤ 50 µg / kg, **** max. level: ≤ 5 µg / kg
→ cereals

Muesli ingredients may contain the following → mycotoxins:
→ aflatoxins
incidence: 2/7, conc. range: 1-5 µg / kg, country: UK
→ deoxynivalenol
incidence: 1/4, conc.: traces, country: UK
→ ochratoxin A
incidence: 2/7, conc. range: 0.2-1.49 µg / kg, country: Germany
incidence: 3/7, conc. range: < 10 µg / kg, country: UK
→ trichothecenes
incidence: 1*/1, conc.: nc, country: UK
*max. level: ≤ 5 µg / kg

Muffin → Maize muffin

Muffin mix may contain the following → mycotoxins:
→ fumonisin B_1
incidence: 1/2*, conc.: 80 µg / kg, country: USA, *→ maize based
→ fumonisin B_2
incidence: 1/2*, conc.: 10 µg / kg, country: USA, *maize based

→ fumonisins
incidence: 1/3*, conc.: nc, country: USA
incidence: 6/6*, conc. range: 450-1450 µg / kg, country: USA,
*maize based
→ zearalenone
incidence: 1/5, conc.: 3.1 µg / kg, country: USA

Mung beans → Beans

mutagenic is a biological, chemical or physical agent which increases the degree of mutation

Mycophenolic acid is a 6-(4-hydroxy-6-methoxy-7-methyl-3-oxo-5-phthalanyl)-4-methyl-4-hexenoic acid (→ mycotoxins) which was first isolated in 1896 from → Penicillium brevicompactum Dierckx (see Figure Mycophenolic acid).

CHEMICAL DATA
Empirical formula: $C_{17}H_{20}O_6$, molecular weight: 320

FUNGAL SOURCES
P. brevicompactum, *P. raciborskii*, → Penicillium roquefortii Thom chemotype I and II.

NATURAL OCCURRENCE
→ cheese, → cheese, Bleu des Causses, → cheese, Blue, → cheese, Gorgonzola, → cheese, Roquefort
Generally, blue veined cheeses are very good substrates for mycophenolic acid production and may contain relatively high concentrations but Roquefort cheese is particularly suitable for the formation of mycophenolic acid.

TOXICITY
relatively less toxic, showing antibiotic, antitumor, and → mutagenic activity, chromosome aberrations occurred in mice
LD_{50} (po): 2500 mg / kg bw mice

DETECTION
mainyl TLC

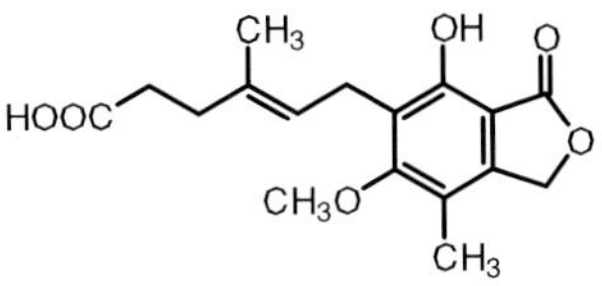

Mycophenolic acid

FURTHER COMMENTS
Although some adverse clinical reactions such as diarrhea, cramps and nausea occurred after a daily application of 2.4 - 7.2 g for 52-104 weeks, this substance seems to be a promising drug for the treatment of psoriasis.

Mycosphaerellaceae → Dothideales

Mycotoxic nephropathy → Mycotoxic porcine nephropathy

Mycotoxic porcine nephropathy (Syn.: Mycotoxic nephropathy) was first discovered in 1928 in Denmark while epidemics occurred in 1963 and 1971, associated with high-moisture grain (→ grains) due to unusual climatic conditions. The major causal agent is → ochratoxin A but other substances like → citrinin and → viomellein (quinone) may also be involved. These nephrotoxic → mycotoxins have been isolated from → barley associated with mycotoxin porcine nephropathy. They mainly act on the → renal tuber system, especially on the proximal tubules.
→ Polydypsia and / or → polyuria are the most characteristic renal alterations in domestic animals like pigs and horses due to feeding of moldy grain or hay. In pigs and → poultry chronic ochratoxicosis may be manifestated by retarded growth rates. Since renal damages are easily overlooked they are usually detected only during inspection in slaughterhouses.
Experimental studies showed that in pigs, the corresponding symptoms occurred

after feeding (i) grain infected with a citrinin / ochratoxin producer, (ii) this mold directly, (iii) application of citrinin and / or OTA as pure substances. 0.01-0.08% of slaughtered pigs from slaughterhouses in Denmark showed prevalence rates of porcine nephropathy. In most cases the affected kidneys (10 μg OTA / kg < 25 μg OTA / kg) are condemned but the remaining carcass is accepted for consumption. In different European countries 25-39% of affected kidneys contained 2-100 μg OTA / kg. These kidneys are swollen and pale with a mottled surface. Histopathological abnormalities may include periglomerular → fibrosis, degeneration of the proximal tubules, followed by atrophy of the tubular epithelium, hyalinization of some glomeruli, and interstitial fibriosis in the → cortex. In a later stage, enlargement of the kidneys to several times the normal size may occur, associated with pronounced gross changes in texture and color, → jaundice, and advanced cellular damage. Renal functions are deeply impaired. Depressed weight gains and decreased performance may also occur after feeding higher toxin levels. If uremia is developed, the whole carcass is condemned at → meat inspection in Denmark (≥ 25 μg OTA in the kidneys / kg).
OTA contaminated → meat and organs of pigs (kidneys, liver) may be a source for human OTA intake mainly due to the consumption of contaminated → sausages.

Mycotoxicosis Toxic syndromes resulting from the ingestion of → foods or feeds contaminated with fungal toxins by man and animals are known as mycotoxicosis. These, often seasonally occurring, disorders are primarily found in climatic regions with high rainfall, high relative humidity, and high temperatures. In some cases drought, insect damage and / or cracked kernels during harvesting enhance fungal growth (e.g → Aspergillus flavus Link, → Aspergillus parasiticus Speare / → peanuts) and subsequent mycotoxin contamination. The development of the fungi is further promoted by the presence of excessive chaff in the harvested → grains and seeds although the molds may even no longer be present in contaminated grain.
It is suggested that about 50 fungal metabolites are involved in man and animal mycotoxicosis while ten of these → mycotoxins such as → aflatoxin B_1, → ochratoxin A, → sterigmatocystin and → fumonisin B_1 are carcinogenic (→ carcinoma) (see Table Mycotoxicosis).
Substantial difficulties arise when making the right diagnose of a mycotoxicosis because (i) mycotoxins, especially at low dosis, or unknown toxins are difficult to detect in food and feed, (ii) contaminated food or feed are often disposed before a mycotoxicosis is suspected, (iii) clinical signs and / or symptoms are often not precise and of an acute nature, (iv) physicians and veterinarians are often not familar with the symptomatology of mycotoxicosis, (v) the "dose-response" principle is difficult to apply to the diagnosis of a mycotoxin-related disease, (vi) the interaction between individual mycotoxins and their effects on man and animals is yet not well elucidated.
Because of animals due to lower feed quality and the way in which they are fed, animals are more likely to be exposed to mycotoxins than humans and a higher incidence of mycotoxicosis in animals has been reported.
Certain common features for a mycotoxicosis are presented below:
- true cause is not immediately identified
- outbreaks often are seasonal
- disease is food or feed-related (e.g. → peanuts, → maize, → rice)*
- no significant pathogenic microorganisms are present*

- fungal activity is obvious in the suspected food / feed
- treatment with drugs or antibiotics is not effective
- disease is not contagious, neither transmissible nor infectious*
- age, sex, and nutritional status are often decisive for severity of the disease
- withdrawal of suspected food / feed leads to signs of improvement*
- isolation of sufficient amounts of the mycotoxin(s) from the food / feed or man / animals to cause the disease*
- mycotoxin(s) isolated are known to produce the typical symptoms of the disease*
- feeding trials with the suspect ration reproduce the disease
  *criteria for a true mycotoxicosis

Possible Mycotoxicosis
→ acute cardiac beri-beri, → aflatoxicosis, AIDS, → akakabi byo disease, → alimentary toxic aleukia, → Balkan endemic nephropathy, → equine leukoencephalomalacia, → ergotism, → indian childhood cirrhosis, → Kashin-Beck disease, → kodua poisoning, → Kwashiorkor, → onyalai, → Pellagra, → porcine pulmonary edema → premature thelarche, → Reye's syndrom

Mycotoxicosis. Possible involvement of food relevant mycotoxins in human mycotoxicosis

Mycotoxicosis	Involved mycotoxin(s)	Involved foodstuff
Acute cardiac beriberi	Citreoviridin	rice
Aflatoxicosis (acute)	Aflatoxins	maize, peanuts
Akakabi byo disease	Trichothecenes (e.g. deoxynivalenol, fusarenon X)	maize, wheat
Alimentary toxic aleukia	Trichothecenes (e.g. diacetoxyscirpenol, HT-2 toxin, T-2 toxin, nivalenol)	cereals, mainly proso millet & wheat but also barley, rye, oats, buckwheat
Arthrinium sugarcane poisoning	β-Nitropropionic acid, fumonisins?	sugarcane
Balkan endemic nephropathy	Ochratoxin A, citrinin	cereals, cereal products, beans, pig products
Deoxynivalenol toxicosis	deoxynivalenol, nivalenol, acetyldeoxynivalenol, T-2 toxin	grains, flours
Ergotism	Ergot alkaloids	cereals, mainyl rye
Esophageal cancer	Fumonisins	maize, maize products
Fusariotoxicosis	Fusarium toxins, e.g. deoxynivalenol, zearalenoene	maize, wheat
Indian childhood cirrhosis	Aflatoxins	rice, peanut oil, human breast milk
Kashin-Beck disease	fusarochromanone, T-2 toxin	cereals, mainly maize, wheat
Kodua poisoning	Cyclopiazonic acid	grains (e.g. rice), bread
Kwashiorkor	Aflatoxins	starchy and low in protein (e.g. rice, maize, plantains
Mseleni Joint disease		maize, peanuts
Reye´s syndrome	*Fusarium* toxins and others	different kinds of foodstuff e.g. milk, peanuts, rice
Onyalai	Aflatoxins	millet, sorghum
Pellagra	Tenuazonic acid, moniliformin fumonisins, kojic acid, trichothecenes, zearalenone	maize
Sago hemolysis	?	sago

Mycotoxin control Prevention (→ mycotoxin prevention) of mycotoxin contamination by "good farm management practice" is the most effective measure in the production of mycotoxin free or low contaminated → foods and feeds. This includes moisture and temperature control which have a crucial effect on fungal growth and mycotoxin formation.

Mycotoxin degradation
physical: adsorption, heat, irradiation
chemical: acids, bases, bisulfite, oxidizing agents, vitamin C; however, many of these compounds are not in practical use since they may render the products unsafe
biological: various fungi (e.g. → Aspergillus *niger*, *Rhizopus stolonifer*), bacteria (e.g. → Flavobacterium aurantiacum, lacto-bacteria)

Mycotoxin detection is carried out by different techniques e.g.
physicochemical: gas chromatography (GC), high performance liquid chromatography (HPLC), thin layer chromatography (TLC)
immunoassays: enzyme-linked immunosorbent assay (ELISA), radio-immunoassay (RAI)
bioassays: animals, cells and tissue cultures, microorganisms
In seeds (e.g. → peanuts) and fruits (e.g. → figs) detection of → mycotoxins is difficult since there is an uneven distribution in these kinds of substrates. However, in processed → foods and drinks mycotoxins seem to be distributed in a more homogenous manner.

Mycotoxin legislation Currently, 77 countries are known for their mycotoxin regulations (see Table Mycotoxin legislation in the Appendix).

Mycotoxin prevention is achieved by the following factors:
control of the **moisture** content of stored products: cereal → grains: < 13%, → soybeans: < 12%, seeds with a high lipid content: 7%
control of the **temperature** in storage: in general, temperatures below 4 °C will prevent mycotoxin production
control of the **atmosphere** in storage: mold growth / mycotoxin production is depressed by low oxygen and / or high concentration of other gases. Inhibition of aflatoxin (→ aflatoxins) formation occurred at 1% O_2 while the production of → sterigmatocystin and → patulin was completely depressed at 0.2% O_2. Only small amounts of sterigmatocystin were produced at 90% CO_2, patulin could not be detected.
microbial competition: different microorganisms such as → Aspergillus *niger*, *Rhizopus stolonifer* or lactic bacteria, decreased / inhibited aflatoxin production. Little to no aflatoxin contamination occurred in grain invaded by a mixture of fungi, including → Aspergillus flavus Link.
antimycotic agents: growth of mycotoxigenic fungi is inhibited by sufficient amounts of e.g. acetic acid, benzoic acid, propionic acid, sorbic acid or natamycin. In addition, production and manufacture of low fungal contaminated raw material, pasteurisation and sterilization of intermediate- and endproducts, suitable packaging, use of preservatives, suitable cooling, freezing and drying techniques, feeding of mycotoxin-free feed, and the use of non-toxic starter cultures in the fermentation industry enables the production of non-contaminated foods.

Mycotoxin producers Almost 350 mold species, mainly mitosporic fungi and only a few ascomycota (e.g. → Claviceps *purpurea*), are known for their mycotoxi-

genic potential (see Table Mycotoxin producers). The most common and important mycotoxin producers which cause mycotoxin contamination of plants (e.g seeds, → fruits) in the field belong to the genera → Alternaria (to a minor degree), → Aspergillus and → Fusarium. The main mycotoxigenic species which attack plant products after harvest, during transport or when in storage are *Aspergillus* and → Penicillium (for further information see also the listed species belonging to each single genus).
Correct fungal identification based on internationally agreed criteria is necessary since there is a close relationship between fungal species and the secondary metabolites they produce. Mycotoxin production is not uniform throughout all strains of a species. Even in the case of a producing strain, mycotoxin formation depends on environmental and nutritional conditions. These phenomena enable strains of potentially mycotoxigenic fungi to be used in food manufacture i.e. *Aspergillus flavus* group (e.g. → aflatoxins) / koji, → Fusarium graminearum Schwabe (e.g. → zearalenone) / microbial protein, → Penicillium roquefortii Thom (e.g. → PR toxin) / Blue cheese (→ cheese, Blue).
→ mycotoxins

Mycotoxin production Since → mycotoxins are secondary metabolites, they are usually produced in the late exponential or early stationary phase. Production is mainly influenced by the following factors:
Moisture: High humidity and a high → a_w favor mycotoxin production. Synthesis of → aflatoxins starts at $a_w > 0.83$, → citrinin: a_w 0.83, → ochratoxin A: a_w 0.83, → patulin: a_w 0.85, → penicillic acid: a_w 0.80. In → grains maximum amounts of → mycotoxins are produced at moisture contents between 20-25%.
Temperature: → Aspergillus spp. aflatoxins: 9-42 °C, → sterigmatocystin: refrigeration temperature
→ Penicillium spp. → cyclopiazonic acid: 4 °C, ochratoxin A, penicillic acid: 4-31 °C, patulin: 0-24 °C, pentirem A (→ penitrems): 6 °C
→ Fusarium spp. → trichothecenes: at and below 10 °C (→ Fusarium sporotrichioides Sherb.: 1.5 to 4 °C optimal production).
Temperatures well below 0 °C will prevent mycotoxin formation. There is a close link between moisture and temperature in mycotoxin production.
Oxygen levels: In general fungi need adequate oxygen concentrations to grow but some species of the genera, e.g. *Mucor*,

Mycotoxin producers. Mycotoxigenic fungal genera

Acremonium	*Dichotomomyces***	*Myrothecium*	*Rosellinia***
*Alternaria**	*Diplodia*	*Microdochium*	*Sclerotinia***
*Aspergillus**	*Drechslera*	*Monographella***	*Spacelia*
Bipolaris	*Epichloe***	*Nigrosabulum***	*Stachybotrys*
Botryodiplodia	*Epicoccum*	*Nigrospora*	*Talaromyces***
Byssochlamys ***	*Fusarium**	*Paecilomyces*	*Thielavia**
*Ceratocystis***	*Gibberella* ***	*Penicillium**	*Trichoderma*
*Chaetomium***	*Gliocladium*	*Periconia*	*Trichothecium*
*Cladosporium**	*Gloeotinia***	*Phoma**	*Verticillium*
Claviceps ***	*Khuskia***	*Phomopsis*	*Verticimonosporium*
Colletotrichum	*Metarhizium*	*Pithomyces*	*Zygosporium*
Curvularia			

* important in food
** teleomorphic state

Rhizopus and *Fusarium* are able to develop under anaerobic conditions (→ atmosphere)

Substrate: Mycotoxin formation is enhanced by carbohydrates (e.g. glucose, saccharose), certain amino acids (e.g. asparagin, glycin), fatty acids and zinc (aflatoxins). Generally, plant-derived-products characterized by a high carbohydrate content are more likely to be prone to mycotoxin contamination than animal products (high protein content). The only important exception is → milk.

Damage, plant stress: Damage of plants (mechanical and / or insects) and / or drought stress in e.g. → peanuts or → maize facilitate invasion of aflatoxigenic fungi and subsequent aflatoxin formation.

In addition, mycotoxin formation is influenced by the availability of trace elements, genetic strain variation and and / or competition with other organisms.

Mycotoxin stability In general, → mycotoxins are quite (heat) stable in most → food products but there are some exceptions; see e.g. → fusarin C, → patulin, and → penicillic acid. (For further information see each single mycotoxin and the contaminated food items.)

Mycotoxins are structurally diverse complex organic compounds of low molecular weight (MW generally lower than 700; → fumonisins, e.g. FB_1 = 721) which belong to the large and diverse group of secondary fungal metabolites. They are not all necessarily → mycotoxins such as the antibiotic penicillin. Based on the inherent toxic effects in higher organisms, a chemical might be called a mycotoxin.

Mycotoxins are found in different chemical groups e.g. pyrones, anthrachinones, coumarins, macrolides, steroids and cyclic polypeptides. Formation usually occurs during the late exponential or early stationary phase of fungal development. These non-antigenic organic compounds are produced by a wide range of fungi. At least 15 different mycotoxins are synthesized by some species of these genera. They are further characterized by their frequent specificity with regard to the taxonomy of the producing fungi. It is estimated that approximately 400 toxic fungal compounds do exist.

Almost all plant products may serve for mold growth and mycotoxin production. To a minor degree animal products such as → milk and → meat may be contaminated. Humans are exposed to mycotoxins mainly through the consumption of → foods directly contaminated by mycotoxin-producers and their mycotoxins (e.g. → aflatoxins, → trichothecenes, → patulin) or by ingestion of residue containing → meat (e.g. → ochratoxin A) or → milk (e.g. → aflatoxin M_1).

FUNGAL SOURCES

Although ca. 350 different fungal species are known to be mycotoxin producers, fungi of the genera → Aspergillus, → Fusarium, → Penicillium and → Alternaria (to a minor degree) are the most important. Worldwide at least 100 mycotoxigenic fungal species are associated with naturally occurring diseases in animals and humans. → mycotoxin producers

NATURAL OCCURRENCE

→ Cereals and → oil seeds (→ nuts) and products derived from them are most likely to be contaminated by mycotoxigenic fungi / → mycotoxins. Several factors like area of crop growth, climate, conditions during growth, harvesting and storage are decisive for mycotoxin contamination of the crop. The warm and moist weather in tropical and subtropical countries favors the rapid growth of (aflatoxigenic) fungi and subsequent mycotoxin contamination (especially → aflatoxins) in such → seeds. Crops

grown in temperate regions are less prone to mycotoxin contamination. Here, → trichothecenes and → ochratoxin A predominate. Worldwide 25% of the annually produced food crops are contaminated with detectable amounts of mycotoxins (FAO 1985) resulting in economic losses of billions of dollars / year. The mycotoxins most commonly found in → food and feedstuff are aflatoxins, → fumonisins, ochratoxin A, patulin, trichothecenes and → zearalenone.
To minimize mycotoxin exposure to man almost 80 countries possess legal or recommended limits for mycotoxins such as aflatoxins, chaetomin, → deoxynivalenol, → diacetoxyscirpenol, → fumonisin B_1, → fumonisin B_2, ochratoxin A, patulin, phomopsin, stachybotryotoxin, → T-2 toxin, and zearalenone.

Toxicity

Often a substrate is contaminated by different mycotoxins which may act synergistically or additively. This fact limits the value of the administration of a pure crystalline mycotoxin, e.g. turkey "X" disease / aflatoxins and → cyclopiazonic acid. Several factors such as molecular structure, dosage, duration of intake, species, age, sex, condition and nutrient status of the affected organism are decisive for the detrimental effect of a mycotoxin. Its toxicity may be limited to only one or a few species, but another mycotoxin may affect a wide range of organisms. Mycotoxins are carcinogenic (e.g. → aflatoxin B_1, fumonisin B_1, → fusarenon X, griseofulvin, → sterigmatocystin), cardiotoxic (e.g. → ergot alkaloids, → penicillic acid), dermatotoxic (e.g. trichothecenes such as → HT-2 toxin), emetic (e.g. deoxynivalenol, T2- toxin), hemorrhagic (e.g. → byssochlamic acid, patulin), hepatotoxic (e.g. → islanditoxin, → luteoskyrin, → rubratoxins, → rugulosin), → immunosuppressive (e.g. ochratoxin A, trichothecenes), mutagenic (e.g. aflatoxins, → alternariol methyl ether, → altertoxin I-III), nephrotoxic (e.g. citrinin, ochratoxin A, penicillic acid, → viomellein, → xanthomegnin), estrogenic (zearalenone), neurotoxic (e.g. → citreoviridin, cyclopiazonic acid, ergot alkaloids, → penitrems), teratogenic (e.g. aflatoxins, → alternariol) and / or tremorgenic (e.g. → tremorgenic mycotoxins) (for further information see also each single mycotoxin). Chronic effects are merely the inhibition of protein synthesis and / or growth. At least some mycotoxins probably have synergistic effects in vivo (see Table Mycotoxins 1).

Mycotoxins 1. Toxicological effects of mycotoxins (Pohland 1993, modified)

Mycotoxin	Mutagenic	Teratogenic	Carcinogenic
Aflatoxin	+++	+++	+++
Citrinin	−+	+	+
Cyclochlorotine			+
Fumonisin B_1			+
Fusarenon X		+	+
Luteoskyrin	−		+
Ochratoxin A	+	+	+
Patulin	+	+	+
Penicillic acid	+	−	+
Rugulosin	−+		+
Sterigmactocystin	+	+	+++
T-2 toxin	−	+	+
Zearalenone	+	+	+

Mycotoxins 2. Possible routes for mycotoxin contamination of human foods (Jarvis 1976, modified)

1.	Mold damaged foodstuffs of plant origin	
	a) Agricultural products	e.g. cereals, fruits, oilseeds (mainyl nuts), pulses, spices
	b) Consumer foods	
2.	Residues in tissues and products of animal origin due to mold contaminated feedstuff	e.g. meat (mainly kidneys, liver) & meat products (mainly sausages), milk, dairy products (mainly cheese)
3.	Mold-ripened foods	e.g. cheeses (mainly Roquefort & Camembert cheese), meat products
4.	Fermentation products	e.g. enzymes, microbial proteins, organic acids, other food additives

Testing different mycotoxins from *Aspergillus* spp. and *Penicillium* spp. the toxicity decreased between 0.001-100 µg/embryo as follows: aflatoxin B_1, ochratoxin A, → PR toxin, → aflatoxin B_2, aflatoxin M_1, sterigmatocystin, → aflatoxin G_2, patulin, rubratoxin B (→ rubratoxins), secalonic acid D (→ secalonic acids), → mycophenolic acid, α-cyclopiazonic acid, penicillic acid, citrinin, brevianamide A and griseofulvin.

However, it is very difficult to assess the present-day risk to human health because quantifying exposure of mycotoxins in the diet is problematic (see Table Mycotoxins 2).
→ extracellular mycotoxins, → intracellular mycotoxins

Myocin (Syn.: → patulin)

N

Neosartorya → Trichocomaceae, anamorph → Aspergillus fumigatus group
N. fischeri possesses heat-resistant ascospores which cause spoilage of → fruit juices and other heated (pasteurized) fruit-based products. *N. fischeri* may produce → mycotoxins such as avenaciolide, fumitremorgins, terrein, verruculogen.

Neosolaniol (Syn.: solaniol, 8α-hydroxydiacetoxyscirpenol) belongs to the group of naturally-occurring → trichothecenes (4β,15-diacetoxy-3α,8α-dihydroxy-12,13-epoxytrichothec-9-ene), which was first isolated from → Fusarium sporotrichioides Sherb. in 1971 (see Figure Neosolaniol). The previous name solaniol was changed by Ueno in 1972 to neosolaniol.

CHEMICAL DATA
Empirical formula: $C_{19}H_{26}O_8$, molecular weight: 382

FUNGAL SOURCES
F. acuminatum, → Fusarium avenaceum (Fr.) Sacc.?, → Fusarium culmorum (W. G. Smith) Sacc.?, → Fusarium equiseti (Corda) Sacc. sensu Gordon, → Fusarium graminearum Schwabe, → Fusarium oxysporum Schlecht. emend. Snyd. & Hansen, → Fusarium poae (Peck) Wollenw., → Fusarium sambucinum Fuckel, → Fusarium semitectum Berk. & Rav.?, *F. sporotrichioides*

NATURAL OCCURRENCE
→ barley, → curry, → ginger, → maize, → oats, → wheat

TOXICITY
cellular degeneration, karyorrhexis in actively dividing cells of thymus, lymph nodes, spleen, bone marrow, intestine, and testes, dermatotoxic
LD_{50} (ip): 14.5 mg / kg bw mice

DETECTION
GC, MS, spectroscopy, TLC

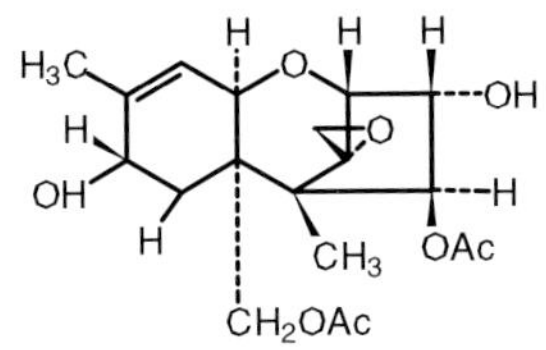

Neosolaniol

FURTHER COMMENTS
Neosolaniol produced by *F. sporotrichioides* may be associated with outbreaks of → ATA and → bean hull poisoning.

Nephritis inflammation of the kidney

Nephropathy → Mycotoxic porcine nephropathy, → Balkan endemic nephropathy

Nephrotoxin e.g. a mycotoxin which damages the kidney tissue

β-Nitropropionic acid (Syn.: bovinocidin, hiptagenic acid, 3-nitropropionic acid) was first isolated from the root bark of Hiptage in 1920 but later it was reported as being a metabolite of → Penicillium spp. and *Streptomyces* spp. (see Figure ß-Nitropropionic acid). As a toxic metabolite of different → mitosporic fungi it is probably involved in a Chinese → mycotoxicosis (→ mycotoxins).

CHEMICAL DATA
Empirical formula: $C_3H_5NO_4$, molecular weight: 119

FUNGAL SOURCES
Arthrinium ssp. (*A. sacchari*, *A. sacchari-cola*), → Aspergillus spp. (e.g. possibly → Aspergillus flavus Link, → Aspergillus oryzae (Ahlburg) Cohn, → Aspergillus parasiticus Speare), → Penicillium spp. (e.g. *P. atrovenetum*).

NATURAL OCCURRENCE
→ cheese, sugarcane ?,

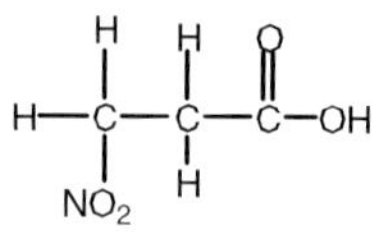

β-Nitropropionic acid

Toxicity
clinical signs: rapid respiration with subsequent apnea, incoordination, marked dilation (subcutaneous and visceral blood vessels), mottled liver
LD_{50} (po): 110 and 68.1 mg / kg bw male and female mice, respectively

Detection
TLC

Possible Mycotoxicosis
→ Arthrinium sugarcane poisoning

Nivalenol belongs to the group of naturally-occurring → trichothecenes (3α,4β,7α,15-tetrahydroxy-12,13-epoxy-trichothec-9-en-8-one) which was first isolated from → Fusarium sporotrichioides Sherb. in 1967 (see Figure Nivalenol). The first report on natural occurrence (Japanese scabby → barley) dates from 1972 (together with → deoxynivalenol).

Chemical Data
Empirical formula: $C_{15}H_{20}O_7$, molecular weight: 312

Fungal Sources
→ Fusarium equiseti (Corda) Sacc. sensu Gordon, → Fusarium graminearum Schwabe, → Fusarium sambucinum Fuckel (?), → Fusarium semitectum Berk. & Rav. (?), F. sporotrichioides

Natural Occurrence
barley, → barley flour, → barley malt, → beer, → bread, → chapatti, → chilli sauce, → curry, → curry paste, → flour, → foods, → garlic, → ginger, → grains, → job's tears, → maize, → millet, → millet meal, → noodles, → oats, → rice, → rye, → rye flour, → sesame seeds, → sorghum, → soybeans → tandoori, → wheat, → wheat bran

In contrast to deoxynivalenol, nivalenol is a less common contaminant of grains.

Toxicity
Similar to deoxynivalenol although DON has a greater acute toxicity.
dermatotoxic, emetic, inhibition of DNA synthesis follows inhibition of protein synthesis
LD_{50} (ip): 4.1 mg / kg bw mice

Detection
GC, HPLC, spectroscopy, TLC

Further Comments
Nivalenol is often co-occurring with → deoxynivalenol.
It may be produced from → fusarenon X by a chemical or enzymatic deacetylation reaction (→ Fusarium nivale (Fr.) Ces.) because nivalenol lacks one acetyl group which is characteristic for fusarenon X.

Noodles During Chinese noodle making losses of → deoxynivalenol and → nivalenol amounted to ≈ 30-40%. In these noodles no → diacetoxyscirpenol, → neosolaniol, → T-2 toxin and → fusarenon X could be detected after manufacturing (artifical contamination). Losses of the afore mentioned → mycotoxins during processing of Japanese noodles were in the range of ≈ 40-70%.
Noodles may contain the following mycotoxins:
→ aflatoxins (AFB_1, AFB_2, AFG_1, AFG_2)
incidence: 1/4*, conc.: 0.4 μg / kg, country: UK, *→ wheat
→ deoxynivalenol
incidence: 2/2*, conc. range: 2500-2720 μg / kg, ∅ conc.: 2610 μg / kg, country: Canada, *Japanese noodles

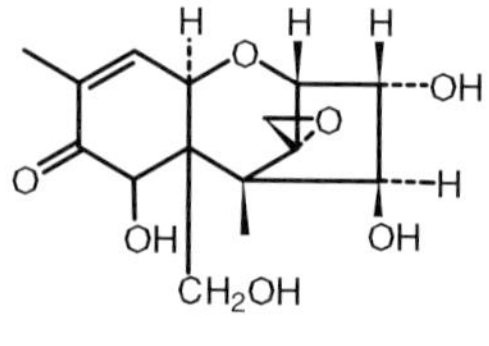

Nivalenol

incidence: 2/2*, conc. range: 2120-2310 μg / kg, ∅ conc.: 2215 μg / kg, country: Canada, *Chinese noodles
incidence: nc/4*, conc. range: 11-92 μg / kg, country: UK, *wheat
→ fumonisins (FB_1, FB_2)
incidence: 1/4*, conc.: 26 μg / kg, country: UK, *wheat
→ nivalenol
incidence: nc/4*, conc. range: 14-26 μg / kg, country: UK, *wheat
→ ochratoxin A
incidence: 97/106, conc. range: $\leq$ 4.9-5.3 μg / kg, country: Germany
incidence: 1/4*, conc.: 0.2 μg / kg, country: UK, *wheat

Nutmeg may contain the following → mycotoxins:
→ aflatoxin B_1
incidence: 3/5*, conc. range: 2.5-5.5 μg / kg, country: Canada, *imported
incidence: 4/13, conc. range: 5-37.5 μg / kg, country: West India
incidence: 11/28, conc. range: tr-7.7 μg / kg, country: Germany
incidence: 3*/13, conc. range: 5-15 μg / kg, country: Germany, *moldy
incidence: 2/3, conc. range: 0.4-0.6 μg / kg, ∅ conc.: 0.5 μg / kg, country: Japan
incidence: 29/67, conc. range: 0.2-16 μg / kg, country: Japan
incidence: 25/56, conc. range: 0.2-60.3 μg / kg, country: Japan
incidence: 30/32, conc. range: 1-23.2 μg / kg, country: The Netherlands
→ aflatoxin B_2
incidence: 3/5*, conc. range: 0.75-1.1 μg / kg, country: Canada, *imported
incidence: 1/3, conc.: 0.2 μg / kg, country: Japan
incidence: 8/67, conc. range: tr-0.6 μg / kg, country: Japan
incidence: 25/56, conc. range: 0.1-0.2 μg / kg, country: Japan
→ aflatoxin G_1
incidence: 1/3, conc.: 0.2 μg / kg, country: Japan
incidence: 1/67, conc.: 0.3 μg / kg, country: Japan
incidence: 25/56, conc. range: 0.2-1.4 μg / kg, country: Japan
→ aflatoxin G_2
incidence: 25/56, conc. range: 0.3 μg / kg, country: Japan
→ aflatoxins (no specification)
incidence: 30*/32, conc. range: 2.7-36.5 μg / kg, country: The Netherlands, *AFB_1, AFB_2, AFG_1, AFG_2
incidence: 5/5*, conc. range: $\leq$ 20 μg / kg, ∅ conc.: 13 μg / kg, country: USA, *imported
→ spices

Nuts (no specification)
Almost all types of nuts grown in different geographic regions are prone to mycotoxin contamination, mainly → aflatoxins. The degree of contamination as well as the observed levels are subject to significant variation. Although some kind of mycotoxin contamination has been established, nuts like → almonds, → cashew nuts, → hazelnuts, and → walnuts generally show a lower degree of contamination due to shell texture and harvesting methods.
→ Peanuts and their derived products are most frequently and heavily contaminated. Individual → Brazil nuts and → pistachio nuts may contain levels of aflatoxins as high as several micrograms per gram. A blue fluorescence under ultraviolet light in e.g. Brazil nuts, peanuts, → pecans, and pistachio nuts may indicate an aflatoxin contamination. Mechanical and electronic sorting leads to a significant reduction in aflatoxin contamination. Pneumatic separation is also used to remove contaminated nuts because fungal infected nuts are often lighter than healthy ones.

However, it is not possible to detect fungal and mycotoxin contamination of in-shell nuts during manufacture. Suspected individual nuts have to be handsorted and removed by the consumer.
Compared with retail marketed whole nuts such as peanuts and almonds, processed nuts (chopped, sliced, grind etc.) usually show a higher degree of mycotoxin contamination. It seems that the aflatoxins are more evenly distributed in these comminuted and mixed samples. These brands give a better reflection of the true toxin concentrations compared to whole nuts due to inadequate sampling techniques.
According to Frisvad (1988) nuts may be contaminated with the following mycotoxins: aflatoxins, → citrinin, → cyclopiazonic acid, emodin, roquefortine A (→ roquefortine A & B), pentirem A (→ penitrems), rugulovasine A, → secalonic acid D, → sterigmatocystin, wentilacton.
Nuts may contain the following → mycotoxins:
aflatoxins
incidence: 3/5, conc. range: 1-8900 µg / kg, country: UK
→ ochratoxin A
incidence: 1/5, conc.: traces, country: UK
incidence: 3/5, conc.: ca. 1 µg / kg, country: UK
almonds, Brazil nuts, cashew nuts, → coconut, hazelnuts, → marzipan, peanuts, pecans, pistachio nuts, walnuts

Nuts (mixed)
may contain the following → mycotoxins:
→ aflatoxin B_1
incidence: 2/10*, conc. range: 10-93 µg / kg, Ø conc.: 51.5 µg / kg, country: Finland, *imported
→ aflatoxin B_2
incidence: 1/10*, conc. range: 29 µg / kg, country: Finland, *imported
incidence: 1/16*, conc.: traces, country: Norway, *imported
→ aflatoxin G_1
incidence: 1/16*, conc.: traces, country: Norway, *imported
→ aflatoxin G_2
incidence: 1/16*, conc.: traces, country: Norway, *imported
→ aflatoxins
incidence: 1/3, conc.: 7 µg / kg, country: USA

O

Oat bran may contain the following → mycotoxins:
→ ochratoxin A
incidence: 1/14, conc.: 0.1 µg / kg, country: Germany
incidence: 5/13, conc.: ≤ 4.9 µg / kg, country: Sweden
→ bran

Oat flakes may contain the following → mycotoxins:
→ aflatoxin B_1
incidence: 7/65, conc.: ≤ 1.5 µg / kg, country: Germany
→ ochratoxin A
incidence: 4/92, conc. range: 1.2-2.0 µg / kg, country: Germany
→ corn flakes, → maize flakes, → cereal flakes

Oats may contain the following → mycotoxins:
3-acetoxynivalenol
incidence: 2/6, conc. range: < 200 µg / kg, country: Finland
→ 3-acetyldeoxynivalenol
incidence: 12/21, conc. range: 6-219 µg / kg, ∅ conc.: 67 µg / kg, country: Finland
→ aflatoxin B_1
incidence: 3/304, ∅ conc.: 6 µg / kg, country: USA
→ alternariol
incidence: 8/339, conc.: nc, country: Austria
incidence: 24/139, conc. range: 3-64 µg / kg, ∅ conc.: 15.6 µg / kg, country: Germany
incidence: 2/10, conc. range: 270-900 µg / kg, ∅ conc.: 596 µg / kg, country: Poland
→ alternariol methyl ether
incidence: 105/339, conc.: nc, country: Austria
incidence: 33/156, conc. range: 5-95 µg / kg, ∅ conc.: 28.9 µg / kg, country: Germany
incidence: 5/10, conc. range: 450-750 µg / kg, ∅ conc.: 437 µg / kg, country: Poland
→ deoxynivalenol
incidence: 6/6, conc. range: 1-6300 µg / kg, country: Finland
incidence: 21/21, conc. range: 7-861 µg / kg, ∅ conc.: 168 µg / kg, country: Finland
incidence: 3/8, ∅ conc.: 60 µg / kg, country: Germany
incidence: 1/2, conc.: 365 µg / kg, country: Germany
incidence: 3/37, conc. range: 200-700 µg / kg, ∅ conc.: 500 µg / kg, country: Germany
incidence: 7/7*, conc. range: 70-90 µg / kg, ∅ conc.: 270 µg / kg, country: Germany, *moldy
incidence: 4/10, ∅ conc.: 200 µg / kg, country: Germany
incidence: 11/72, conc. range: 20-500 µg / kg, country: Germany
incidence: 2/3, conc. range: ≤ 80 µg / kg, country: New Zealand
incidence: 3/6, conc. range: 420-520 µg / kg, ∅ conc.: 470 µg / kg, country: Sweden
incidence: 11/32, conc. range: 40-260 µg / kg, ∅ conc.: 140 µg / kg, country: Sweden
incidence: 14/45, conc. range: 40-500 µg / kg, ∅ conc.: 200 µg / kg, country: Sweden
incidence: 1/6, conc.: < 100 µg / kg, country: UK
incidence: 1/1, ∅ conc.: 31 µg / kg, country: USSR
→ diacetoxyscirpenol
incidence: 3/6, conc. range: 10-1700 µg / kg, country: Finland
→ fusarenon X
incidence: 1/6, conc.: < 40 µg / kg, country: Finland

→ HT-2 toxin
incidence: 1/6, conc.: < 80 μg / kg, country: Finland
incidence: 2/21, conc. range: 33-44 μg / kg, ∅ conc.: 39 μg / kg, country: Finland
incidence: 18/68, conc. range: ≤ 700 μg / kg, country: Germany
incidence: 10/107, conc. range: 300-900 μg / kg, country: Germany
incidence: 18/59, conc. range: 100-700 μg / kg, country: Germany
→ nivalenol
incidence: 2/6, conc. range: < 1000 μg / kg, country: Finland
incidence: 3/21, conc. range: 48-83 μg / kg, ∅ conc.: 70 μg / kg, country: Finland
incidence: 1/8, conc.: 1464 μg / kg, country: Germany
incidence: 4/7, conc. range: 16 μg / kg, country: Nepal
incidence: 2/3, conc. range: ≤ 610 μg / kg, country: New Zealand
incidence: 1/1, conc.: 1100 μg / kg, country: USSR
→ moniliformin
incidence: 3*/3, conc. range: 15,700-38,300 μg / kg, ∅ conc.: 24,060 μg / kg, country: Poland, * hand-selected, visible fungal damage
→ monoacetoxyscirpenol
incidence: 1/107, conc.: 50 μg / kg, country: Germany
→ neosolaniol
incidence: 2/107, conc. range: 300-400 μg / kg, ∅ conc.: 350 μg / kg, country: Germany
→ ochratoxin A
incidence: 23/48, conc. range: 5-1000 μg / kg, country: Austria
incidence: 1/1, conc.: ca. 1700 μg / kg, country: Canada
incidence: 1/19, conc.: 1-2 μg / kg, country: Czechoslovakia
incidence: 21/50*, conc. range: 0.05-4.9 μg / kg (20 samples), 5.6 μg / kg (1 sa), ∅ conc.: 0.5 μg / kg, country: Denmark, *conventional
incidence: 6/17*, conc. range: 0.05-4.2 μg / kg, ∅ conc.: 0.3 μg / kg, country: Denmark, *ecological
incidence: 13/25*, conc. range: 0.05-4.6 μg / kg, ∅ conc.: 0.5 μg / kg, country: Denmark, *conventional, imported
incidence: 12/93, conc. range: 1-58.8 μg / kg, ∅ conc.: 9.5 μg / kg, country: Germany
incidence: 2/34, conc. range: 1.4-56.6 μg / kg, ∅ conc.: 29 μg / kg, country: Italy
incidence: 2/14, conc.: ≤ 2.4 μg / kg, country: The Netherlands
incidence: 4/18, conc. range: 0.1-2.4 μg / kg, country: The Netherlands
incidence: 17/24, conc.: ≤ 3.8 μg / kg, ∅ conc.: 0.95 μg / kg, country: UK
incidence: 1/46, conc.: 80 μg / kg, country: UK
incidence: 2/28, conc.: 52-110 μg / kg, ∅ conc.: 81 μg / kg, country: USA
→ T-2 toxin
incidence: 1/6, conc.: < 24 μg / kg, country: Finland
incidence: 11/19, conc. range: 1-160 μg / kg, ∅ conc.: 27 μg / kg, country: Finland
incidence: 2/21, conc. range: 45-73 μg / kg, ∅ conc.: 59 μg / kg, country: Finland
incidence: ?, conc. range: 10-90 μg / kg, country: Germany
incidence: 16/49, conc. range: 10-50 μg / kg, ∅ conc.: 300 μg / kg, country: Germany
incidence: 7/10, conc. range: 13-500 μg / kg, ∅ conc: 220 μg / kg, country: Germany
incidence: 15/82, conc. range: 70-300 μg / kg, country: Germany
T-2 triol
incidence: 3/66, conc. range: 100-300 μg / kg, country: Germany

→ zearalenone
incidence: 3/21, conc. range: 30-86 µg / kg, ∅ conc.: 63 µg / kg, country: Finland
incidence: 3/8, ∅ conc.: 49 µg / kg, country: Germany
incidence: 1/2, conc.: 41 µg / kg, country: Germany
incidence: 22/144, conc. range: 1-150 µg / kg, country: Germany
incidence: 6/7*, conc. range: $\leq$ 8 µg / kg, ∅ conc.: 3 µg / kg, country: Germany, *damaged kernels
incidence: 4/10, ∅ conc.: 50 µg / kg, country: Germany
incidence: 17/80, conc. range: 10-440 µg / kg, country: Germany
incidence: 1/5, ∅ conc.: 2 µg / kg, country: Italy
incidence: 5/7, ∅ conc.: 6 µg / kg, country: Nepal
incidence: 10/29, conc. range: $\leq$ 90 µg / kg, country: New Zealand
→ cereals

Ochratoxicosis This worldwild-occurring → mycotoxicosis, due to the intake of → ochratoxin A, is primarily a problem in temperate climates of such countries as Canada, Denmark, Ireland, Norway, Sweden, and the US. Pigs and → poultry are mainly affected but humans may also suffer from this disease (→ Balkan endemic nephropathy).
The occurrence of the → mycotoxic porcine nephropathy in Denmark is linked with apparently "extreme climatic conditions" such as high moisture and relatively high temperatures (about 25 °C). These conditions favor the growth of ochratoxin producing fungi like → Aspergillus ochraceus group and → Penicillium verrucosum Dierckx.
→ nephropathy

Ochratoxin A (Abbr.: OTA) is a N-[[(3R)-5-chloro-3,4-dihydro-8-hydroxy-3-methyl-1-oxo-1H-2-benzopyran-7-yl]carbonyl]-L-phenylalanine which belongs to the isocumarins having an amide linkage to L-phenylalanine (→ mycotoxins). Similar to → aflatoxin B_1 it contains a lactone group but the Cl-atom is striking for a natural substance (see Figure 1 Ochratoxin A). It was first isolated from → Aspergillus ochraceus K. Wilh. in 1965 by African scientists during laboratory screening for toxigenic fungi. OTA was found to occur naturally for the first time in an US → maize sample in 1969. In contrast to South Africa this mycotoxin caused economically important animal diseases and possibly also a human disease in other, northern parts of the world due to the contamination of → food and feedstuff.

Chemical Data
Empirical formula: $C_{20}H_{18}O_6NCl$, molecular weight: 403

Fungal Sources
In tropical and semitropical regions OTA is mainly produced by members of the → Aspergillus ochraceus group. → Aspergillus alutaceus var. alutaceus Berkely & Curtis (formerly → A. ochraceus K. Wilh.) is the best known ochratoxin producer of the genus → Aspergillus, e.g. *A. melleus*, *A. sclerotiorum* and *A. sulphureus* are of minor importance. Their incidence in food is rare. In temperate regions → Penicillium verrucosum Dierckx predominates. Further producers: *A. niger* group, → Eurotium *herbariorum*, → Penicillium spp. (e.g. *P. purpurescens*), → Petromyces *alliaceus*

Natural Occurrence
→ almonds, → baby food, → bacon, → bakery products, → barley, → barley malt, → beans, → beer, → beer, wheat, → biscuits, → bran, → bread, → breakfast cereals, → breakfast drinks, → buckwheat, → cardamom, greater, → cassava flour, → cereal flakes, → cereal food, → cereal products, → cereals, → chapatti, → cheese, → cheese, Bhutanese,

→ cheese, Cheddar, → cheese, Chesire, → cheese, Double Gloucester, → cheese, Edam, → cheese, Emmental, → cheese, Leicester, → cheese, Wensleydale, → cheese, cake, → cheese trimmings, → chicken, → chicken, yolk, → chilli pickles, → chilli powder, → chilli sauce, → cocoa beans, → cocoa nibs, → cocoa presscake, → cocoa products, → coffee beans, → coffee, → confectionery, → copra, → coriander, → corn flakes, → cow peas, → curry, → curry paste, → duck, → fennel, → figs, → fish, → flour, → foods, → garlic pickle, → ginger, → goose, → grains, → grape juice, → ham, → hazelnuts, → human breast milk, → kulen, → lentils, → maize, → maize flour, → maize grits, → maize products, → majoran, → malt, → meat, → milk, → millet, → muesli, → muesli ingredients, → noodles, → nuts, → oats, → oat bran, → oat flakes, → olive oil, → olives, → paprika, → peanuts, → peas, → pepper, → pig blood, → pig kidneys, → pig liver, → pig serum, → pop corn, → pork, → porridge, → poultry, → rice, → rice bran, → rice cake, → rye,→ rye bran, → rye flour, → rye grits, → sausages, → sesame seeds, → sesame oil, → snack food, → soybean, → soybean concentrate, → spelt, → spices, → sunflower seeds, → tandoori, → tapioca, → triticale, → turkey, → vegetables, → wheat, → wheat grits, → wheat products, → wine, → zwieback

OTA occurs widely in plants and plant products but most frequently in cereal grains infected with *P. verrucosum*, particularly in north temperate growing areas. Compared to pre-harvest production, post-harvest OTA formation is regarded as the predominant factor in the contamination of insufficiently dried starch-rich foodstuffs (cereals and derived products). It seems that the distribution of OTA in food and / or crops resembles that of aflatoxin with respect to inhomogenicity.

OTA contamination of wines (up to 0.4 μg / l) from southern parts of Europe may be responsible for increased OTA levels found in the blood of males in southern Switzerland . Grape juice samples may also be contaminated ($\approx$ 0.2 μg OTA / l). Although infection of → meat and → fish with *P. verrucosum* (and possible mycotoxin formation) has been reported, contamination of meat products is more usually due to the → carry over of OTA from contaminated animal feed into blood, kidneys and muscles.

Since OTA is extensively metabolized in the forestomachs by protozoan and bacterial enzymes to nontoxic metabolites, tissues of ruminants are not contaminated to any significant extent. Even at higher concentrations the rapid hydrolysis of OTA greatly impedes absorption and may cause only a transient suppression of → milk production in cattle. It was estimated that the application of at least 1.66 mg OTA / kg bw for four days is necessary to detect any residues of OTA in the milk. Therefore, OTA levels commonly found in *P. verrucosum*-contaminated feeds do not represent a substantial health risk to these animals.

However, significant contamination in a number of tissues of single-stomach food animals (e.g. pigs and poultry), especially the kidneys, due to carry over from feed is possible. These animals belong to the group of susceptible monogastric livestock showing nephropathy. → Pork and → bacon as well as pork-derived meat products (e.g. → sausages, black pudding) may contain higher amounts of ochratoxin. A high incidence of ochratoxin A in swine blood samples was related to a high moisture content in barley (main ingredient of swine feed). In poultry muscles 29 μg OTA / kg have been found at slaughter under natural conditions but in general liver and kidney typically contain the highest residues. However, OTA residues in tissues

decrease rapidly following removal of the contaminated diet.
It is suggested that for humans the bioavailability for OTA residues is higher in cereals than in meats, as in the latter OTA is bound to proteins.
In the blood, ochratoxin A is present bound to serum albumin and in its free form. Particularly in humans, → cattle and pigs, OTA is strongly bound to serum albumin. → Human breast milk may also be contaminated with OTA (see Figure 2 Ochratoxin A).

TOXICITY
Clinical symptoms: emetic, strong nephrotoxic, hepatotoxic, → immunosuppressive, → teratogenic, → mutagenic, cancerogenic
LD_{50} (po): 20-22 mg / kg bw rats
In all the mammalian species tested, the kidney is the major target for the toxicity of OTA. Besides the → Balkan endemic nephropathy elevated exposure to OTA should also be associated with human nephropathies in Algeria and Tunisia. Furthermore, in rural Scandinavian populations high kidney failure rates have been observed which may be due to the ingestion of pig meat contaminated with excessive amounts of OTA. The fact that the half-life of OTA in humans is 8-12 times longer than in rats is important for risk assessment. Since this mycotoxin is fat soluble and not readily excreted, accumulation in fatty tissues occurs. Decreased weight gains in swine and poultry as well as losses in egg production occur at levels higher than 2000 µg / kg. Higher dosis are often fatal. However, cattle are resistant to the OTA levels found naturally in feed.
The primary source of excretion is the urine (rats) but faecal excretion also occurs to some extent.
In experimental animals a synergistic effect between OTA and citrinin as well as → penicillic acid has been observed.

DETECTION
ELISA, HPLC (fluorescence detection), LC-MS, RIA, TLC

POSSIBLE MYCOTOXICOSIS
→ Mycotoxic porcine nephropathy, Balkan endemic nephropathy

FURTHER COMMENTS
It seems that cereals and cereal products are the main contributors to OTA intake in northern climates (Europe). There are no large uncertainties about the consumption data of this kind of food since they are regularly eaten by most people. In addition, roasted coffee, beer, pig meat, blood products, wine, and pulses may contribute to the intake of OTA. Because of the particularly strong binding of ochratoxin A to serum albumin of pig blood, products like black pudding are most likely to contain ochratoxin A. Estimations revealed the following mean daily OA intakes for adults: Germany = 1.2 and 1.3 ng / kg bm for women and men, respectively, Sweden = 0.4 ng / kg bm, Swiss = 0.7 ng / kg bm men (residing north of the Alpes), Canada = 1.1 ng / kg bm for males (12-19 years).
Due to the fact that OTA occurs in blood at a much higher rate than the frequency that nephropathy has detected, the analysis of swine blood residue levels may be a more suitable indicator of low amounts, or early exposition to the toxin. The use of this analytic technique allows the essential elimination of OTA residues from the kidneys and / or carcasses that have been fed on an OTA-free diet for a period of four weeks before slaughter. In this way the entry of contaminated meat into the food chain may be prevented.
OTA possesses a relatively long half-life in certain edible animal species as well as in humans (35 days in serum). This explains the high incidence of OTA in human blood sera (but generally at low levels). It further documents the potential widespread occurrence of OTA in Eur-

opean food, particularly in whole-grain breads, → pork and pig-blood-based products.

OTA is often accompanied by → citrinin and the naphthoquinones viomellein and xanthomegnin (all nephrotoxic) which are products of → Penicillium aurantiogriseum Dierckx.

Production: Minimum → a_w for ochratoxin production is a_w 0.85 *A. ochraceus* and a_w 0.83 *P. verrucosum* with an optimum at > a_w 0.97. At the optimum a_w OTA is produced in a temperature range of 12-37 °C (optimum 25 °C) for *A. ochraceus* and 4-31 °C (optimum 24 °C) for *P. verrucosum*. On bread (pH 5.6) the minimum a_w for OTA prodcution amounted to a_w 0.80, the optimum was a_w 0.92 (→ Penicillium viridicatum Westling ? = *P. verrucosum*). The optimum pH for ochratoxin A production under *in vitro* conditions is ≈ pH 5.6 which is the same for both species (*A. sulphureus* = pH 6.0-6.3).

In general the medium composition, especially the presence or absence of micronutrients such as metal ions, more strongly influenced biosynthesis of OTA than growth. Groundnuts and soybeans were the optimal substrata for OTA production of *A. alutaceus* whereas *P. verrucosum* produced highest yields on maize and wheat.

Levels of nitrogen applied to growing barley increased the protein content as well as the ochratoxin production of *A. ochraceus* and *P. verrucosum* on barley postharvest.

Subinhibitory concentrations of phosphine may increase the levels of ochratoxin produced whereas ochratoxin biosynthesis is inhibited by dichlorvos at concentrations which have relatively little effect on fungal growth. Growth and ochratoxin production by *A. sulphureus* and *P. verrucosum* are inhibited due to antimicrobial food additives like methyl paraben, sodium propionate, and potassium sorbate.

Irradation (≤ 10 kGy) enhanced OTA production by *A. ochraceus*.

Stability: Compared to → patulin or penicillic acid OTA appears to be more stable in foods but it is probably somewhat less stable than → aflatoxins. Once ochratoxin A has been formed in a food this moderatly stable mycotoxin survives most food processing stages (such as cooking, roasting, fermenting) to quite appreciable degree. Even temperatures as high as 250 °C are not sufficient for complete degradation of OTA. Losses of ≈ 20% occurred during frying of blood-pudding, kidneys, and muscular tissue from pigs. No losses were observed in adipose tissue after frying.

Frying (150-160 °C) of certain pig products such as ground muscle, fat, diced kidneys, and sliced blood pudding caused losses in total toxin of about 20% but in frozen pig kidney a high stability of OTA was observed. Cooking of animal products was less effective in the destruction of OTA.

No destruction of OTA occurred during bread baking but OTA levels partially decreased after biscuit baking. Soaking, blanching, cooking or canning of beans generally resulted in only small losses of OTA (10-34%). Cooking of wheat was also ineffective.

Regarding the stability of OTA, the roasting of coffee gave the most variable results. Losses were reported in the range of 0-100%. This may be due to several factors such as roasting conditions, inhomogenicity of natural coffee bean contamination, OTA levels, natural contamination versus spiking and analytical method performance. There are different opinions among researchers whether OTA passes into brewed coffee.

The overwhelming majority of more than 600 European coffee samples did not show any OTA contamination while only a few samples contained more than 30 µg / kg.

OTA can slowly break down merely during storage of grains and grain products as is documented by a decrease of more than 60% in naturally contaminated barley over a storage period of 2 years. Depending on the temperature and possibly other factors the moisture can increase or decrease the stability of ochratoxin A during the heating of cereals.

OTA was moderately stable during drying of sausages.

Storage of cheese at room temperature caused a significant decrease in OTA concentration.

It is suggested that the mechanical removal of OTA is probably the most feasible procedure.

Cleaning / milling: Neither cleaning (dry or wet) nor milling did eliminate OTA from naturally-contaminated samples of barley and wheat. After milling similar levels of OTA were found in flour and → bran. However, scouring (removal of the outer layers of the pericarp) as an additional cleaning procedure reduced the OTA concentration in wheat by as much as 50%.

During experimental wet-milling of maize the maize bits (starch, fibre and gluten) contained most of the OTA (51%) of all the maize fractions.

Malting / brewing: There are different results concerning the sources of OTA contamination in beer. OTA appears to be completely destroyed or lost during malting (mainly in the initial steeping stage) of moderately contaminated barley lots. Besides malting (malt mash but not the cooker mash) boiling of the wort with hops, and the final fermentation should also contribute to the destruction of OTA, possibly to ochratoxin α. A transmission rate of 2-28% of the ochratoxin in barley into beer was observed, if heavily contaminated lots are used for malting. However, such severely deteriorated barley brands would probably not be accepted for brewing. Contamination of beer is mainly due to the use of contaminated cereal adjuncts since OTA survives the fermentation step in beer-making. About 20-30% of the original OTA concentration may be found in the finished product.

Control: Proper storage of harvested grains (moisture content and temperature are most important factors) prevents growth of saprophytic storage fungi and subsequent OTA contamination. Since OTA is transmitted into animal tissue, particularly in pigs and poultry, no contaminated feeds should be fed to animals intended for human consumption.

In general, only good practices at all stages of the food chain and approaches based on HACCP concepts will contribute to keep OTA contamination low in foodstuffs.

Regulations: The IARC has classified ochratoxin A as a possible human carcinogen (Group 2B) in 1993 based on sufficient evidence in humans. An acceptable safe level of the tolerable daily intake would fall in the range of 1.5-4.2 ng / kg bw / day. The Word Health Organization / Food and Agricultural Organization Joint Expert Committee on Food Additives (JECFA) recently re-evaluated the toxicity of OTA. A PTWI of 100 ng / kg, bodyweight / week was determined.

Among 77 countries with known mycotoxin regulations eight (Czech Republic, Denmark, France, Greece, Hungary, Sweden, Switzerland, Uruguay) also have specific regulations for ochratoxin A levels in one or more commodities whereas some countries have proposals for ochratoxin A regulations (Austria, Germany, Great Britain, Rumania, The Netherlands). Current (proposed) limits for OTA contamination are as follows: 1-5 µg / kg children and infant foods, 2-50 µg / kg foods, 5-300 µg / kg animal feeds.

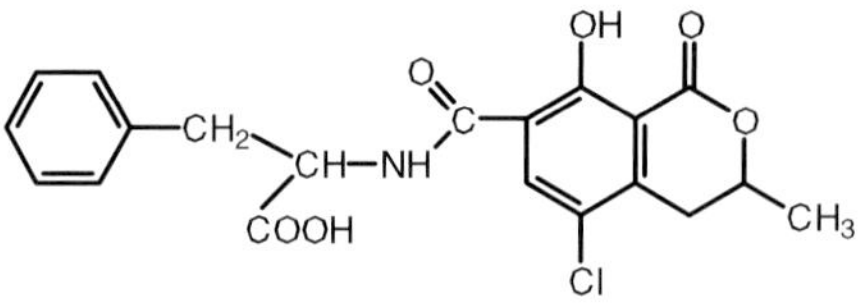

Ochratoxin A (Figure 1)

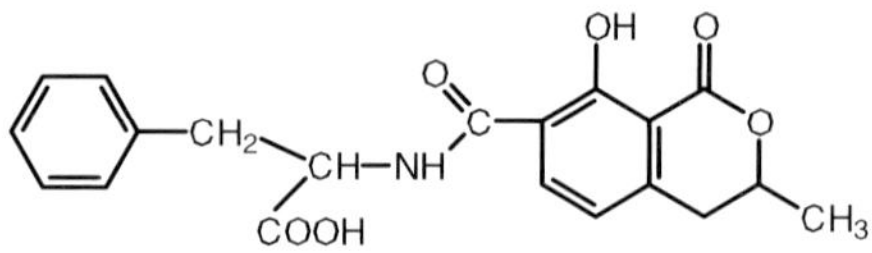

Ochratoxin B

The proposed tolerance levels in the EU are 1 µg / kg infant foods, 5 µg / kg cereals.

Ochratoxin B is the dechloro-analogue of ochratoxin A (N-[[(3R)-3,4-dihydro-8-hydroxy-3-methyl-1-oxo-1H-2-benzopyran-7-yl]carbonyl]-L-phenylalanine) which was first isolated in 1965 from → Aspergillus alutaceus var. alutaceus Berkely & Curtis (→ mycotoxins) (see Figure Ochratoxin B).

CHEMICAL DATA
Empirical formula: $C_{20}H_{19}O_6N$; molecular weight: 369

FUNGAL SOURCES
A. alutaceus var. *alutaceus*

NATURAL OCCURRENCE
→ bread, → maize

TOXICITY
Ochratoxin B is approximately 16 times less toxic to chicks than → ochratoxin A and also less toxic than ochratoxin C. However, similar pathological lesions occurred in chicks and rainbow trout as described for ochratoxin A.

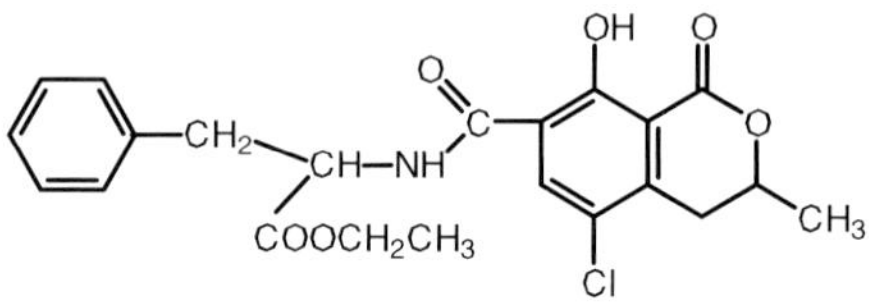

Ochratoxin C

DETECTION
TLC

Ochratoxin C (Syn.: Ochratoxin A ethyl ester) Isolation methods which depend on the free carboxyl group are not successful and therefore the occurrence of this ochratoxin type may be underestimated (N-[[(3R)-5-chloro-3,4-dihydro-8-hydroxy-3-methyl-1-oxo-1H-2-benzopyran-7-yl]carbonyl]-L-phenylalanine; ethyl ester). It is produced by → Aspergillus alutaceus var. alutaceus Berkely & Curtis and was first isolated in 1965. Natural occurrence of ochratoxin C ($C_{22}H_{22}O_6NCl$; mw 431) in → wine has been reported (see Figure Ochratoxin C).

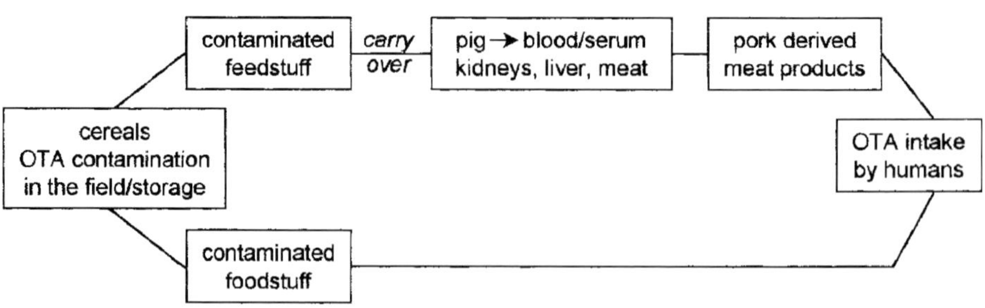

Ochratoxin A (Figure 2). Routes of OTA intake by humans

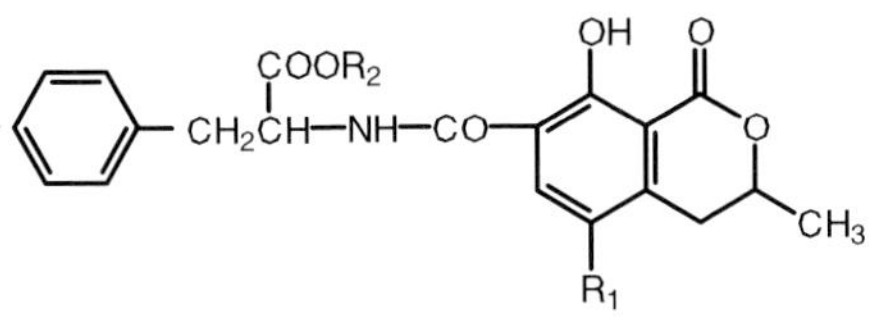

Ochratoxin A: $R_1 = Cl$; $R_2 = H$
Ochratoxin B: $R_1 = H$; $R_2 = H$
Ochratoxin C: $R_1 = Cl$; $R_2 = C_2H_5$
Methylester of Ochratoxin A: $R_1 = Cl$; $R_2 = CH_3$
Methyl or ethyl ester of Ochratoxin B: $R_1 = H$; $R_2 = CH_3$ or C_2H_5

Ochratoxins. Members of the ochratoxin group

Ochratoxins are isocoumarines composed of a 3,4-dihydroy-3-methylisocoumarin moiety linked via the 7-carboxy group to L-β-phenylalanine by an amide bond (→ mycotoxins). The isolation of a chlorine-containing metabolite designated → ochratoxin A succeeded in 1965 when African scientists carried out a screening of toxigenic fungi.
OTA as the major toxic principle in different kinds of food and feedstuff (mainly → grains) is the most important toxic member of a group of nine or more → ochratoxins produced with the highest yield. This group consists of ochratoxin A, its methyl and ethyl esters (all which are toxic), and 4-hydroxyochratoxin A (see Figure Ochratoxins). In contrast, the chlorine free derivative of OTA, ochratoxin B as well as ochratoxin C, is rarely found in → foods and feeds. Contamination of grains with ochratoxins has been reported from e.g. most European countries and North America and is due to → Aspergillus ochraceus group and → Penicillium verrucosum Dierckx. OTA typically co-occurs with low amounts of → citrinin, which is also a → nephrotoxin.

Ogbono is a Nigerian type of foodstuff made from the plant *Irvingia gabunensis*.
Ogbono may contain the following → mycotoxins:
aflatoxin B (→ aflatoxins)
incidence: 1/1, conc.: 168 µg / kg, country: Nigeria

Ogili-ugba is a Nigerian type of foodstuff made from the castor bean, *Riccinus communis*.
Ogili-ugba may contain the following → mycotoxins:
aflatoxin B (→ aflatoxins)
incidence: 1/1, conc.: 362 µg / kg, country: Nigeria

Ogoro is a Nigerian indigenous beverage (palm juice) made from the sap from the stalk of the male inflorescence or the immature shoot of the oil palm (*Elais guinensis*).
Ogoro may contain the following → mycotoxins:
aflatoxin B (→ aflatoxins)
incidence: 2/2, conc. range: 116-118 µg / kg, ∅ conc.: 117 µg / kg, country: Nigeria

Oil If the oil is removed from the → oil seeds, → aflatoxins are mainly found in the oil seed meal. The soap stock as a by-product from the alkali-refining step contains only the low levels that remained in the crude vegetable oil. In general, the refined oil is aflatoxin-free since aflatoxin residues are removed in the bleaching refining steps.
Oil may contain the following → mycotoxins:
aflatoxin (no specification)
incidence: 10/25*, conc. range: ≤ 7 µg / kg, ∅ conc.: 3 µg / kg, country: Philippines, *cooking
aflatoxins (AFB_1, AFB_2, AFG_1, AFG_2)
incidence: 1/4*, conc.: 0.1 µg / kg, country: UK, *chili, almond
→ zearalenone
incidence: 1/4*, conc.: 5.4 µg / kg, country: UK, *chili, almond
→ coconut oil, → olive oil, → peanut oil, → sunflower seed oil

Oil seed rape represents an important agricultural crop which is used as cook-

ing oil and for the production of margarine. Rapeseed meal is also used in cattle concentrates. Weather conditions during harvesting and threshing show extreme variations, enabling the development of different fungi especially if oilseed rape is stored under poor conditions. Subsequent mycotoxin contamination might occur. Oilseed rape may contain the following → mycotoxins:
→ aflatoxin B_1
incidence: 1/20, conc.: 0.25 μg / kg, country: Spain
→ citrinin
incidence: 1/1*, conc.: 4100 μg / kg, country: UK, *moldy
→ sterigmatocystin
incidence: 1/1*, conc.: 40 μg / kg, country: UK, *moldy
→ viomellein
incidence: 1/1*, conc.: traces, country: UK, *moldy
vioxanthin
incidence: 1/1*, conc.: 40 μg / kg, country: UK, *moldy
→ xanthomegnin
incidence: 1/1*, conc.: traces, country: UK, *moldy

Oil seeds (no specification)
may contain the following → mycotoxins:
→ aflatoxin B_1
incidence: 31/73*, conc. range: < 5-2000 μg / kg, country: Natal (Union of South Africa), *includes → peanuts, → sunflower seeds, cottonseeds
→ aflatoxin (no specification)
incidence: 9/80, conc. range: 2-20 μg / kg (7 samples), > 20 μg / kg (2 sa), country: Uruguay
→ patulin
incidence: 8/107*, conc.: nc, country: South Africa, *includes peanuts, sunflower seeds, cottonseeds
→ zearalenone
incidence: 1/107, conc.: nc, country: South Africa
incidence: 6/64, conc. range: 100-200 μg / kg (3 samples), > 200 μg / kg (3 sa), country: Uruguay

Olive oil In some countries farmers sometimes store their → olives for several weeks under conditions that contribute to the growth of molds such as → Aspergillus flavus Link and the → Aspergillus ochraceus group. This may result in aflatoxin and ochratoxin contamination of olives and olive → oil. If the so called "virgin" olive oil is prepared from contaminated crude oil, the refining process which would remove the → aflatoxins is omitted.
Nonchemically treated olive oil made from deteriorated olives may contain low levels of → Alternaria mycotoxins. These low levels should not represent a concern for human health.
The transmission rate from olives into the oil amounted to only 4% → alternariol methyl ether (793.6 μg / kg) and 1.8% → alternariol (285.7 μg / kg). No transmission has been reported for → altenuene and → tenuazonic acid, considering an oil yield of 15% from the processed olives (experimental study).
The results of a limited survey showed that olive oil samples collected from different oil mills did not show any mycotoxin contamination.
However, olive oil may contain the following → mycotoxins:
→ aflatoxin B_1
incidence: 3/46, conc. range: 1-13 μg / kg, ∅ conc.: 11 μg / kg, country: Germany
incidence: 14/16, conc. range: 1-75 μg / kg, ∅ conc.: 361 μg / kg, country: Greece
→ aflatoxin B_2
incidence: 10/16, conc. range: 1-55 μg / kg, ∅ conc.: 185 μg / kg, country: Greece

→ aflatoxin G_1
incidence: 5/16, conc. range: 1-2.5 µg / kg, ∅ conc.: 1.6 µg / kg, country: Greece
→ aflatoxin G_2
incidence: 5/16, conc. range: 1-5 µg / kg, ∅ conc.: 2.2 µg / kg, country: Greece
→ ochratoxin A
incidence: 3/60, conc. range: traces, country: Morocco
→ coconut oil, → oil, → peanut oil

Olives Physical damage of the surface seems to be a prerequisite for → Alternaria mycotoxins to contaminate olives. Such olives are frequently infected with → Alternaria spp., mainly → Alternaria alternata (Fr.) Keissler. Under suitable conditions fungal attack starts with the penetration into the fruit pulp, followed by substantial mycelial growth and subsequent mycotoxin formation. Considerable amounts of *Alternaria* → mycotoxins may be produced in physically damaged (heavily damaged, weathered or moldy) olives in the field before harvesting as well as during storage. Although *Alternaria* spp. could be isolated from sound, undamaged olives properly harvested from the ground in different areas, no mycotoxin contamination could be established.
Olives may contain the following mycotoxins:
→ aflatoxin B_1
incidence: 12/103**, conc. range: 5-37 µg / kg, country: Morocco, **black, Greek-style
→ altenuene
incidence: 1/4*, conc.: 1400 µg / kg, country: Italy
→ alternariol
incidence: 4/4*, conc. range: 109-2320 µg / kg, ∅ conc.: 1120 µg / kg, country: Italy
→ alternariol methyl ether
incidence: 4/4*, conc. range: 30-2870 µg / kg, ∅ conc.: 818 µg / kg, country: Italy
→ ochratoxin A
incidence: 5/103**, conc. range: 40-80 µg / kg, country: Morocco, **black, Greek-style
incidence: 4/7, conc. range: 0.3-46,830 µg / kg, country: Tunesia
→ tenuazonic acid
incidence: 2/4*, conc.: 109-262 µg / kg, ∅ conc. 1865 µg / kg, country: Italy, *samples visibly affected by *Alternaria* rot

Onions → garlic

Ontjom Processing of ontjom might result in detoxication of mycotoxin (→ mycotoxins) contaminated → peanuts.

Onyalai This neither heritable nor infectious disease was first described in Angola (1904) and is widespread in the south of the Sahara. It most frequently occurs in summer. In general, individuals of all ages and both sexes of African races (e.g. Bantu) are almost exclusively affected although some cases in Europeans and Chinese have also been reported.
A distinct feature of this disorder is the sudden appearance of hemorrhagic (→ hemorrhage) bullae (→ bulla) in the mouth and sometimes on the skin. Further symptons are → hematuria, profuse bleeding from the nose, mouth, and conjunctiva. Severe cases are characterized by hemorrhagic shock and cerebral hemorrhages. Death may occur within a few days of the onset of the disease. A mortality rate of 14% in one case study has been reported. However, first-time sufferers may recover spontaneously within two months of the onset. The severity and mortality of this disease depend on the area of its occurrence. Although the cause of onyalai is still unknown, → mycotoxins such as

→ tenuazonic acid (salts) and → moniliformin should be involved. The mycotoxin hypothesis is due to the fact that the disease occurs only among eaters of → millet (*Pennisetum typhoides*). This millet as well as → sorghum was contaminated by → Fusarium spp. and highly toxic → Phoma spp. From *P. sorghina*-inoculated → maize culture calcium-, magnesium-, and sodium-tenuazonate were isolated. Acid treatment resulted in tenuazonic acid, which is also produced by certain → Alternaria species. However, the suggested primary toxic actions of tenuazonic acid (emetic and cardiovascular action) are not consistent with the characteristic hemorrhagic bullae in the oral cavity due to onyalai. In addition, the ability to selectively complex with trace metals *in vivo* is not sufficient to explain the symptoms of onyalai. Different environmental factors or even other mycotoxins may also be involved.
The better nutritional status in the endemic regions is probably one reason for the decrease of onyalai in the last few years.

Oo-hen-mai Japanese: yellow rice (disease)
→ Acute cardiac beriberi

Oranges may contain the following → mycotoxins:
→ aflatoxin B_1
incidence: 1/20*, conc.: 4600 µg / kg, country: Germany
incidence: 4/14*, conc. range: 5-50 µg / kg, country: Germany
→ aflatoxin G_1
incidence: 1/20*, conc.: 21.5 µg / kg, country: Germany
→ aflatoxin G_2
incidence: 1/20*, conc.: 1200 µg / kg, country: Germany
*moldy
→ fruits

Oriental fermentations During the manufacturing of fermentation products like → miso and → shoyu none of the tested industrial used → Aspergillus strains (in Japan) produced → aflatoxins, → ochratoxin A, → patulin, → penicillic acid or → sterigmatocystin. → Cyclopiazonic acid was produced by only a few isolates. Although the strains used for fermentation were able to synthesize aspergillic acid, → kojic acid, nitropropionic acid and oxalic acid, the concentrations were to low to constitue any toxic hazard to humans.

OTA → Ochratoxin A

Oxygen → atmosphere

P

Paecilomyces → mitosporic fungi, teleomorph: → Byssochlamys spp. *Byssochlamys* spp. and *P. variotii* are important producers of → patulin.

Paprika may contain the following → mycotoxins:
→ ochratoxin A
incidence: 1/4, conc.: 40 µg / kg, country: Germany
→ spices

Paralysis Loss of feeling or power to move in any or every part of the body.

Parasiticol (Syn.: → aflatoxin B_3)

paretic incomplete paralysis

Parmesan cheese → cheese, Parmesan

Pasta may contain the following → mycotoxins:
→ citrinin
incidence: 1/2, conc.: 0.5 µg / kg, country: Switzerland
→ ochratoxin A
incidence: 11/21, conc. range: < 5 µg / kg, country: UK

Pasteurized foods According to Frisvad (1988) → patulin may be excreted into → fruit juices and vegetable juices (→ vegetables).

Pastries may contain the following → mycotoxins:
→ aflatoxin B_1
incidence: 1/86, conc.: < 5 µg / kg, country: Germany
→ citrinin
incidence: 1/2, conc.: 0.5 µg / kg, country: Switzerland

Patulin (Syn.: clavacin, clavatin, claviformin, expansine, gigantic aicd, mycoin, penicidin, tercinin, leucopin) This 4-hydroxy-4-*H*-furo-[3,2-c]pyran-2(6H)-one was first isolated from *Penicillium patulum* (= → Penicillium griseofulvum Dierckx) during the search for new antibiotics in 1941 (see Figure Patulin).

CHEMICAL DATA
Empirical formula: $C_7H_6O_4$, molecular weight: 154

FUNGAL SOURCES
→ Aspergillus clavatus Desm., *A. giganteus*, → Aspergillus terreus Thom, → Byssochlamys *nivea*, *B. fulva*, → Eupenicillium spp., → Penicillium expansum Link (most important and the most commonly encountered patulin producer), → Penicillium spp. (e.g. *P. claviforme*, → Penicillium roquefortii Thom chemotype II, *P. melinii*)

NATURAL OCCURRENCE
→ Apples, → apple beverages, → apple butter, → apple flavor, → apple jam, → apple juice, → apple juice concentrate, → apple products, → blueberries, → cereals, → cheese, → cheese, goat, → cider, → cranberries, → fruits, → fruit juices, → fruits products, → grape juice, → jam, → lingonberries, → meat, → oil seeds, → peaches, → pear juice, → pears, → plums, → scented supar, → soft drinks
Apples, apple products, and peaches are excellent substrates for patulin production. In nature patulin is found almost exclusively in apples and apple products but visual inspection will usually identify poor quality items. Patulin contamination of apple juice is an effective indicator of the use of unsound, substandard, *P. expansum* rotted apples in juice manufacture. Such juices may contain up to 1,000,000 µg / l of patulin. Although patulin commonly occurs in rotting apples and the incidence of patulin contamination of apple juice is fairly high, the level of contamination in general is relatively

low (< 100 μg / l). Sporadically very high levels 45,000 μg patulin / l apple juice from a roadside stand in the USA have been detected. Apples and especially apple products such as juice are the major human dietary sources of patulin. Alcoholic fermentation of fruit juices by *Saccharomyces cerevisia* and *S. ellipsoideus* caused almost total destruction (> 99%) of patulin.

Although potential patulin producers are present on foods such as → oranges, oranges juice, wet → maize, durum → wheat, → sorghum, → flour, cheeses, meat and meat products (e.g. → sausages) no or only decreased levels of patulin have been detected. The lack or decrease is mainly attributed to the reaction (binding) with sulfhydryl groups of compounds (cysteine, gluthatione) present in the → foods, although not all contain sulfhydryl groups. This reaction makes patulin chemically undetectable and of lesser toxicity because the binding to functional groups is inhibited.

Since contamination of livestock feeds has not been reported, patulin accumulation in meat and poultry products due to → carry over seems to be unlikely.

Toxicity

antibiotic (e.g. *Mycobacterium tuberculosis*), antifungal, → immunosuppressive, neurotoxic, → teratogenic (?),→ mutagenic, carcinogenic (?)

Gastrointestinal → hyperemia, distension, → hemorrhage and ulceration

LD_{50} (po): 35 mg / kg bw mice

Adducts formed with cysteine possessed a markedly lower toxicity values than patulin itself.

The no-observed-effect level (NOEL) for patulin was considered to be 43 μg patulin / kg bw / day after a three times per week administration. In a guideline, the Joint Food and Agriculture Organization/ World Health Organization Committee on Food Additives (JECFA) lowered the provisional maximum tolerable daily intake (PMTDI) for patulin from a nominal 1 to 0.4 μg / kg bw / day, based on the calculated NOEL and use of a 100-fold safety factor. Based on the fact that the patulin concentrations in apple juices are usually below 50 μg / l, the JECFA estimated maximum intakes to be in the order of 0.2 and 0.1 μg patulin / kg bw / day for children and adults, respectively (WHO 1995).

Detection

HPLC (reverse-phase), TLC

Further Comments

Natural patulin contamination is primarily found in apples and apple products. Two facts are responsible. Besides the inactivation of patulin by distinct compounds in certain foods patulin producing molds represent only a low percentage of the total fungal strains isolated from most of the food (1% of the penicillia from flour and bread, 1.42% of the total fungi from European-style dry sausages, 0.9% of the total fungi isolated from corn meal). However, almost 70% (*P. expansum*) of the isolated fungi from naturally rotted apples produce patulin. Storage of fruits under a controlled atmosphere reduced patulin formation significantly but after evacuation *P. expansum*-infected fruits show a rapid increase in patulin levels. Diffusion of the toxin into the surrounding tissue has been reported for peaches, pears, and tomatoes but not for apples (up to 1 cm).

Patulin contamination is mainly seen as an indicator of bad manufacturing practices (use of rotten raw materials) although it seems to be only a minor threat to human and animal health.

Although patulin exhibits strong antibacterial activity it was too toxic for all test animals (carcinogenic, mutagenic) to have been used therapeutically.

Production: Patulin production of *Penicillium* spp. occurs in a temperature

range from 0 °C to 31 °C. The limiting a_w amounted to a_w 0.95 (*P. griseofulvum*) and in excess of a_w 0.95 for *P. expansum* in a synthetic medium. The pH optimum for patulin production is between pH 3-6.5 whereas the optimum temperature ranges from 20-25 °C (*P. expansum*). Irradiation (15 kGy) increased patulin formation of *P. griseofulvum*.
If the headspace O_2 levels in cans or jars of grape juice are below 0.5%, growth of *Byssochlamys* spp. is significantly reduced and no substantial patulin production can be expected.
Patulin production (up to 50 mg / kg) has been reported in soil under certain circumstances.
Reduction / elimination: An overall 24% decrease in patulin concentration has been observed in pressed apple juice during "down-line" technological production of concentrates. In addition, various chemicals like ascorbic acid, charcoal, sulfur dioxide, vitamin B_1 as well as irradiation are suitable for reducing or destroying patulin during "down-line" processing. Besides the inactivation of patulin by sulfhydryl compounds this mycotoxin is also unstable in the presence of alkali. Patulin is more stable at acidic pH whereas temperatures up to 80 °C do not cause a significant reduction.

Peaches may contain the following → mycotoxins:
→ aflatoxin B_1
incidence: 9/20*, conc. range: 5-15 µg / kg, country: Germany, *moldy
→ patulin
incidence: 2/4*, conc. range: 200-400 µg / kg, Ø conc.: 300 µg / kg, country: Germany, *stewed, moldy

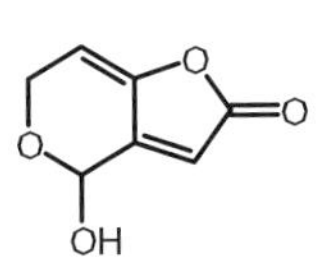

Patulin

incidence: 1/8, conc.: 6 µg / kg, country: Sweden
→ fruits

Peach kernels may contain the following → mycotoxins:
→ aflatoxins (AFB_1, AFB_2, AFG_1)
incidence: nc, conc.: ≤ 10 µg / kg, country: Germany

Peanut brittle may contain the following → mycotoxins:
→ aflatoxin B_1
incidence: 8/19, conc. range: 0.5-5 µg / kg (5 samples), 6-10 µg / kg (1 sa), 11-30 µg / kg (1 sa), 142 µg / kg (1 sa), country: UK
→ aflatoxins (AFB_1, AFB_2, AFG_1, AFG_2)
incidence: 8/19, conc. range: 6-10 µg / kg (6 samples), 31-100 µg / kg (1 sa), 190 µg / kg (1 sa), country: UK

Peanut butter The use of crushed → peanuts which are lower in grade than whole peanuts contributes to the aflatoxin contamination of peanut butter. High peak exposure to → aflatoxins is reduced by mixing and blending processes. However, average exposure to → aflatoxin B_1 which is 60% of the total aflatoxins in peanuts remains the same.
Highly effective automatic (electronic) peanut selectors are used in the peanut butter manufacturing process to remove poor-quality nuts. A sorting machine measuring near-infrared transmission spectra allows the dectection of molds in the inner part of shelled peanuts covered with inner skin.
Peanut butter may contain the following → mycotoxins:
aflatoxin B_1
incidence: 64/111, conc. range: < 5 µg / kg (36 samples), 10-662 µg / kg (28 sa), country: Germany
incidence: 44/182, Ø conc.: 46 µg / kg, country: Germany

incidence: 4/4, conc. range: 147-208 μg / kg, country: Germany
incidence: 1/1, conc.: 233 μg / kg, country: Germany
incidence: 2/2, conc. range: 3.5-5.2 μg / kg, ∅ conc.: 4.4 μg / kg, country: Germany
incidence: 3/4, conc. range: 0.6-1.4 μg / kg, ∅ conc.: 1.3 μg / kg, country: Japan
incidence: 3/6, conc. range: 0.6-2.4 μg / kg, country: Japan
incidence: 31/32, conc. range: < 10 μg / kg, country: UK
incidence: 10/63, conc. range: 2-20 μg / kg, ∅ conc.: 7 μg / kg, country: USA
→ aflatoxin B_2
incidence: 2/2, conc. range: 0.5-0.6 μg / kg, ∅ conc.: 0.55 μg / kg, country: Germany
incidence: 3/4, conc. range: 0.1-0.3, ∅ conc.: 0.2 μg / kg, country: Japan
incidence: 3/6, conc. range: 0.4 μg / kg, country: Japan
→ aflatoxin G_1
incidence: 2/2, conc. range: 3.5-5.2 μg / kg, ∅ conc.: 4.4 μg / kg, country: Germany
incidence: 2/4, conc. range: 0.3 μg / kg, ∅ conc.: 0.3 μg / kg, country: Japan
incidence: 3/6, conc. range: 0.1-0.4 μg / kg, country: Japan
→ aflatoxin G_2
incidence: 2/2, conc. range: 1.3-1.7 μg / kg, ∅ conc.: 1.5 μg / kg, country: Germany
aflatoxins (AFB_1, AFB_2, AFG_1, AFG_2)
incidence: 64/111, conc. range: < 5 μg / kg (36 samples), 7-362 μg / kg (28 sa), country: Germany
incidence: 44/182, ∅ conc.: 59 μg / kg, country: Germany
incidence: 1/1, conc.: 278 μg / kg, country: Germany
incidence: 98/2092, conc. range: 5-19.9 μg / kg (95 samples), > 25 μg / kg (3 sa), country: Canada
aflatoxins (no specification)
incidence: 25/2477, conc. range: 15-30 μg / kg (18 samples), 31-60 μg / kg (3 sa), 61-90 μg / kg (2 sa), 90 μg / kg (2 sa), country: Canada
incidence: 29/29, conc. range: 30-8600 μg / kg, country: Philippines
incidence: 5/522, conc. range: ≤ 6600 μg / kg, ∅ conc.: 186 μg / kg, country: Philippines
incidence: 17/104, conc. range: ≤ 27 μg / kg, ∅ conc.: 14 μg / kg, country: USA
incidence: 1/3*, conc.: 43 μg / kg, country: USA, *imported

Peanut butter (crunchy) may contain the following → mycotoxins:
→ aflatoxin B_1
incidence: 5/16*, conc. range: 2-5 μg / kg (3 samples), 6-10 μg / kg (1 sa), 12 μg / kg (1 sa), country: UK, *regular
incidence: 10/14**, conc. range: 2-5 μg / kg (2 samples), 6-10 μg / kg (1 sa), 11-30 μg / kg (1 sa), 31-100 μg / kg (1 sa), > 100 ≤ 318 μg / kg (5 sa), country: UK
incidence: 7/9**, conc. range: 2-5 μg / kg (5 samples), 6-10 μg / kg (1 sa), 58 μg / kg (1 sa), country: UK
incidence: 7/15**, conc. range: 6-10 μg / kg (1 sa), 11-30 μg / kg (3 sa), 31 ≤ 73 μg / kg (3 sa), country: UK
**health food
→ aflatoxins (AFB_1, AFB_2, AFG_1, AFG_2)
incidence: 47/59, conc. range: 1-50 μg / kg (32 samples), 51-100 μg / kg (4 sa), > 100 μg / kg (11 sa), country: UK
incidence: 6/16*, conc. range: 2-5 μg / kg (4 samples), 6-10 μg / kg (1 sa), 14 μg / kg (1 sa), country: UK, *regular
incidence: 10/14**, conc. range: 2-5 μg / kg (1 sample), 6-10 μg / kg (2 sa), 11-30 μg / kg (1 sa), > 100 ≤ 345 μg / kg (5 sa), country: UK
incidence: 7/9**, conc. range: 2-5 μg / kg (3 samples), 6-10 μg / kg (2 sa), 11-30 μg / kg (1 sa), 211 μg / kg (1 sa), country: UK

incidence: 7/15**, conc. range: 6-10 μg / kg (1 sample), 11-30 μg / kg (1 sa), 31-100 μg / kg (2 sa), > 100 ≤ 147 μg / kg (3 sa), country: UK
**health food

Peanut butter (smooth) may contain the following → mycotoxins:
→ aflatoxin B_1
incidence: 7/16*, conc. range: 2-5 μg / kg (6 samples), 7 μg / kg (1 sa), country: UK, *regular
incidence: 4/11**, conc. range: 6-10 μg / kg (2 samples), 31 ≤ 49 μg / kg (2 sa), country: UK
incidence: 5/6**, conc. range: 11-30 μg / kg (3 samples), 31 ≤ 76 μg / kg (2 sa), country: UK
incidence: 1/4**, conc: 13 μg / kg, country: UK
**health food
→ aflatoxins (AFB_1, AFB_2, AFG_1, AFG_2)
incidence: 31/33, conc. range: 1-50 μg / kg (25 samples), 51-100 μg / kg (4 sa), > 100 μg / kg (2 sa), country: UK
incidence: 7/16*, conc. range: 2-5 μg / kg (6 samples), 8 μg / kg (1 sa), country: UK, *regular
incidence: 6/11**, conc. range: 2-5 μg / kg (1 sample), 6-10 μg / kg (2 sa), 11-30 μg / kg (1 sa), 31 ≤ 85 μg / kg (2 sa), country: UK
incidence: 6/6**, conc. range: 6-10 μg / kg (1 sample), 11-30 μg / kg (1 sa), 31-100 μg / kg (3 sa), 175 μg / kg (1 sa), country: UK
incidence: 1/4**, conc.: 27 μg / kg, country: UK
**health food

Peanut candy may contain the following → mycotoxins:
→ aflatoxins
incidence: 10/18, conc. range: ≤ 20 μg / kg, ∅ conc: 10 μg / kg, country: USA

Peanut mix may contain the following → mycotoxins:
→ aflatoxins
incidence: 1/1, conc.: 302 μg / kg, country: USA

Peanut oil In general, peanut → oil does not contain any significant amounts of → aflatoxins. These → mycotoxins are removed during processing due to the use of solvents or they are destroyed by conventional alkali washing as a part of the refining process. Subsequent bleaching operation further contributes to the elimination of the aflatoxins.
Low aflatoxin amounts have been found in crude oils which are not suitable for human consumption. They are obtained by solvent extraction or by hydraulic pressing of ground moldy peanuts. The corresponding meals contained the major portion of the aflatoxins.
Peanut oil may contain the following → mycotoxins:
→ aflatoxin B_1
incidence: 4/6, conc. range: ≤ 0.7 μg / kg, country: India
→ aflatoxin B_2
incidence: 4/6, conc. range: ≤ 0.1 μg / kg, country: Japan
→ aflatoxin G_1
incidence: 4/6, conc. range: ≤ 0.1 μg / kg, country: Japan
aflatoxin (no specification)
incidence: 5/8*, conc. range: ≤ 310 μg / kg, ∅ conc.: 246 μg / kg, country: USA, *crude peanut oil
aflatoxins (no specification)
incidence: 544/1209, conc. range: 71-5000 μg / kg, country: India
→ coconut oil, → oil, → olive oil, → peanuts, → sunflower seed oil

Peanut paste may contain the following → mycotoxins:

→ aflatoxins
incidence: 3/4*, conc. range: 11 μg / kg, ∅ conc.: 9 μg / kg, country: USA, *imported
→ peanuts

Peanut products (no specification)
may contain the following → mycotoxins:
→ aflatoxin B_1
incidence: 13/20*, conc. range: 15-138 μg / kg, ∅ conc.: 64.9 μg / kg, country: Germany
→ aflatoxin B_2
incidence: 2/20*, conc. range: 3-24 μg / kg, ∅ conc.: 13.5 μg / kg, country: Germany
→ aflatoxin G_1
incidence: 8/20*, conc. range: 9-44 μg / kg, ∅ conc.: 28 μg / kg, country: Germany
→ aflatoxin G_2
incidence: 2/20*, conc. range: 4-18 μg / kg, ∅ conc.: 11 μg / kg, country: Germany, *suspected
aflatoxin (no specification)
incidence: 1/6, conc.: 2 μg / kg, country: USA
→ aflatoxins (no specification)
incidence: 11/32, conc. range: > 30-≤ 220 μg / kg, country: Philippines
→ peanuts

Peanut sauce may contain the following → mycotoxins:
→ aflatoxins (AFB_1, AFB_2, AFG_1, AFG_2)
incidence: 18/20, conc. range: 18-943 μg / kg, ∅ conc.: 162 μg / kg, country: Gambia
→ peanuts

Peanuts (no specification)
From all types of → nuts peanuts which grow in the soil are most susceptible to mycotoxin (→ aflatoxins) contamination. Contamination mainly occurs in the field during the harvest while the nuts are being dried. When harvesting is associated with wet weather conditions a higher contamination rate was established.
Pre-harvesting, harvesting, handling and storage conditions greatly influence the degree of aflatoxin contamination. A soil temperature of between 25.7-27 °C and drought stress during the latter part of the growing season should contribute to aflatoxin contamination in the field before harvest. Although the surface of the pods easily comes into contact with the soil borne → Aspergillus flavus Link, pods of intact kernels are difficult to penetrate by the fungus. However, damaged kernels, especially mature ones which contain 30-60% water at the time of harvest, are very susceptible to *A. flavus* infection. Damages are due to various biotic and abiotic factors: insects (e.g. termites), fungi (e.g. *Macrophomina phaseoli*, *Sclerotium rolfsii*), nematodes (e.g. *Meloidogyne arenaria*), very rapid growth of the peanuts, over-advanced maturity and direct mechanical damage. In addition, pods harvested during the rains showed a high infestation rate while pods harvested during the dry season showed only little infection.
Growth of *A. flavus* in infected peanuts immediately starts after lifting. The optimum moisture content for fungal growth in peanuts is between 10(15)-(25)30% but growth occurs in the range from 9-35%. The minimum → a_w for aflatoxin production in immature broken peanuts is a_w 0.83.
Contamination has been observed before digging, after digging and before combining, between combining and drying as well as in storage. The avoidance of pre-harvest stress in combination with effective drying techniques (moisture content < 9-10%) and storage conditions (e.g. 32 °C / 50% relative humidity, adequate ventilation) immediately after harvest lower or even prevent aflatoxin contamination. During improper storage the total amount of the produced aflatoxins and the ratio of different aflatoxin types is influenced by the temperature. The ratio AFB_1 :

AFG_1 is smaller at higher temperatures (35 °C) than at lower temperatures (20 °C). Approximately 60% of the total aflatoxins found in peanuts is aflatoxin B_1. Stored in-shell peanuts having a moisture content > 11% in combination with a relative humidity of at least 84% allow the development of aflatoxin producing fungi.
However, according to the FDA aflatoxin contamination mainly occurs prior to harvest of the peanuts whereas very high kernel moistures may prevent aflatoxin production.
In Brazil the protein-enriched peanut meal and husks as by-products of peanut oil processing are fed to animals. Mycotoxin contamination of these products is harmful and may result in the contamination of → meat and → milk (→ carry over).
Wrinkled kernels seem to contain higher aflatoxin concentrations (up to 70 times) than the dark kernels while most of sound mature peanuts do not contain aflatoxin.
There are different ways for decontaminating contaminated peanuts. Roasting reduces (50-70%) but does not eliminate aflatoxin contamination whereas boiling and baking are less effective (20-30%). Microwave and oven roasting caused destructions of ≈ 55% AFB_1 and ≈ 36% AFG_1.
Peanuts may contain the following → mycotoxins:
→ aflatoxin B_1
incidence: 1/1, conc.: 625 µg / kg, country: Angola
incidence: 1/88, conc.: 5 µg / kg, country: China
incidence: 3/40, conc. range: 98-1056 µg / kg, country: Egypt
incidence: 2/6, conc. range: 3.6-5.4 µg / kg, country: Egypt
incidence: 9/104*, conc. range: ≤ 1-954 µg / kg, country: Finland, *imported
incidence: 7/8, conc. range: 22-2222 µg / kg, ∅ conc.: 682 µg / kg, country: Finland
incidence: 3/6, conc. range: 15-240 µg / kg, country: Gambia
incidence: 42/1038, ∅ conc.: 97 µg / kg, country: Germany
incidence: 46/62*, conc. range: 20-28,000 µg / kg, country: Germany, *moldy
incidence: 39/40, conc. range: 1.3-1600 µg / kg, country: Germany
incidence: 9/19, conc. range: 5-15 µg / kg, country: India
incidence: 926*/2062, conc. range: ≤ 833 µg / kg, contry: India, *exceeded 5 µg / kg
incidence: 1/2, conc.: 5 µg / kg, country: Nigeria
incidence: 7/40*, conc. range: tr-400 µg / kg, country: Norway, *imported
incidence: 48/1962, conc. range: 5-200 µg / kg, country: South Africa
incidence: 10/553, conc. range: 5-20 µg / kg, country: South Africa
incidence: 1/259, conc.: 20 µg / kg, country: South Africa
incidence: 97/157, conc. range: 5-3000 µg / kg, country: Sudan
incidence: 1/20, conc.: 40 µg / kg, country: Sudan
incidence: 2/14, conc. range: 5-20 µg / kg, country: Sudan
incidence: 106*/216, ∅ conc.: 1530 µg / kg, country: Thailand, *total: ∅ conc.: 12,256 µg / kg AFB_1, AFB_2, AFG_1, AFG_2
incidence: 4/65, conc. range: 6-46 µg / kg, country: Tunisia
incidence: 59/605, conc. range: 5-625 µg / kg, country: USA
incidence: 2/56, conc. range: 10-125 µg / kg, country: USA
incidence: 10/63, conc. range: ≤ 5 - > 5 µg / kg, country: USA
→ aflatoxin B_2
incidence: 1/1, conc.: 180 µg / kg, country: Angola

incidence: 2/6, conc. range: 1.8-2.6 µg / kg, country: Egypt
incidence: 9/104*, conc. range: ≤ 1-568 µg / kg, country: Finland, *imported
incidence: 3/8, conc. range: 167-1111 µg / kg, ∅ conc.: 482 µg / kg, country: Finland
incidence: 39/40, conc. range: 1.5-744 µg / kg, country: Germany
incidence: 7/40*, conc. range: 2-50 µg / kg, country: Norway, *imported
→ aflatoxin G_1
incidence: 1/1, conc.: 315 µg / kg, country: Angola
incidence: 2/109*, conc. range: 3-136 µg / kg, country: Finland, *imported
incidence: 4/8, conc. range: 333-556 µg / kg, ∅ conc.: 500 µg / kg, country: Finland
incidence: 39/40, conc. range: 1-1540 µg / kg, country: Germany
incidence: 7/40*, conc. range: tr-350 µg / kg, country: Norway, *imported
incidence: 4/65, conc. range: ≤ 0.38 µg / kg, country: Tunisia
→ aflatoxin G_2
incidence: 1/1, conc.: 40 µg / kg, country: Angola
incidence: 1/109*, conc.: 34 µg / kg, country: Finland, *imported
incidence: 2/8, conc.: 167 µg / kg, ∅ conc.: 167 µg / kg, country: Finland
incidence: 39/40, conc. range: 1-548 µg / kg, country: Germany
incidence: 7/40*, conc. range: tr-30 µg / kg, country: Norway, *imported
→ aflatoxins (no specification)
incidence: 2/2, conc. range: 31-50 µg / kg, country: Brazil
incidence: 284*/1679, conc. range: > 5-24.9 µg / kg (186 samples), > 25 µg / kg (98 sa), country: Canada, *AFB_1, AFB_2, AFG_1, AFG_2
incidence: 1/2, conc. range: 51-100 µg / kg, country: Egypt
incidence: 5/5, conc. range: 1-440 µg / kg, country: Gambia
incidence: 42/1038, ∅ conc.: 141 µg / kg, country: Germany
incidence: 505/8081*, conc. range: nc, country: Germany, *peanuts and peanut products
incidence: 17/35, conc. range: 1-410 µg / kg, country: India
incidence: 93/160, conc. range: tr-5850 µg / kg, country: India
incidence: 20/20*, conc. range: 126-1603 µg / kg, country: Indonesia, *from local farmers
incidence: 80/80*, conc. range: 81-14,565 µg / kg, country: Indonesia, *from the market
incidence: 26/53, conc. range: 1-300 µg / kg, country: Malawi
incidence: 5/67, conc. range: nc, country: Mocambique
incidence: 5/71, conc.: > 30- ≤ 100 µg / kg, country: Philippines
incidence: 27*/152, conc. range: 1-100 µg / kg (11 samples), 100-1000 µg / kg (8 sa), > 1000 µg / kg (8 sa), country: Uganda, *24 samples contained AFB_1, 16 AFB_2, 17 AFG_1, 7 AFG_2
incidence: 13/56, conc. range: 1-200 µg / kg, country: USA
incidence: 50/50, conc. range: 3-22,000 µg / kg, ∅ conc.: 1685 µg / kg, country: USA
→ citrinin
incidence: 16/160, conc. range: tr-1200 µg / kg, country: India
→ cyclopiazonic acid
incidence: 1/6, conc. range: traces, country: USA
incidence: 45/50, conc. range: < 50-2900 µg / kg, ∅ conc.: 460 µg / kg, country: USA
incidence: 21/27* conc. range: 32-6525 µg / kg, country: USA, *loose-shell kernel fractions
incidence: 4/21* conc. range: 32-130 µg / kg, country: USA, *sound mature kernel fractions

→ ochratoxin A
incidence: 1/1*, conc.: 4900 μg / kg, country: Canada, *visible moldy
→ nuts

Peanuts (boiled) may contain the following → mycotoxins:
aflatoxin (no specification) (→ aflatoxins)
incidence: 8/8, conc. range: ≤ 103 μg / kg, ∅ conc.: 24 μg / kg, country: Philippines

Peanuts (chocolate-coated) may contain the following → mycotoxins:
→ aflatoxin B_1
incidence: 3/17, conc. range: 0.5 ≤ 3 μg / kg, country: UK
→ aflatoxins (AFB_1, AFB_2, AFG_1, AFG_2)
incidence: 3/17, conc. range: 0.5-5 μg / kg, country: UK

Peanuts (dry roasted) may contain the following → mycotoxins:
→ aflatoxin B_1
incidence: 3/14, conc. range: 0.5 - ≤ 5 μg / kg, country: UK
→ aflatoxins (AFB_1, AFB_2, AFG_1, AFG_2)
incidence: 3/14, conc. range: 0.5 - ≤ 5 μg / kg, country: UK

Peanuts (fresh, raw) may contain the following → mycotoxins:
aflatoxin (no specification) (→ aflatoxins)
incidence: 110/169, conc. range: ≤ 885 μg / kg, ∅ conc.: 58 μg / kg, country: Philippines

Peanuts (in-shell) may contain the following → mycotoxins:
→ aflatoxins (no specification)
incidence: 15/26, conc. range: 0.5-10 μg / kg (10 samples), 11-50 μg / kg (2 sa), > 50 μg / kg (3 sa), country: UK

Peanuts (in-shell, raw) may contain the following → mycotoxins:
→ aflatoxin B_1
incidence: 12/12, conc. range: 05-5 μg / kg (9 samples), 11-30 μg / kg (2 sa), 2520 μg / kg (1 sa), country: UK
→ aflatoxins
incidence: 13*/24, conc. range: 0.5-5 μg / kg (8 samples), 6-30 μg / kg (2 sa), 31-100 μg / kg (2 sa), 4920 μg / kg (1 sa), country: UK, *AFB_1, AFB_2, AFG_1, AFG_2
incidence: 1/4, conc.: 273 μg / kg, country: USA

Peanuts (in-shell, roasted) may contain the following → mycotoxins:
→ aflatoxin B_1
incidence: 5/13, conc. range: 0.5-5 μg / kg (4 samples), 9 μg / kg (1 sa), country: UK
→ aflatoxins (AFB_1, AFB_2, AFG_1, AFG_2)
incidence: 5/10, conc. range: 0.5-10 μg / kg (4 samples), 11-50 μg / kg (1 sa), country: UK
incidence: 7/13, conc. range: 0.5-5 μg / kg (3 samples), 6-10 μg / kg (2 sa), 11-28 μg / kg (2 sa), country: UK

Peanuts (processed) may contain the following → mycotoxins:
→ aflatoxins (AFB_1, AFB_2, AFG_1, AFG_2)
incidence: 7/150, conc. range: 5-14.9 μg / kg (5 samples), > 25 μg / kg (2 sa), country: Canada

Peanuts (roasted) Since the roasting process destroys → aflatoxins in contaminated peanuts to a varying degree some manufactures might use low-quality raw materials (see also sliced and crushed → peanuts). The use of aflatoxin contaminated peanut oil in roasting further contributes to the contamination of the → nuts (absorption). In Finland 6.6% and in Sweden 37% of imported roasted peanuts have been found to contain aflatoxins.
Roasted peanuts may contain the following → mycotoxins:

→ aflatoxin B_1
incidence: 3/17, conc. range: 0.5-5 μg / kg (2 samples), 6 μg / kg (1 sa), country: UK
→ aflatoxins (AFB_1, AFB_2, AFG_1, AFG_2)
incidence: 4/17, conc. range: 0.5-5 μg / kg (3 samples), 7 μg / kg (1 sa), country: UK

Peanuts (shelled) may contain the following → mycotoxins:
→ aflatoxins (AFB_1, AFB_2, AFG_1, AFG_2)
incidence: 3/8, conc. range: 0.5-10 μg / kg, country: UK

Peanuts (shelled, raw) may contain the following → mycotoxins:
→ aflatoxin B_1
incidence: 1/2, conc.: 2.7 μg / kg, country: Syria
incidence: 5/8, conc. range: 0.5-5 μg / kg (4 sa), 88 μg / kg (1 sa), country: UK
→ aflatoxins (AFB_1, AFB_2, AFG_1, AFG_2)
incidence: 5/8, conc. range: 0.5-5 μg / kg (4 sa), 182 μg / kg (1 sa), country: UK

Peanuts (shelled, roasted) may contain the following → mycotoxins:
→ aflatoxin B_1
incidence: 8/121*, conc. range: 3-716 μg / kg, ∅ conc.: 160 μg / kg, country: Finland, *imported
incidence: 1/26, conc.: 0.1 μg / kg, country: Japan
incidence: 2/3, conc. range: 0.4-1.9 μg / kg, country: Syria
incidence: 3/14, conc. range: 0.5-10 μg / kg, country: UK
→ aflatoxin B_2
incidence: 8/121*, conc. range: 1-89 μg / kg, ∅ conc.: 21.6 μg / kg, country: Finland, *imported
incidence: 2/3, conc. range: 0.3-0.6 μg / kg, country: Syria
→ aflatoxin G_1
incidence: 2/108*, conc. range: 12-20 μg / kg, ∅ conc.: 16 μg / kg, country: Finland, *imported
→ aflatoxins (no specification)
incidence: 6/55, conc. range: ≤ 329 μg / kg, ∅ conc.: 68 μg / kg, country: USA
incidence: 1/1, conc.: 4 μg / kg, country: USA

Peanuts (shelled, roasted, salted) may contain the following → mycotoxins:
→ aflatoxins
incidence: 4*/17, conc. range: 0.5-10 μg / kg, country: UK, *AFB_1, AFB_2, AFG_1, AFG_2
incidence: 5/3, conc. range: nc, country: UK

Peanuts (shelled, roasted, unsalted) may contain the following → mycotoxins:
→ aflatoxins (no specification)
incidence: 5/12, conc. range: nc, country: UK

Peanuts (sliced) may contain the following → mycotoxins:
→ aflatoxin B_1
incidence: 144/718, conc. range: 5-665 μg / kg, ∅ conc. 134 μg / kg, country: Germany

Peanuts (sliced and crushed): In a Finnish study sliced and crushed peanut samples were frequently contaminated with → aflatoxins. It was suggested that a more even distribution of aflatoxin in these lots and / or the use of low-quality material in the preparation of the corresponding lots are responsible.
Sliced and crushed peanuts may contain the following → mycotoxins:
→ aflatoxin B_1
incidence: 20/68*, conc. range: ≤ 1-716 μg / kg, country: Finland, *imported
→ aflatoxin B_2
incidence: 18/68*, conc. range: ≤ 1-76 μg / kg, country: Finland, *imported
→ aflatoxin G_1
incidence: 6/68*, conc. range: ≤ 1-91 μg / kg, country: Finland, *imported

→ aflatoxin G_2
incidence: 4/68*, conc. range: ≤ 1-14 µg / kg, country: Finland, *imported

Pear juice may contain the following → mycotoxins:
→ patulin
incidence: 1/4, conc.: 24 µg / kg, country: Germany

Pears may contain the following → mycotoxins:
→ patulin
incidence: 8/24, conc. range: 0.9-10 µg / kg, country: Spain
→ apples

Peas may contain the following → mycotoxins:
→ aflatoxin B_1
incidence: 1/35, conc.: 25 µg / kg, country: Tunisia
→ aflatoxin G_1
incidence: 1/35, conc.: 42 µg / kg, country: Tunisia
→ aflatoxins
incidence: 3*/19, conc. range: 1-100 µg / kg, country: Uganda
* 2 samples contained AFB_1, 1 AFB_2, 1 AFG_1
aflatoxin (no specification)
incidence: 3/8, conc. range: ≤ 40 µg / kg, ∅ conc.: 13 µg / kg, country: Philippines
→ ochratoxin A
incidence: 2/72, conc.: 10 µg / kg, country: Sweden
incidence: 6/71, conc. range: 10-442 µg / kg, country: Sweden
→ beans, → cabbage, → cowpeas → lentils, → pigeon peas, → soybeans, → vegetables

Pecans Since → aflatoxins have been detected in damaged as well as in non-visibly damaged kernels the major cause of contamination is not clear. Nevertheless, weevil-damaged and late-harvested → nuts (shell integrity) may be more susceptible to mold invasion. The prevailing orchard temperatures during the latter part of the harvest season greatly influence the degree of contamination. In addition, nuts falling to the ground in pastures, especially on wet soil, are more likely to mold than those falling in non-pasture orchards.
Besides other mycotoxins → alternariol and → alternariol methyl ether have been detected in pecans. These → mycotoxins only occurred in discolored kernels which were removed from shelled pecans during processing. They would probably be rejected by consumers of in-shell pecans. In addition, per capita consumption of pecans is very low which further reduces the risk of intake of → Alternaria mycotoxins.
Pecans may contain the following → mycotoxins:
→ aflatoxin B_1
incidence: 3/48, conc. range: tr-25 µg / kg, country: USA
→ aflatoxin G_2
incidence: 3/48, conc. range: traces, country: USA
→ aflatoxins
incidence: 1*/55, conc. range: 5-9.9 µg / kg, country: Canada, *AFB_1, AFB_2, AFG_1, AFG_2
incidence: 39/575, conc. range: ≤ 172 µg / kg, ∅ conc.: 86 µg / kg, country: USA
incidence: 1/229, conc.: 40 µg / kg, country: USA
incidence: 3/17, conc. range: ≤ 334 µg / kg, ∅ conc.: 135 µg / kg, country: USA
→ alternariol*
incidence: nc/50, conc. range: nc, country: USA
→ alternariol methyl ether*
incidence: nc/50, conc. range: nc, country: USA
*in discolored pecans ("pickouts")
→ citreoviridin
incidence: 1/1*, conc.: nc, country: USA, *moldy fragments

→ sterigmatocystin
incidence: 1/20, conc.: 20,000 µg / kg, country: USA
→ nuts

Pellagra This human disease is characterized by the insufficient intake / failure to absorb the B complex vitamin niacin or its amide. People consuming deteriorated → maize as a staple food are most frequently affected. Maize contains only low levels of niacin in an available form and the concentration of certain niacin precursors is also low. These compounds are essential for the activity of certain enzymes which are involved in detoxification processes of the → mycotoxins ingested via contaminated maize. It is suggested, that the effects of this malnutrition are enhanced by certain mycotoxins such as → trichothecenes, → fumonisins, → koji acid and → zearalenone. Pellagra is more common in spring time and it is concluded that storing maize under cool humid conditions in winter promotes trichothecene (especially → T-2 toxin) production and contamination. However, up to now the real cause of Pellagra remains unresolved.

Penicidin (Syn.: → patulin)

Penicillic acid This 3-methoxy-5-methyl-4-oxo-2,5-hexadienoic acid or 2-keto-β-methoxy-δ-methylene-Δα-hexenoic acid (→ mycotoxins) was one of the first metabolites isolated (→ Penicillium *puberulum*, 1913). It was recognized as a toxic fungal metabolite possibly as early as 1896 (see Figure Penicillic acid).

Chemical Data
Empirical formula $C_8H_{10}O_4$, molecular weight: 170

Fungal Sources
important producers: → Penicillium aurantiogriseum Dierckx and varieties, → Penicillium roquefortii Thom chemotype II (only a few isolates), *P. janczewskii*, → Eupenicillium spp., → Petromyces alliaceus Malloch & Cain, → Aspergillus alutaceus var. alutaceus Berkely & Curtis, → Aspergillus quercinus (Bain.) Thom & Church, *A. sclerotiorum*.

Natural Occurrence
→ apples, → barley, → beans, → cereals, → cheese, → cheese, Blue, → cheese, goat, → cheese, Swiss, → maize, → rye
Certain strains of *P. roquefortii* which have been used in the cheese industry produced penicillic acid. Penicillic acid is not stable in foods containing reactive amino acids.

Toxicity
nephrotoxic, → mutangenic, carcinogenic,
LD_{50} (po) : 35-600 mg / kg bw mice
A potentiated effect in the nephrotoxic action of penicillic acid and → ochratoxin A was observed. Furthermore, a synergistic effect between → patulin and penicillic acid is evident.
Although the adducts of penicillic acid with cysteine or glutathione should be biologically inactive they retained some toxicity to the chick embryo.

Detection
GC, TLC

Further Comments
Stability: The inactivation of penicillic acid by SH-compounds is due to a reaction with the isolated rather than the conjugated double bond. In aqueous solution the reaction product formed with cysteine derived from the open-chain form of penicillic acid or from the lactone form.
Compared to the aflatoxins, the penicillic acid like → patulin seems to be less stable in certain foods. Thiol compounds should be responsible for the instability in orange juice and → flour but stability was observed in → grape juice and → apple juice. No penicillic acid could be

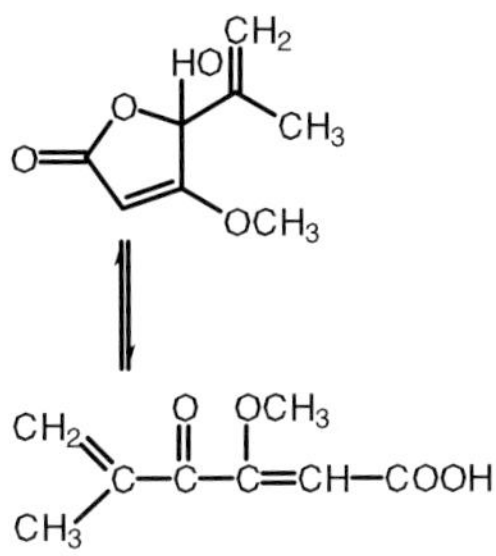

Penicillic acid

detected in bologna 48 h, storage temperature 5 °C. Only low levels (< 10%) of this mycotoxin persisted in Swiss cheese after one week at 5 °C. A rapid loss of penicillic acid has been reported during the grinding of maize. The formation of penicillic acid during the aging of meats (→ meat) will result in non-toxic products since this mycotoxin reacts with amino acids.
Penicillic acid is not stable in stored → wheat flour. After 2 weeks only 10% remained at 22 °C. Within 3 days a complete decomposition was observed after treament with 2% aqueous ammonia.

Penicillium anamorphic → Trichocomaceae, teleomorphs → Eupenicillium, *Talaromyces*
This ubiquitous distributed genus is more common in temperate climatic regions of the world such as Canada and northern Europe. Although generally accepted as storage fungi some species occur on cereal → grains before harvest. *Penicillium* spp. are usually the dominant organisms of the blue and green molds associated with the spoilage of → foods, especially → fruits (citrusfruits) and → vegetables. Cool storage (-2 to 5 °C) of damp grain (→ a_w > 0.90) causes the "blue eye" disease.
85 *Penicillium* species are known to be toxigenic but most → mycotoxins in this genus are produced by a small, well defined range of species. Within each species a wide range of mycotoxins with an extreme diverse molecular composition is synthesized. At least 27 *Penicillium* metabolites are known to be toxic to man and animals which are produced by 32 species. Although the toxicity of these mycotoxins is also very diverse, most toxins either affect liver and kidney function or they are neurotoxins.
A water activity of a_w 0.80-0.82 is sufficient for the growth of *P. aurantiogriseum* and *P. verrucosum* whereas a_w levels between 0.86-0.89 are necessary for mycotoxin production. Important mycotoxin producers are e.g. → Penicillium aurantiogriseum Dierckx, → Penicillium citroenigrum Dierckx, → Penicillium expansum Link, → Penicillium islandicum Sopp, → Penicillium roquefortii Thom, → Penicillium verrucosum Dierckx. Important mycotoxins are e.g. → citrinin, → ochratoxin A, → penicillic acid, and → xanthomegnin.
In general *Penicillium* spp. are capable of producing mycotoxins at lower temperatures than are → Aspergillus spp.

Penicillium aurantiogriseum Dierckx (Syn.: *P. cyclopium*) is of ubiquitous distribution and found on different kinds of food such as → bread, → cereals, → cheese, → coffee beans, → grains, frozen → meat, → nuts, → sausages, → shrimps. *P. aurantiogriseum* is the most important member of all Penicillia in stored → cereals. There is a broad temperature range for → ochratoxin A? and → penicillic acid production (4-31 °C). The minimum a_w for penicillic acid production is a_w 0.97-0.99.
P. aurantiogriseum may produce → mycotoxins such as penicillic acid, penitrem A (→ penitrems), terrestric acids, verrucosidin, → viomellein, viridicatins, xanthomegnins (→ xanthomegnin).

Penicillium camembertii Thom (Syn.: e.g. *P. candidum*, *P. caseicola*) is a white growing mold used for the manufacturing of Camembert cheese (→ cheese, Camembert). Surface growth of this mold prevents (i) undesirable fungal infections and causes (ii) proteolytic degradation of casein and (iii) hydrolysis of triglycerides. However, it seems that *P. camembertii* is a consistent producer of → cyclopiazonic acid whereas the minimum temperature for production is 4 °C.

Penicillium chrysogenum Thom (Syn.: *P. notatum*) is a penicillin producer and common on different types of food such as → almonds, → bread, → cheese, → fish, → flour, → ham, → meat, → nuts, → sausages. In some countries (e.g. Canada) it is frequently isolated from → cereals. → Roquefortine C might occur naturally in cereals infected with *P. chrysogenum*. → Ochratoxin A production of this fungus could not be confirmed (see Figure *Penicillium chrysogenum* Thom).
P. chrysogenum may produce → mycotoxins such as → PR toxin, roquefortine C and D, xanthocillins.

Penicillium citreonigrum Dierckx (Syn.: *Penicillium citreo-viride*) although widely distributed is not a commonly isolated species. → Rice seems to be the best substrate whereas growth starts soon after the → grains become wet under improper storage conditions. The lower temperatures and shorter hours of daylight in the more temperate rice-growing areas favor its growth. At a moisture content of 14.6% *P. citreonigrum* starts growing in stored rice. It is overgrown by other fungi if the moisture content reaches 15.6%. This fungus is involved in the → Yellow rice disease / → acute cardiac beriberi.
P. citreonigrum may produce → mycotoxins such as → citreoviridin.

Penicillium citrinum Thom as an ubiquitous fungus is a contaminant of nearly every kind of foodstuff but is found predominately on subtropical and tropical → cereals. Besides cereal → grains and → flour (the most common sources) this mold has been isolated from e.g. → almonds, → bread, → cheese, → coffee beans, → fish, → fruit juices, → meat, → nuts, → spices (see Figure *Penicillium citrinum* Thom). *P. citrinum* is a consistent producer of → citrinin although in the presence of → Aspergillus niger and / or *Trichoderma viride* toxin production is inhibited.
P. citrinum may produce → mycotoxins such as citrinins.

Penicillium commune Thom may produce the following → mycotoxins: cyclopaldic acid, → cyclopiazonic acid, → roquefortine A & B (*P. commune* chemotype II), rugulovasines.

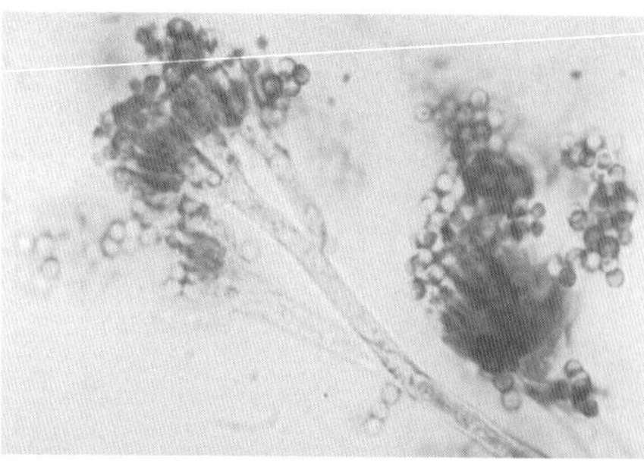

Penicillium chrysogenum Thom

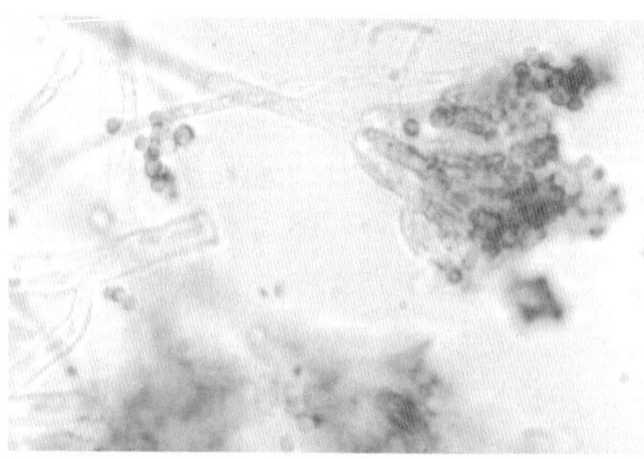

Penicillium citrinum Thom

Penicillium crustosum Thom is an ubiquitous spoilage fungus, preferring products with a high content of lipids and proteins. → Cereals are less often affected. Occurrence has also been reported for → almonds, → cheese, → flour, → fruit juices, → maize (high lipid content), → meat (processed), and → nuts. Furthermore, it is a weak pathogen on pomaceous → fruits and cucurbits. Nearly all isolates produce the tremorgenic penitrem A and therefore *P. crustosum* is by far the most important source of this mycotoxin.
P. crustosum may produce → mycotoxins such as penitrem A (→ penitrems), → roquefortine A, B, C, terrestric acid, viridicatin, → xanthomegnin.

Penicillium expansum Link is a common storage mold in → apples and → pears. Strains of *P. expansum* tolerate low oxygen levels as well as high CO_2 tensions (see Figure *Penicillium expansum* Link). Since *P. expansum* is the most important → patulin producer, infection is usually associated with patulin contamination of the → fruits (Golden Delicious: 2-100 μg / g). Conventional CO_2 and O_2 tensions in CA storage inhibit the growth of this fungus. A minimum → a_w of 0.99 is needed for patulin production (temperature 0-24 °C).
P. expansum may produce → mycotoxins such as chaetoglobosins, → citrinin, → patulin, → roquefortine C.

Penicillium griseofulvum Dierckx as a ubiquitous species could be isolated from different kinds of → foods such as → cereals and → meat. The minimum → a_w that allows → patulin production is ≈ a_w 0.94 (temperature 30 °C). Temperatures that enabled patulin production were in the range of 4-31 °C (see Figure *Penicillium griseofulvum* Dierckx).
P. griseofulvum may produce → mycotoxins such as → cyclopiazonic acid, griseofulvins, → patulin, → roquefortine C.

Possible Mycotoxicosis
Patulin (malt) and cyclopiazonic acid (cereals) are involved in mycotoxicoxsis.

Penicillium islandicum Sopp is widely distributed but occurs infrequently in nature. Classified as a storage fungus or a member of the mycoflora of soils, this species is uncommon at least in the temperate zones. As a more or less frequent contaminant of → rice, *P. islandicum* is involved in the → yellow rice disease (see Figure *Penicillium islandicum* Sopp). It represents an important problem for rice consuming peoples in most Asiatic and African countries with high temperatures and a humid climate. In rice mycotoxin production is favored by moisture con-

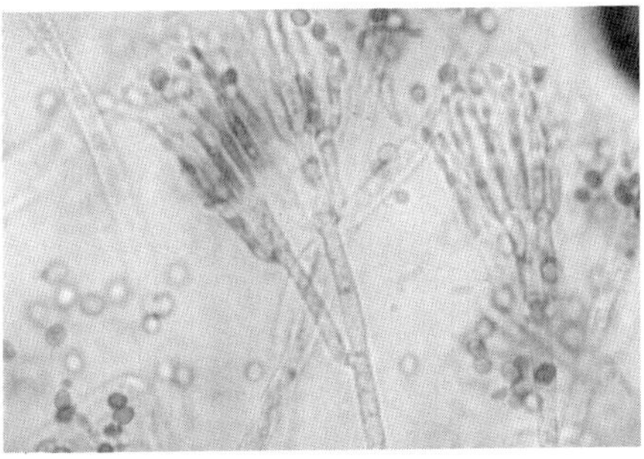

Penicillium expansum Link

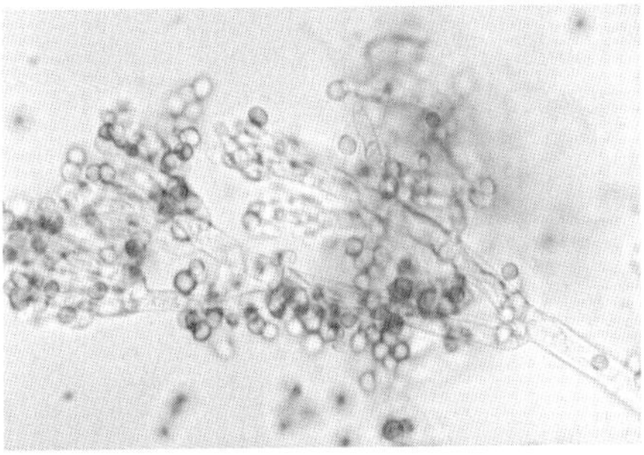

Penicillium griseofulvum Dierckx

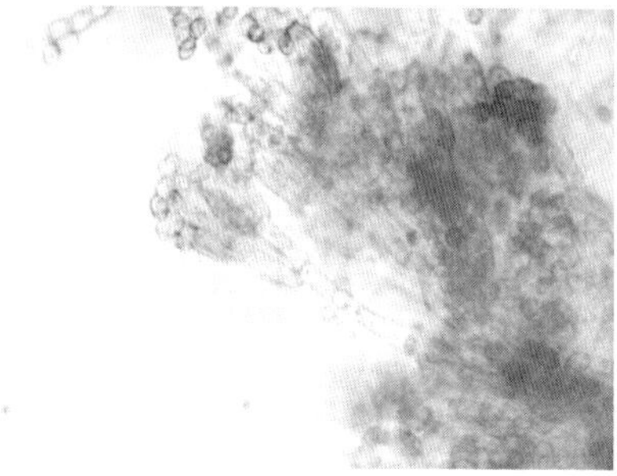

Penicillium islandicum Sopp

tents > 16% in combination with high temperatures (≈ 33 °C optimum). After the first isolation by Sopp (1912) on the Island Skyr (Norway) it became obvious that *P. islandicum* produces a series of very hepatotoxic substances. They cause acute liver atrophy, liver → cirrhosis and liver tumors.
P. islandicum may produce → mycotoxins such as emodin, erythroskyrin, islandic acid, → islanditoxin, → luteoskyrin, → rugulosin, skyrin.

Penicillium roquefortii Thom (Syn.: *P. casei, P. biourgei, P. gorgonzolae, P. stilton, P. vesiculosum*) is frequently found in → cereals stored under controlled → atmosphere and silage. Starter cultures of this fungus are used in the cheese industry for the preparation of blue veined cheese. *P. roquefortii* prevents (i) undesirable fungal infections and causes (ii) proteolytic degradation of casein as well as (iii) hydrolysis of triglycerides. However, this fungus produces a variety of toxic metabolites. Of these, → mycophenolic acid, → penicillic acid, → roquefortine C and → roquefortine A & B have been detected in naturally contaminated → cheeses. The natural contamination of blue veined cheese with these mycotoxins as well as their toxicological properties do not represent a risk for human health.
P. roquefortii may produce → mycotoxins such as → cyclopiazonic acid, mycophenolic acid, → patulin, penicillic acid, PR-toxins (PR-toxin *P. roquefortii* chemotype I only), roquefortine A, B, C (*P. roquefortii* Chemotype I and II) & D.

Penicillium verrucosum Dierckx is very common in temperate regions of the world (especially northern parts of Europe) where it almost exclusively occurs (see Figure *Penicillium verrucosum* Dierckx). Cereal → grains (→ maize, → wheat, → barley) are most frequently affected resulting in a possible OTA contamination. Infestation of some kernels from anthesis and surface contamination is common at harvest. The absolute amount of pre-harvest infection is influenced by site and season. During combine harvesting, conidia of *P. verrucosum* are disseminated resulting in the contamination of other grains. In addition, refrigerated → meat and → cheese products in subtropical areas as well as → fish may also contaminated.
P. verrucosum is the only known and confirmed producer of → ochratoxin A within the genus → Penicillium. Formation of this mycotoxin is enhanced by the amino acids proline and glutamic acid. A positive correlation between the protein concentration of → barley and the production of OTA was established.
P. verrusocum is the causal microorganism of → Mycotoxic porcine nephropathy in pigs in Denmark, Sweden and Hungary. It is suggested that this disorder due to ochratoxin A may be enhanced by → citrinin and oxalic acid. Particularly at lower temperatures *P. verrucosum* causes citrinin-contamination of cereals whereas no citrinin is produced on oilseeds crops. Similarly, wheat gave better OTA yields than corn or the oilseed crops.
P. verrucosum may produce → mycotoxins such as → citrinin (*P. verrucosum* Chemotype II), → ochratoxin A (*P. verrucosum* Chemotype I and II).

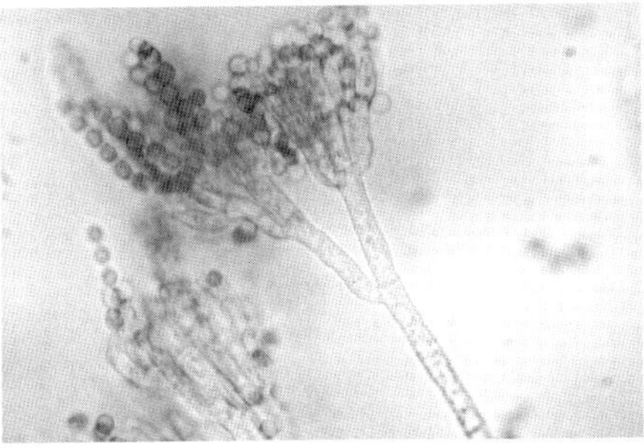

Penicillium verrucosum Dierckx

Penicillium viridicatum Westling should be involved in the → Mycotoxic porcine nephropathy of Danish pigs but it could be shown, after correct identification, that → Penicillium verrucosum Dierckx was the causal organism. *P. viridicatum* has been isolated from e.g. → almonds, → cereals, → cereal products, → fish, → meat products, → nuts, → shrimps (see Figure *Penicillium viridicatum* Westling).
P. viridicatum may produce → mycotoxins such as → penicillic acid, → viomellein, viridicatins, → xanthomegnin.

Penitrems are indole derivatives (→ mycotoxins) which contain only one nitrogen per molecule (see Figure Penitrems). Penitrem A, B, C are produced by → Penicillium spp. even at low (refrigeration) temperature.

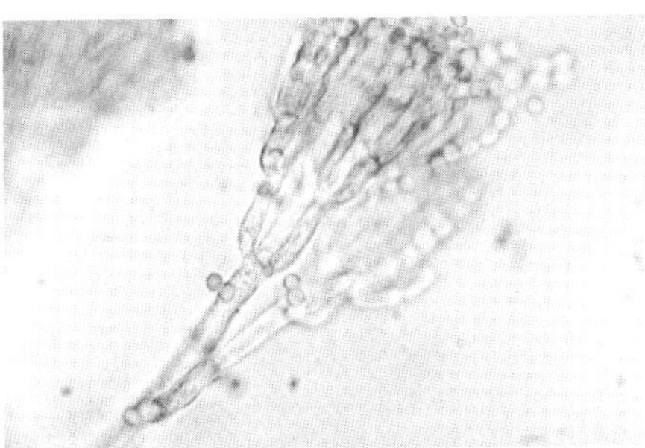

Penicillium viridicatum Westling

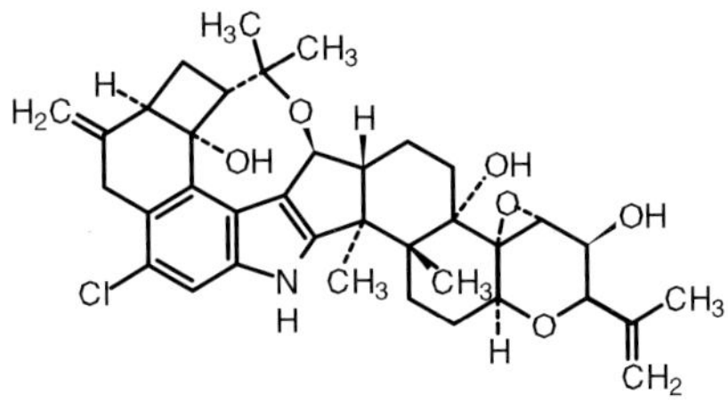

Penitrems. Penitrem A

CHEMICAL DATA
Empirical formula: $C_{37}H_{44}O_6NCl$, molecular weight: 633 (penitrem A)

FUNGAL SOURCES
→ Penicillium crustosum Thom, *P. clavigenum*, *P. glandicola*

NATURAL OCCURRENCE
→ cheese, cream, → walnuts

TOXICITY
neurological (tremors, → convulsions) and → renal effects
LD_{50} (ip): 1.05 mg / kg bw mice (penitrem A)
The mammalian toxicity of penitrem C is unknown.
In humans dizziness and vomiting may be caused by the intake of penitrems but patients recovered completely in all cases.

DETECTION
HPLC, MS, spectroscopy, TLC

FURTHER COMMENTS
In nature compounds causing sustained trembling are rare, whereas most of them are synthesized by molds. The intoxication of dogs consuming moldy cream cheese was the first definitive natural occurrence of penitrem A toxicosis.

Pepper may contain the following → mycotoxins:
→ alternariol
incidence: 1/1*, conc.: 640 µg / kg, country: Italy
→ alternariol methyl ether
incidence: 1/1*, conc.: 49 µg / kg, country: Italy

→ ochratoxin A
incidence: 11/11, conc. range: ≤ 4.9-8 µg / kg, country: Austria
incidence: 1/4, conc.: 40 µg / kg, country: Germany
→ tenuazonic acid
incidence: 1/1*, conc.: 54 µg / kg, country: Italy
*sample was visibly affected by → Alternaria rot
→ spices

Pepper (black) may contain the following → mycotoxins:
→ aflatoxin B_1
incidence: 4/15, Ø conc.: 35 µg / kg, country: Egypt
incidence: 5/8, conc. range: 17-190 µg / kg, country: India
→ aflatoxin B_2
incidence: 5/8, conc. range: 12-150 µg / kg, country: India
→ aflatoxin G_1
incidence: 2/20*, conc. range: 1.72-3.18 µg / kg, Ø conc.: 2.45 µg / kg, country: Egypt, *different → spices
incidence: 3/7*, conc. range: 1.8-3.7 µg / kg, country: Canada, *imported
incidence: nc/137*, conc.: 1.1 µg / kg, country: Canada, *imported
incidence: 5/8, conc. range: 15-75 µg / kg, country: India
→ aflatoxin G_2
incidence: 5/8, conc. range: 12-76 µg / kg, country: India
→ citrinin
incidence: 1/8, conc.: 50 µg / kg, country: India
→ sterigmatocystin
incidence: 2/8, conc. range: 105-125 µg / kg, Ø conc.: 115 µg / kg, country: India
→ zearalenone
incidence: 1/8, conc.: nc, country: India

Pepper (red): It was suggested that aflatoxin production in red peppers starts during eight to ten days they were spread out to dry after harvesting.
Red pepper may contain the following → mycotoxins:
→ aflatoxin B_1
incidence: 11/22, conc. range: tr-24 µg / kg, country: Germany
incidence: 6/6, conc. range: tr-6 µg / kg, country: India
incidence: 4/9, conc. range: 15-146 µg / kg, country: India
incidence: 1/2*, conc.: 0.8 µg / kg, country: Japan, *imported
incidence: nc, conc. range: ≤ 700 µg / kg, country: Nigeria
incidence: 12*/106**, Ø conc.: 125 µg / kg, country: Thailand
*total: Ø conc.: 966 µg / kg AFB_1, AFB_2, AFG_1, AFG_2, **chili peppers
incidence: 7/15, conc. range: 0.2-32.9 µg / kg, Ø conc.: 9.21 µg / kg, country: USA
→ aflatoxin B_2
incidence: 4/9, conc. range: 11-88 µg / kg, country: India
incidence: 7/15, conc. range: 0.1-1.5 µg / kg, Ø conc.: 0.43 µg / kg, country: USA
→ aflatoxin G_1
incidence: 4/9, conc. range: 8-58 µg / kg, country: India
incidence: 4/15, conc. range: 0.7-28.4 µg / kg, Ø conc.: 9.07 µg / kg, country: USA
→ aflatoxin G_2
incidence: 4/9, conc. range: 6-40 µg / kg, country: India
incidence: 1/15, conc.: 1.1 µg / kg, country: USA
→ aflatoxins
incidence: 18/50*, conc. range: 1-3.9 µg** / kg (7 samples), 4-50 µg** / kg (11 sa), country: UK, *imported, **AFB_1, AFB_2, AFG_1, AFG_2 (total)
incidence: 9/14*, conc. range: 1-3.9 µg** / kg (5 samples), 4- > 50 µg** / kg (4 sa), country: UK, *imported, port samples, **AFB_1, AFB_2, AFG_1, AFG_2 (total)

incidence: 9/12*, conc. range: ≤ 30 µg / kg, ∅ conc.: 10 µg / kg, country: USA, *imported
→ ochratoxin A
incidence: 13/18, conc. range: ≤ 4.9-38 µg / kg, country: Austria
incidence: 4/4, conc. range: ≤ 4.9-50.4 µg / kg, country: UK
→ zearalenone
incidence: 1/9, conc.: nc, country: India
→ spices

Pepper (white) may contain the following → mycotoxins:
→ aflatoxin B_1
incidence: 1/7, conc.: 0.3 µg / kg, country: USA
incidence: 4/15, ∅ conc. range: ≤ 22 µg / kg, country: Egypt
incidence: 1/13*, conc.: 0.6 µg / kg, country: Japan, *imported
incidence: 7/24, conc. range: 0.6-2.3 µg / kg, country: Japan
incidence: 1/7, conc.: 0.3 µg / kg, country: USA
→ aflatoxin B_2
incidence: 7/24, conc. range: 0.1-0.2 µg / kg, country: Japan
→ aflatoxin G_1
incidence: 7/24, conc. range: 0.2-1.4 µg / kg, country: Japan
→ spices

Pepper cheese → cheese, pepper→

Persipan (apricot seed paste)
Blanched peach and apricot seeds for persipan manufacture should be processed immediately after blanching because aflatoxin contamination may occur very rapidly.
Persipan may contain the following → mycotoxins:
→ aflatoxin B_1
incidence: 6/16, conc. range: tr-5 µg / kg, country: Germany
→ aflatoxin B_2
incidence: 3/16, conc.: traces, country: Germany
→ aflatoxin G_1
incidence: 2/16, conc.: tr-3 µg / kg, country: Germany
→ marzipan

Petromyces → Trichocomaceae, anamorph → Aspergillus (*ochraceus* group)
P. alliaceus belongs to the genus *Aspergillus*, subgenus Circumdati, section Circumdati. *P. alliaceus* is a known → ochratoxin A producer.

Pheasants may contain the following → mycotoxins:
→ aflatoxin B_1
incidence: 56/94*, conc. range: 0.3-0.985 µg / kg, ∅ conc.: 0.329 µg / kg, country: Czechoslovakia, *liver
incidence: 79/94*, conc. range: 0.3-1.67 µg / kg, ∅ conc.: 0.679 µg / kg, country: Czechoslovakia, *kidney
→ meat

Phoma anamorphic Pleosporaceae, teleomorph *Pleospora*

Pig blood may contain the following → mycotoxins:
→ ochratoxin A
incidence: 910/1200, conc. range: 5-20 µg / l (861 samples), 20-100 µg / l (44 sa), 100-229 µg / l (5 sa), country: Canada
incidence: 574/1169, ∅ conc.: ca. 1 µg / l, country: Germany
incidence: 178/216, conc. range: > 5 µg / l, country: Norway
incidence: 36/195, conc. range: 3-270 µg / l, country: Poland
incidence: 47/279, conc. range: 2-187 µg / l, ∅ conc.: 15.7 µg / l, country: Sweden
incidence: 6/76, ∅ conc. range: 36-37 µg / l, country: Yugoslavia
→ meat, → pork

Pig kidneys (normal) may contain the following → mycotoxins:
→ citrinin
incidence: 9/125, conc. range: 0.1- > 10 µg / kg, country: UK
→ ochratoxin A
incidence: 9/95, conc. range: 0.2- > 80 µg / kg, country: Belgium
incidence: 1/63, conc.: 1-5 µg / kg, country: Czechoslovakia
incidence: 4403/7639, conc. range: > 25 µg / kg (4293 samples), > 150 µg / kg (110 sa), country: Denmark
incidence: 137/686, conc. range: 2-67 µg / kg, country: Denmark
incidence: 5/25, conc. range: > 25 µg / kg, country: Denmark
incidence: 20/20, conc. range: 0.5-1955 µg / kg, country: Denmark
incidence: 10/193*, conc. range: < 0.1-5 µg / kg, country: Finland, *normal and suspected
incidence: 47/354, ∅ conc.: 1.4 µg / kg, country: Germany
incidence: 42/300, conc. range: 0.5-10.2 µg / kg, country: Germany
incidence: 15/100, conc. range: 0.5-16.4 µg / kg, country: Germany
incidence: 48/122, conc. range: 2-100 µg / kg, country: Hungary
incidence: 2/131, conc. range: 7-10 µg / kg, country: Norway
incidence: 32/129, conc. range: 2-104 µg / kg, country: Sweden
incidence: 24/90, conc. range: 2-88 µg / kg, country: Sweden
incidence: 12/36, conc. range: 0.1-0.2 µg / kg (11 samples), 0,3 µg / kg (1 sa), country: Switzerland
incidence: 7/12, conc.: ≤ 1.0 µg / kg, country: The Netherlands
incidence: 1/6, conc. range: 0.2-0.8 µg / kg, country: The Netherlands
incidence: 15/104, conc. range: ≤ 4.9-9.3 µg / kg, ∅ conc.: 0.84 µg / kg, country: UK
incidence: 242/378, conc. range: 0.5- > 10 µg / kg, country: UK
incidence: 43/278, conc. range: 1-10 µg / kg (41 samples), 22-44 µg / kg (2 sa), country: UK
incidence: 4/76*, ∅ conc.: 21 µg / kg, country: Yugoslavia, *partly suspected
→ meat, → pork

Pig kidneys (suspected) may contain the following → mycotoxins:
→ ochratoxin A
incidence: 69/104, ∅ conc.: 0.75 µg / kg, country: Austria
incidence: 28/95, conc. range: 0.2-9.99 µg / kg, country. Belgium
incidence: 68/385, conc. range: 0.2-12 µg / kg, country: Belgium
incidence: 76/96, conc. range: 1-20 µg / kg, country: Czechoslovakia
incidence: 21/60, conc. range: 2-68 µg / kg, country: Denmark
incidence: 20/20, conc. range: 0.2-1965 µg / kg, ∅ conc.: 34.2 µg / kg, country: Denmark
incidence: 3/38, conc. range: ≤ 4.9 µg / kg, ∅ conc.: 0.7 µg / kg, country: Finland
incidence: 22/104, conc. range: 0.1-1.8 µg / kg, ∅ conc.: 0.45 µg / kg, country: Germany
incidence:48/122, conc. range: 10-7100 µg / kg, country: Hungary
incidence: 77/197, conc. range: 5-100 µg / kg, country: Hungary
incidence: 27/113, conc. range: tr-23 µg / kg, country: Poland
incidence: 33/73, conc. range: 2-23 µg / kg, country: Poland
incidence: 32/129, conc. range: 2- < 5 µg / kg (25 samples), 5- < 10 µg / kg (2 sa), 10- ≤ 104 µg / kg (5 sa), country: Sweden
incidence: 35/75, conc. range: ≤ 2.0 µg / kg, country: The Netherlands
incidence: 33/46, conc. range: 0.2-2 µg / kg, country: The Netherlands
incidence: 6/6, conc. range: 0.2-1 µg / kg, country: The Netherlands

incidence: 2/29, conc. range: 0.2-0.4 µg / kg, country: The Netherlands
incidence: 17/24*, conc. range: 0.2-240 µg / kg, country: The Netherlands, *originating from Denmark
incidence: 112/303*, conc. range: 0.5- < 5 µg / kg (104 samples), 5- < 10 µg / kg (6 sa), 11.5-12.4 µg / kg (2 sa), country: UK, *unsuitable for human consumption
→ meat, → pork

Pig liver may contain the following → mycotoxins:
→ aflatoxin B_1
incidence: 5/13, conc. range: < 5 µg / kg, country: Germany
→ ochratoxin A
incidence: 4/76*, Ø conc.: 21 µg / kg, country: Yugoslavia, *partly suspected
→ meat, → pork

Pig serum Certain → sausages, e.g. frankfurter-type, are produced with pig serum (plasma) and may therefore be contaminated with ochratoxin A.
Pig serum may contain the following → mycotoxins:
→ ochratoxin A
incidence: 32/1445, Ø conc.: 12.6 µg / l, country: Canada
incidence: 72/143, Ø conc.: 21 µg / l, country: Canada
incidence: 146/283, Ø conc.: 1.2 µg / l, country: Germany
incidence: 93/191, conc. range: 0.1-67.3 µg / l, Ø conc.: 5.8 µg / l, country: Germany
→ meat, → pork

Pigeon peas may contain the following → mycotoxins:
aflatoxin (no specification) (→ aflatoxins)
incidence: 5/9, conc. range: ≤ 23 µg / kg, Ø conc.: 7 µg / kg, country: Philippines
→ beans, → cabbage, → cowpeas, → lentils, → peas, → soybeans, → vegetables

Pine nuts Pudding prepared from contaminated → nuts contained more than 83% of the original amount of aflatoxin. Pine nuts may contain the following → mycotoxins:
→ aflatoxin B_1
incidence: 26/50, conc. range: 25-2080 µg / kg, country: Tunisia
→ aflatoxin G_1
incidence: 26/50, conc. range: 56-4570 µg / kg, country: Tunisia
→ aflatoxins (AFB_1, AFB_2, AFG_1, AFG_2)
incidence: 26/50, conc. range: 95-7550 µg / kg, country: Tunisia

Piper betle (medicinal seeds) may contain the following → mycotoxins:
→ aflatoxin B_1
incidence: nc/nc, conc. range: 20-1000 µg / kg, country: India
→ citrinin
incidence: nc/nc, conc. range: 10-720 µg / kg, country: India

Pipian paste may contain the following → mycotoxins:
→ aflatoxins (no specification)
incidence: 3/3*, conc. range: ≤ 78 µg / kg, Ø conc.: 53 µg / kg, country: USA, *imported

Pistachio candy may contain the following → mycotoxins:
→ aflatoxins
incidence: 1/1*, conc.: 78 µg / kg, country: USA, *imported

Pistachio nuts As in the case of → peanuts an uneven distribution of → aflatoxins has been established in pistachio nuts samples. Only a few nuts contained high aflatoxin concentrations (≤ 1.4 g / kg). The highest contamination occurred only in brown, brown spotted or fluorescent pistachio kernels. Using an automatic sorter, which removes → nuts with fluorescent shells, the aflatoxin content could

be reduced by ca. 50%. However, non-fluorescent nuts (shells) which may also contain significant levels of aflatoxin, escape this control measure.
Aflatoxin contamination of pistachio nuts occurs after soaking to remove the hulls from the shells and / or during improper storage. Aflatoxin producers gain entry to the nut along the vascular system. This tissue connects the kernel with the shell.
Pistachios may contain the following → mycotoxins:
→ aflatoxicol
incidence: 5/54, conc. range: 0.2-13.9 µg / kg, ∅ conc.: 3.62 µg / kg, country: Thailand
incidence: 51/247, ∅ conc.: 27 µg / kg, country: Germany
→ aflatoxin B_1
incidence: 6/54, conc. range: 7.9-1830 µg / kg, ∅ conc.: 585 µg / kg, country: Japan
incidence: 51/247, ∅ conc.: 21 µg / kg, country: Germany
incidence: 67/140, conc. range: < 5 µg / kg (40 samples), 11-35 µg / kg (27 sa), country: Germany
incidence: 1/19, conc.: 22 µg / kg, country: Tunisia
→ aflatoxin B_2
incidence: 6/54, conc. range: 1.5-235 µg / kg, ∅ conc.: 86 µg / kg, country: Japan
→ aflatoxin M_1
incidence: 5/54, conc. range: 0.9-51.8 µg / kg, ∅ conc.: 21.7 µg / kg, country: Japan
→ aflatoxins (no specification)
incidence: 19*/175, conc. range: 5-24.9 µg / kg (12 samples), > 25 µg / kg (7 sa), country: Canada, *AFB_1, AFB_2, AFG_1, AFG_2
incidence: 61/993, conc. range: nc, country: Germany
incidence: 7/22, conc. range: ≤ 252 µg / kg, ∅ conc.: 58 µg / kg, country: USA
incidence: 10/21, conc. range: ≤ 133 µg / kg, ∅ conc.: 41 µg / kg, country: USA
nuts

Pito → beer, pito

Pleosporaceae → Pleosporales

Pleosporales → Dothideales

Plums may contain the following → mycotoxins:
→ patulin
incidence: 1/6, conc.: 4 µg / kg, country: Sweden

Polenta → maize grits

Polydypsia excessive thirst

Polyuria excessive urination

Popcorn may contain the following → mycotoxins:
→ aflatoxin B_1
incidence: 5/15, conc. range: 20-47 µg / kg, ∅ conc.: 35 µg / kg, country: Brazil
incidence: 3/28, conc. range: ≤ 1.5 µg / kg, country: Germany
→ aflatoxin G_1
incidence: 1/15, conc.: 18 µg / kg, country: Brazil
→ aflatoxin G_2
incidence: 1/15, conc.: 8 µg / kg, country: Brazil
→ deoxynivalenol
incidence: 2/12*, conc. range: 12-250 µg / kg, country: Japan, *import from USA
incidence: 7/7, conc. range: 12-250 µg / kg, country: USA
incidence: 1/1, conc.: 30 µg / kg, country: USA

→ fumonisin B_1
incidence: 4/6, conc. range: 10-60 µg / kg, ∅ conc.: 28.3 µg / kg, country: Italy
incidence: 6/6, conc. range: ca. < 10-122 µg / kg, ∅ conc.: 70 µg / kg, country: Germany
incidence: 13/29, conc. range: < 10-160 µg / kg, country: Germany
incidence: 7/22, conc. range: ≤ 1003 µg / kg, ∅ conc.: 347 µg / kg, country: Thailand
incidence: 5/5, conc. range: < 100-500 µg / kg, ∅ conc.: 100 µg / kg, country: USA
incidence: 2/2, conc. range: 10-60 µg / kg, ∅ conc.: 35 µg / kg, country: USA
→ fumonisin B_2
incidence: 1/6, conc.: 20 µg / kg, country: Italy
incidence: 7/22, conc. range: ≤ 273 µg / kg, ∅ conc.: 116 µg / kg, country: Thailand
→ fumonisins
incidence: 5/5, conc. range: < 10-100 µg / kg (HPLC), country: Germany
incidence: 6/13*, conc. range: 14-784 µg / kg, ∅ conc.: 83 µg / kg, country: UK, *popping and microwaveable corn
incidence: 1/1, conc.: 250 µg / kg, country: USA
→ ochratoxin A
incidence: 1/29, conc.: 1.4 µg / kg, country: Germany
→ zearalenone
incidence: 4/7, conc. range: 2.5-130 µg / kg, ∅ conc.: 38 µg / kg, country: USA
→ maize

Poppadoms may contain the following → mycotoxins:
→ aflatoxins (AFB_1, AFB_2, AFG_1, AFG_2)
incidence: nc/4, conc. range: 0.6-2 µg / kg, country: UK

Porcine nephropathy → Mycotoxic porcine nephropathy

Porcine pulmonary edema (Abbr.: PPE) This lethal disorder in swine due to the ingestion of fumonisin B_1 and FB_2 (→ fumonisins) causes severe lung → edema and hydrothorax. Rapid death occurs after an acute onset of → dyspnea, weakness, and cyanosis. Oral as well as intravenous administration induced the disease.

Pork Feed to tissue ratios of less than 100 (→ ochratoxin A / → pig kidneys) indicate an extensive → carry over for this mycotoxin. The biological halflife of OTA in swine tissue is 4.5 days (oral administration). A several week withdrawal period would be necessary to eliminate all OTA residues after exposure to a contaminated dietary.
Since 1978 in Denmark the kidneys of all slaughtered pigs have been examined for macroscopic changes. Suspected kidneys are chemically analysed. The level for rejection of the entire carcass is 25 µg OTA / kg pig kidney. This level ensures that the concentration in → meat does not exceed 10 µg OTA / kg because it could be shown that pig meat contains only ca. 40% of the OTA found in pig kidney.
→ Aflatoxin B_1 feeding studies show that the kidneys (followed by the liver) of pigs accumulate most aflatoxin residues, mainly aflatoxin M_1 and to a lesser extent AFB_1 and → aflatoxicol. Minor levels were found in muscle. In comparison to → cattle (ca. 18 days) pigs might require a shorter withdrawal period (ca. 7 days).
Pork may contain the following → mycotoxins:
→ ochratoxin A
incidence: 64/76*, conc. range: ≤ 1.3 µg / kg, ∅ conc.: 0.11 µg / kg, country: Denmark, *produced conventionally
incidence: 4/7*, conc. range: ≤ 0.12 µg / kg, ∅ conc.: 0.05 µg / kg, country: Denmark, *produced ecologically

incidence: 1/12, conc.: 5 µg / kg, country: Yugoslavia
→ pig blood, → pig kidneys, → pig liver, → pig serum

Porridge may contain the following → mycotoxins:
→ ochratoxin A
incidence: 3/6*, conc.: ≤ 0.3 µg / kg, ∅ conc.: 0.10 µg / kg, country: Germany, *ready made
incidence: 4/92*, conc.: ≤ 2 µg / kg, ∅ conc. 0.10 µg / kg, country: Germany, *→ oats
→ cereals

Port wine → Wine

Potatoes Since artificial inoculation with *Fusarium sambucinum* or *F. sulphureum* resulted in the production of → monoacetoxyscirpenol and → diacetoxyscirpenol (≤ 5 µg / g rot fresh weight) → trichothecenes might be found in moldy potato → tubers.
Potatoes may contain the following → mycotoxins:
→ deoxynivalenol
incidence: 4/17, conc.: nc, country: Canada
→ sambutoxin
incidence: 9/21*, conc. range: 15.8-78.1 µg / kg, ∅ conc.: 49.2 µg / kg, country: Korea, *rotten

Poultry Poultry tolerate relatively high levels of → trichothecenes in their diet but only very small traces are transmitted into → meat and eggs. Residues quickly decline to negligible levels if the contaminated diet is removed.
Poultry meat may contain the following → mycotoxins:
→ ochratoxin A
incidence: 62/113, conc. range: ≤ 0.18 µg / kg, country: Denmark

PPE → Porcine pulmonary edema

PR toxin (Abbr.: PRT) is a 2-(acetyloxy)-2,3,3a,4,6,7b-hexahydro-3,3´,3a-trimethyl-6-oxo-spiro[naphth[1,2-b]oxirene-5(1aH),2´-oxirane]-3´-carboxaldehyde which was first isolated from → Penicillium roquefortii Thom chemotype I in 1973 (see Figure PR toxin).

CHEMICAL DATA
Empirical formula: $C_{17}H_{20}O_6$, molecular weight: 320

FUNGAL SOURCES
Penicillium roquefortii

NATURAL OCCURRENCE
→ cheese, Blue

TOXICITY
causes degenerative changes in liver and kidney of rat
LD_{50} (po): 58-100 mg kg / bw mice

DETECTION
HPLC, spectroscopy, TLC

FURTHER COMMENTS
If neutral and basic amino acids are present, PR-imines are formed. Compared to PR toxin, the toxicity of PR-imines is much lower. Further degradation products of PR toxin are PR-amide and the eremofortins A, B, C. The three latter ones are probably non-toxic.

Premature thelarche This → mycotoxicosis may be induced by → zearalenone.

Primary hepatocellular carcinoma (Abbr.: PHC) In various areas of central and southern Africa, Thailand, and Indonesia a high incidence of PHC in humans has been found which might be due to the ingestion of → aflatoxins in the diet. A linear dose-response relationship between

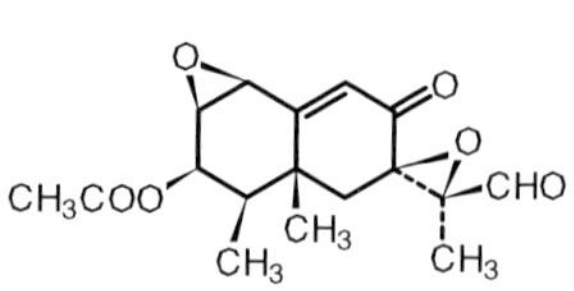

PR toxin

the consumption of aflatoxins (0.003-0.222 µg / kg bw) and human liver cancer has been demonstrated in several countries like Kenya, Mozambique, Uganda and Thailand.

Processed cheese → cheese, processed

Proteinuria resulted in increased serum protein levels in the urine.

pulmonary pertaining the lung

Pulses Pulses may show a high contamination with → ochratoxin A. They may be regarded as a possible contributor to OTA intake if they are regularly ingested. → beans, → cowpeas, → lentils, → peas, → pigeon peas, → soybeans

Pumpkin seeds may contain the following mycotoxins:
→ aflatoxins
incidence: 31/130, conc. range: nc, country: Germany

R

Ragi (*Eleusine coracan* (L.) Gaertn.) Fluctuation of temperature, change in relative humidity and excessive rainfall may contribute to → Alternaria mycotoxin contamination under field conditions.
Ragi may contain the following → mycotoxins: → altenuene
incidence: 1/8, conc.: 30 µg / kg, country: India
→ alternariol methyl ether
incidence: 2/8, conc. range: 800-1400 µg / kg, ∅ conc.: 1100 µg / kg, country: India
→ tenuazonic acid
incidence: 3/8, conc. range: 2030-5700 µg / kg, ∅ conc.: 3843 µg / kg, country: India

Rape → oilseed rape

Rd-toxin (Syn.: → deoxynivalenol)

Red mold toxicosis (Syn.: akakabi byo disease, red mold disease, red mold poisoning, scab disease) It takes its name from the reddish coloration of the predominately infected → wheat and → barley kernels. → Fusarium graminearum Schwabe (*Gibberella zeae*) and other species like *F. heterosporum*, → Fusarium nivale (Fr.) Ces., → Fusarium poae (Peck) Wollenw., and → Fusarium oxysporum Schlecht. emend. Snyd. & Hans. are mainly responsible for infection. Excessive rainfall and low temperatures throughout the ripening and harvest seasons favor the rate of invasion. In the severe development of this disease, more than 90% of the annual yield was damaged.
Typically, people who ingested the discolored and shrivelled → grains became ill from 5 to 30 min (2 h) after consumption. The following symptoms have been described: nausea, vomiting, diarrhea, feedrefusal, congestion or → hemorrhage in the lung, adrenals, intestine, uterus, vagina and brain, and destruction of the bone marrow. It was concluded that toxic metabolites of fusaria, especially → deoxynivalenol, are responsible for this disease.
Red mold disease due to DON contamination has occurred sporadically during the years from 1946 to 1963 in several northern districts of rural Japan and in a southern area of Korea. The disease is akin to the wheat scab which has frequently been recorded in the USA. It became particularly widespread in Canada between 1980 and 1982. In India (Kashmir Valley) an outbreak of the disease for the last time in the 1980s was reported after the consumption of DON-contaminated wheat and → wheat products (ca. 10,000 µg → trichothecenes / kg). From 1961 to 1985 the disease affected at least 7818 victims and was attributed to consumption of scabby wheat and moldy → maize; no deaths were reported. In China over a 20-year period ca. 10,000 cases of acute trichothecene toxicosis were reported but it is estimated that the real rate is significantly higher due to the difficulties of acquiring and evaluating such information from rural China and India.

renal pertaining to the kidney

Reye's syndrome A disease originally described in Australia by Reye but it is also known for children in Czechoslovakia, New Zealand, Thailand, and the U.S. In these cases → aflatoxins have been implicated. Livers and blood serum from patients with Reye's syndrome contained aflatoxins. Children who suffer from an infection with influenza virus type A or B or with varicella were mainly (exclusively) affected. It seems that RS is a complex disease caused by a combination of factors including viral infection and

xenobiotic compounds possibly aflatoxins. The disease is characterized by clinical signs such as vomiting, convulsions, coma, and death within 24 h to 48 h after onset. Histopathological changes show acute encephalopathy, fatty degeneration of the liver (and kidney), pale, slightly widened → renal cortexes (→ cortex), associated with elevated serum transaminase levels. A mortality rate of 81% of the originally diagnosed cases (21) has been reported. In Thailand the typical histopathological changes have been reproduced experimentally by feeding aflatoxin to Macaque monkeys.
The involvement of aflatoxins in the Reye's syndrome was linked to the seasonal and geographic distribution of the disease (Thailand). Especially in rural areas there was a high incidence of death among children, which may be correlated with the aflatoxins detected in the food. There was also an increasing incidence of this disorder towards the latter part of the rainy season. This is typical for a → mycotoxicosis (acute → aflatoxicosis). From 1963 to 1974 more than 250 cases were reported in the US; 139 occurred in Thailand. In the following years (1973-1981) the RS incidence ranged from 0.37-0.88 per 100,000 per year, with a value of 2-4 during influenza epidemics (USA). Although a lower RS incidence was established in Britain during a five year study, a higher mortality rate (59%) occurred. According to the Center for Disease Control (USA) the following criteria are used to diagnose RS: (i) acute onset of encephalopathy, (ii) hepatic involvement (enlargement) of the liver or elevated serum transaminase levels (glutamic: pyruvic acid transaminase / oxalacetic acid transaminase), (iii) the disorder cannot be explained in any other way.

Rice Approximately one third of US-strains of → Aspergillus flavus Link isolated from rice produces significant amounts of → aflatoxins. In addition, these aflatoxin producers make up a significant part of the normal mycoflora. Toxin formation probably results from rapid growth of the mold although limited penetration of the endosperm has been observed. Approximately 95% of the toxin is found in the → bran layer.
During ordinary → milling procedures much of the aflatoxin in a contaminated kernel (rough rice) is removed. In consequence, the bran fractions contained ca. 10 times more aflatoxin than the milled rice fractions. Naturally contaminated rice may contain aflatoxins at levels of < 4-50 µg / kg. Since artificial drying is an efficient and effective operation, and since the toxin level in contaminated rice is greatly reduced by the milling process, contamination of rice with aflatoxins is not a serious problem at this time (USA). White rice, which is most widely used in human diets, did not contain any detectable levels of → fumonisins. However, rough rice and rice hulls (feeding diets) contained fumonisin concentrations above 5 µg / kg, indicating that fumonisins are localized primarily in the hulls and bran. However, since fumonisins are heat-stable they would probably not be destroyed by rice cooking methods and, to assure maximum safety, the rice should be monitored for fumonisin contamination (US). Studies documenting the fumonisin contamination of rice grown in other geographical areas are necessary.
Rice may contain the following → mycotoxins:
→ aflatoxin B_1
incidence: 2/52*, conc. range: 26-38 µg / kg, ∅ conc.: 32 µg / kg, country: Brazil, *polished
incidence: 1/1, conc.: 8 µg / kg, country: Egypt

incidence: 1/50, conc.: 28 µg / kg, country: Italy
incidence: 6/8, conc. range: < 2.5-15 µg / kg, country: Nepal
incidence: 4/4*, conc. range: < 2.5-12.5 µg / kg, country: Nepal, *parboiled
incidence: 7/364, ∅ conc.: 20 µg / kg, country: Thailand
incidence: 9/9*, conc. range: ≤ 600 µg / kg, ∅ conc.: < 1-2 µg / kg, country: Thailand, *total: ∅ conc.: 98 µg / kg AFB_1, AFB_2, AFG_1, AFG_2
incidence: 1/182, conc.: 5 µg / kg, country: USA
→ aflatoxin B_2
incidence: 1/52*, conc.: 15 µg / kg, country: Brazil, *polished
incidence: 1/1, conc.: 2 µg / kg, country: Egypt
incidence: 1/4*, conc. range: 1.8 µg / kg, country: Nepal, *parboiled
→ aflatoxin G_1
incidence: 1/52*, conc.: 20 µg / kg, country: Brazil, *polished
incidence: 2/84, conc. range: 73.1-77.5 µg / kg, ∅ conc.: 75.3 µg / kg, country: Malaysia
→ aflatoxin G_2
incidence: 3/84, conc. range: 3.7-96.3 µg / kg, ∅ conc.: 45.6 µg / kg, country: Malaysia
aflatoxin (no specification)
incidence: 3/15*, conc. range: ≤ 38 µg / kg, ∅ conc.: 16 µg / kg, country: Philippines, *rice bran
incidence: 17/82*, conc. range: ≤ 43 µg / kg, ∅ conc.: 12 µg / kg, country: Philippines, *milled
incidence:1/6*, conc.: ≤ 3 µg / kg, ∅ conc.: 3 µg / kg, country: Philippines, *pop
incidence: 3/10*, conc. range: ≤ 18 µg / kg, ∅ conc.: 15 µg / kg, country: Philippines, *rough
→ aflatoxins (AFB_1, AFB_2, AFG_1, AFG_2)
incidence: 13/30, conc. range: 22-317 µg AFB_1 / kg, 15-125 µg AFB_2 / kg, 14-107 µg AFG_1 / kg, 20-98 µg AFG_2 / kg, country: India
incidence: nc/4*, conc. range: 0.1-2.4 µg / kg, country: UK, *Basmati rice
aflatoxins (no specification)
incidence: 14/20, conc. range: 2-19 µg / kg, ∅ conc.: 7.9 µg / kg, country: Gambia
incidence: 12/80*, conc. range: tr-430 µg / kg, country: India, *cyclone-affected
incidence: 23/81*, conc. range: 30-1130 µg / kg, country: India, *cyclone-affected
incidence: 32/43*, conc. range: 30-130 µg / kg, country: India, *parboiled
incidence: 1/23, conc.: 1000 µg / kg, country: Mozambique
incidence: 16/72, conc. range: ≤ 33 µg / kg, ∅ conc.: 16 µg / kg, country: Philippines
→ citrinin
incidence: 4/30, conc. range: 49-92, country: India
incidence: 2*/2, conc. range: 700-1130 µg / kg, country: Japan
→ deoxynivalenol
incidence: 1/1*, conc.: 90 µg / kg, country: Papua New Guinea, *imported, brown trukai
incidence: nc/4*, conc. range: 4-6 µg / kg, country: UK, *Basmati rice
incidence: nc/4*, conc. range: 4-7 µg / kg, country: UK, *Chinese rice
→ fumonisin B_1
incidence: 8/20, conc. range: ≤ 4300 µg / kg, country: USA
→ fumonisin B_2
incidence: 6/20, conc. range: ≤ 1200 µg / kg, country: USA
→ fumonisin B_3
incidence: 5/20, conc. range: ≤ 600 µg / kg, country: USA
→ fumonisins (FB_1, FB_2)
incidence: 1/4*, conc.: 28 µg / kg, country: UK, *Basmati rice
→ nivalenol
incidence: 2/9, ∅ conc.: 22 µg / kg, country: Nepal

incidence: 1/1*, conc.: 63 μg / kg, country: Papua New Guinea, *imported
incidence: nc/4*, conc. range: 4-11 μg / kg, country: UK, *Basmati rice
→ ochratoxin A
incidence: 1/3*, conc.: 533 μg / kg, country: Egypt, *rice germ
incidence: 2/36, conc. range: ≤ 0.3 μg / kg, country: Germany
incidence: 2/32, conc. range: 8-25 μg / kg, ∅ conc.: 16.5 μg / kg, country: India
incidence: 2/15, conc. range: 1.7-2.4 μg / kg, country: Indonesia
incidence: 8/15, conc. range: ≤ 1.0 μg / kg, country: Italy
incidence: 1/various → food samples, conc.: 50 μg / kg, country: Japan
incidence: 2*/2, conc. range: 230-430 μg / kg, country: Japan, *deteriorated
→ sterigmatocystin
incidence: 3/30, conc. range: 108-157 μg / kg, country: India
incidence: 2/nc, conc. range: 50-450 μg / kg, country: Japan
incidence: ?/?*, conc. range: 3800-4300 μg / kg, country: Japan, *moldy
incidence: 12/37, conc. range: ≤ 16,300 μg / kg, country: Japan
T2-triol
incidence: 1/4*, conc.: 49 μg / kg, country: UK, *Chinese rice
→ zearalenone
incidence: 1/9, conc.: 8 μg / kg, country: Nepal
incidence: 1/1*, conc.: 3060 μg / kg, country: Papua New Guinea, *imported
incidence: 3/42*, conc. range: > 200 μg / kg, country: Uruguay, *and by-products
Besides the → mycotoxins listed under → cereals rice may additionally be contaminated with → citreoviridin, → islanditoxin, → luteoskyrin (Frisvad 1988).
→ cereals

Rice bran may contain the following → mycotoxins:
→ ochratoxin A
incidence: 1/3, conc.: 9 μg / kg, country: Egypt
→ bran

Rice cake may contain the following → mycotoxins:
→ aflatoxin B_1 & → aflatoxin B_2
incidence: 1*/53**, conc.: ≈ 10 μg / kg, country: Japan, *moldy, **different moldy samples
→ ochratoxin A
incidence: 1/3*, conc.: 4 μg / kg, country: Egypt, *rice germ cake

Roe deer may contain the following → mycotoxins:
→ aflatoxin B_1
incidence: 38/56*, conc. range: 0.3-2.17 μg / kg, ∅ conc.: 0.696 μg / kg, country: Czechoslovakia, liver
incidence: 39/56*, conc. range: 0.3-1.93 μg / kg, ∅ conc.: 0.795 μg / kg, country: Czechoslovakia, kidney
→ meat

Roquefort cheese → cheese, blue; → cheese, Roquefort

Roquefortine (Syn.: roquefortine C) is an indole alkaloid (10b-(1,1-dimethyl-2-propenyl)-6,10b,11,11a-tetrahydro-3-(1H-imidazol-4-ylmethylene)-2H-pyrazinol[1´,2´:1,5]pyrrolo[2,3-b]indole-1,4-(3H,5aH)-dione) which was originally named roquefortine C (→ mycotoxins). It was first isolated from → Penicillium roquefortii Thom in 1975 by Japanese workers (see Figure Roquefortine).

Chemical Data
Epmirical formula: $C_{22}H_{23}N_5O_2$, molecular weight: 389

Fungal Sources
→ Penicillium spp. such as → Penicillium chrysogenum Thom, → Penicillium crustosum Thom, → Penicillium expansum Link,

→ Penicillium griseofulvum Dierckx, *P. roquefortii* chemotype I and II

Natural Occurrence
→ cheese, Blue, → cheese, Blue Castello, → cheese, Danish Blue, → cheese dressing, blue , → cheese, Gorgonzola, → cheese, Roquefort, → cheese, Stilton

Toxicity
LD_{50} (ip): 15-189 mg / kg bw male mice

Detection
Electrochemical detectors, TLC

Further Comments
Roquefortine was detected in the stomach of several dogs. They showed a strychnine-like poisoning.

Roquefortine A & B (Syn.: isofumigaclavine A & B) roquefortine A (9-acetoxy-6,8-dimethylergolin) - roquefortine B (6,8-dimethylergolin-9-ol) is the hydrolysis product - and → roquefortine have been isolated from the mycelium of → Penicillium roquefortii Thom in 1975 (see Figure Roquefortine A & B).

Chemical Data
Empirical formula: $C_{18}H_{20}N_2O_2$, molecular weight: 296 (Roquefrotine A)
Empirical formula: $C_{16}H_{20}N_2O$, molecular weight: 256 (Roquefrotine B)

Fungal Sources
→ Penicillium commune Thom chemotyope II, *P. clavigerum*, *P. roquefortii*,

Natural Occurrence
→ cheese, Blue

Toxicity
Roquefortine A is weakly toxic and possesses neurotoxic properties.

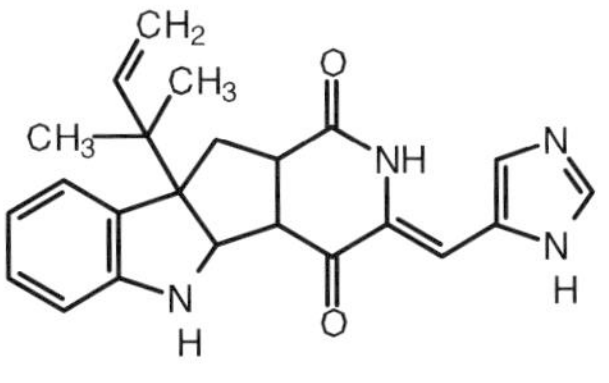

Roquefortine. Roquefortine C

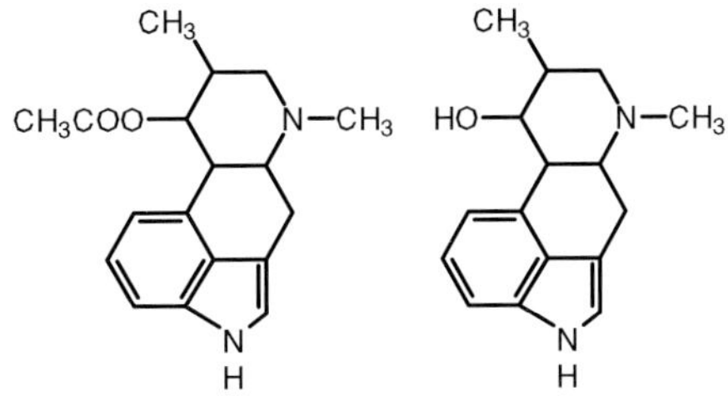

Roquefortine A & B

LD_{50} (ip): 340 mg roquefortine A and 1000 mg roquefortine B / kg bw mice. Weak pharmacological actions (e.g. muscle relactant, antidepressant, and local anaesthetic effects) have been observed.

Detection
TLC

Roquefortine D (Syn.: dihydroroquefortine) a mycotoxin (12,13-dihydroroquefortine) which is a probable precursor of → roquefortine C (→ mycotoxins). It is produced by *P. atramentosum*, → Penicillium chrysogenum Thom, *P. glandicola*, and → Penicillium roquefortii Thom.

Rubratoxins are complex nonarides with anhydride groups (relatively stable) and lactone rings (→ mycotoxins) which were first isolated as pure compounds in 1966 from a culture filtrate of → Penicillium *rubrum* (see Figure Rubratoxins). Since difficulties in isolating the toxic fractions occur, the compounds were named rubratoxin A (10-[(R)-[(2R)-3,6-dihydro-6-oxo-2H-pyran-2-yl]hydroxymethyl]-5,9,10,11-tetrahydro-4-hydroxy-5-[(1S)-1-hydroxyheptyl]-1H-cyclonona[1,2-c:5,6-c′]difuran-1,3,6-trione (4S,5R,10R)) (more easily to isolate) and B. The more prevalent member was named rubratoxin B (10-[(R)-[(2R)-3,6-dihydro-6-oxo-2H-pyran-2-yl]hydroxymethyl]-5,9,10,11-tetrahydro-4-hydroxy-5-[(1S)-1-hydroxyheptyl]-1H-cyclonona[1,2-c:5,6-c′]difuran-1,3,6,8(4H)-tetrone (4S,5R,10R)) and is the more toxic.

Chemical Data
Empirical formula: $C_{26}H_{32}O_{11}$, molecular weight: 520 (Rubratoxin A)
Empirical formula: $C_{26}H_{30}O_{11}$, molecular weight: 518 (Rubratoxin B).
The fact that rubratoxin A is significantly more soluble in ethyl alcohol whereas rubratoxin B is significantly more soluble in ethyl acetate is important in fractionating mixtures of the two toxins.

Fungal Sources
P. purpurogenum, P. rubrum

Natural Occurrence
→ tumeric, → wheat
In addition, rubratoxins have been produced on → maize by *P. purpurogenum* and *P. rubrum.*

Toxicity
Although various effects on animals have been recorded, rubratoxin B is mainly hepatotoxic and nephrotoxic. Rubratoxin A possesses acute toxicity.
LD_{50} (po): 120 mg / kg bw mice

Detection
HPLC, MS, RIA, spectroscopy, TLC

Possible Mycotoxicosis
Rubratoxin B was first implicated in "moldy corn toxicosis" in cattle, pigs, and poultry although their role in natural outbreaks of animals disease is, as yet, not clearly defined.

Further Comments
A synergistic action between → aflatoxin B_1 and rubratoxin B, especially in the case of "hepatitis X", a toxicosis in dogs, is suggested.
Since rubratoxins are excreted into the medium and not retained by the mycelium (*P. rubrum*) they belong to the → extracellular mycotoxins.
Rubratoxins are degraded by *P. puberulum.*

Rubratoxin (Syn.: rubratoxin B (→ rubratoxins))

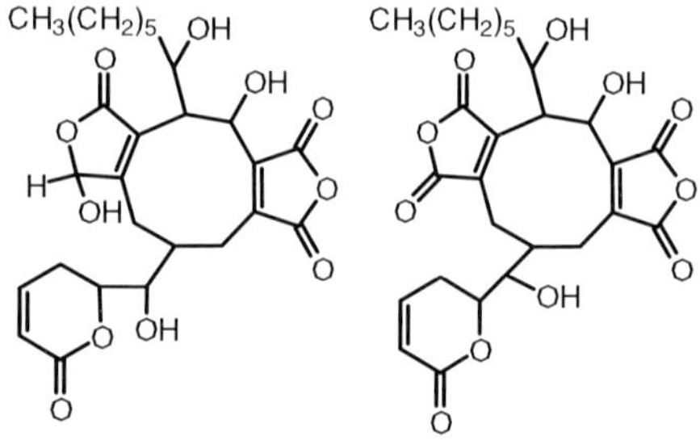

Rubratoxins. Rubratoxin A & B

Rugulosin is a 2,2´,4,4´,5,5´-hexahydroxy-2,2´,3,3´-tetrahydro-7,7´-dimethyl-1,1´-bianthraquinone (→ mycotoxins) (see Figure Rugulosin).

Chemical Data
Empirical formula: $C_{30}H_{22}O_{10}$, molecular weight: 542

Fungal Sources
→ Penicillium spp., e.g. → Penicillium islandicum Sopp (the (-) form), *P. rugulosum, P. variabile, Talaromyces wortmanii*

Natural Occurrence
It might be present in "yellow rice".

Toxicity
antibiotic, hepatotoxic, carcinogenic
LD_{50} (ip) : 83 mg / kg bw mice

Detection
TLC

Possible Mycotoxicosis
Yellow rice disease

Further Comments
Thermal decomposition of rugulosin leads to the mycotoxins emodin and skyrin.
Long term feeding studies demonstrated the hepato-carcinogenicity of rugulosin

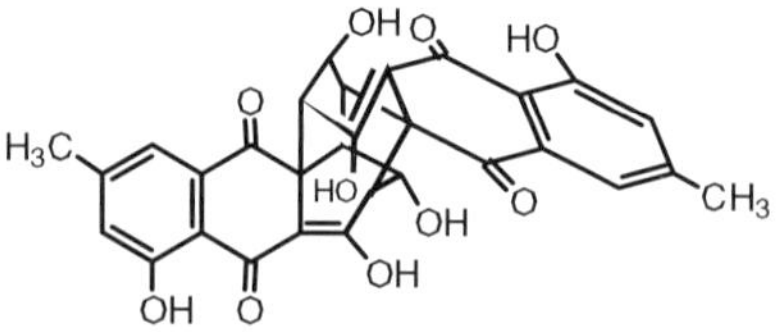

Rugulosin

and → luteoskyrin and caused almost identical clinical signs.

Rye Rye seems to be more contaminated with → ochratoxin A than → wheat.
Rye may contain the following → mycotoxins:
→ 3-acetyldeoxynivalenol
incidence: 4/31, conc. range: 15-38 µg / kg, Ø conc.: 24 µg / kg, country: Finland
→ aflatoxin B_1
incidence: 1/2, conc.: 15 µg / kg, country: Germany
incidence: 2/35, Ø conc. range: traces, country: USA
→ alternariol
incidence: 5/23, conc. range: 20-230 µg / kg, Ø conc.: 85 µg / kg, country: Poland
→ alternariol methyl ether
incidence: 1/8, conc.: 20 µg / kg, country: Germany
incidence: 7/49, conc. range: 20-460 µg / kg, Ø conc.: 117 µg / kg, country: Poland
→ citrinin
incidence: 1/2, conc.: 960 µg / kg, country: Canada
→ deoxynivalenol
incidence: 2/14, conc. range: 420-500 µg / kg, Ø conc.: 460 µg / kg, country: Austria
incidence: 1/1*, conc.: 56 µg / kg, country: Austria, *ecological
incidence: 1/1, conc.: 204 µg / kg, country: Canada
incidence: 8 products analysed, Ø conc.: 100 µg / kg, country: Canada
incidence: 9/10*, conc. range: 10-47 µg / kg, Ø conc.: 31 µg / kg, country: Finland, *imported from Germany, Hungary, Soviet Union, Sweden, USA
incidence: 20/50*, conc. range: ≤ 1250 µg / kg, Ø conc.: 160 µg / kg, country: Germany, *conventional
incidence: 28/50, conc. range: ≤ 500 µg / kg, Ø conc.: 427 µg / kg, country: Germany, *ecological
incidence: 4/22, Ø conc.: 406 µg / kg, country: Germany
incidence: 1/2, conc.: 950 µg / kg, country: Germany
incidence: 24*/31, conc. range: 30-2140 µg / kg, Ø conc.: 330 µg / kg, country: Germany, *moldy
incidence: 1/23, conc.: 100 µg / kg, country: Germany
incidence: 4/7*, conc. range: 31-86 µg / kg, Ø conc.: 53.5 µg / kg, country: Germany, *organic produce
incidence: 24/31, conc. range: 9-93 µg / kg, Ø conc.: 52 µg / kg, country: Finland
incidence: 5/5, Ø conc.: 1 µg / kg, country: Korea
incidence: 4/4, conc. range: 8-384 µg / kg, Ø conc.: 106 µg / kg, country: The Netherlands
→ HT-2 toxin
incidence: 1/31, conc.: 23 µg / kg, country: Finland
incidence: 1/23, conc.: 100 µg / kg, country: Germany
→ nivalenol
incidence: 1/1, conc.: 8 µg / kg, country: Canada
incidence: 4/22, Ø conc.: 12 µg / kg, country: Germany
incidence: 5/5, Ø conc.: 83 µg / kg, country: Korea
incidence: 3/4, conc. range: 10-34 µg / kg, Ø conc.: 21 µg / kg, country: The Netherlands
→ moniliformin
incidence: 3/3*, conc. range: 6100-12,300 µg / kg, Ø conc.: 9030 µg / kg, country: Poland, *hand-selected, visible fungal damage
→ nivalenol
incidence: 1/31, conc.: 33 µg / kg, country: Finland
ochratoxin A
incidence: 1/18, conc.: 2 µg / kg, country: Austria

inicdence: 18/41, conc. range: 5-100 µg / kg, country: Austria
incidence: 1/2, conc.: ca. 480 µg / kg, country: Canada
incidence: 177/503*, conc. range: 0.05-4.9 µg / kg (157 samples), 5-25 µg / kg (16 sa), > 25-121 µg / kg (4 sa), ∅ conc.: 1.2 µg / kg, country: Denmark, *conventional
incidence: 71/91*, conc. range: 0.05-4.9 µg / kg (55 samples), 5-25 µg / kg (12 sa), > 25-120 µg / kg (4 sa), ∅ conc.: 5.4 µg / kg, country: Denmark, *ecological
incidence: 8/22*, conc. range: 0.05-0.7 µg / kg, ∅ conc.: 0.1 µg / kg, country: Denmark, *conventional, imported
incidence: 149/682, conc. range: ≤ 4.9 µg / kg, country: Germany
incidence: 4/45, conc. range: 0.3-4.7 µg / kg, country. Germany
incidence: 5/29, conc. range: 50-800 µg / kg, ∅ conc.: 354 µg / kg, country: Poland
incidence: 62/228, conc. range: 5-2400 µg / kg, country: Poland
incidence: 44/94, conc. range: ≤ 4.9-28 µg / kg, country: Sweden
incidence: 2/12, conc. range: ≤ 16.7 µg / kg, country: The Netherlands
incidence: 5/14*, conc. range: 0.1-16.8 µg / kg, country: The Netherlands, *imported
→ penicillic acid
incidence: 1/29, conc. 2400 µg / kg, country: Poland
→ T-2 toxin
incidence: 1/31, conc.: 17 µg / kg, country: Finland
incidence: 10/25, conc. range: 200-700 µg / kg, country: Germany
→ zearalenone
incidence: 5/14, conc. range: 5-10 µg / kg, ∅ conc.: 9 µg / kg, country: Austria
incidence: 1/1, conc.: 2 µg / kg, country: Canada
incidence: 9/50*, conc. range: ≤ 7 µg / kg, ∅ conc.: 4 µg / kg, country: Germany, *conventional
incidence: 5/50*, conc. range: ≤ 199 µg / kg, ∅ conc.: 51 µg / kg, country: Germany, *ecological
incidence: 3/22, ∅ conc.: 5 µg / kg, country: Germany
incidence: 15/31, conc. range: ≤ 100 µg / kg, ∅ conc.: 17 µg / kg, country: Germany
incidence: 1/26, conc.: < 70 µg / kg, country: Germany
incidence: 3/5*, conc. range: 3-4 µg / kg, country: Korea, *polished
incidence: 1/29, conc.: 2000 µg / kg, country: Poland
incidence: 1/4, conc.: 11 µg / kg, country: The Netherlands
→ cereals

Rye bran may contain the following → mycotoxins:
→ deoxynivalenol
incidence: 1/1, conc.: 150 µg / kg, country: Austria
→ ochratoxin A
incidence: 2/3, conc. range: ≤ 0.6 µg / kg, country: The Netherlands
→ zearalenone
incidence: 1/1, conc.: 30 µg / kg, country: Austria
→ bran

Rye flour may contain the following → mycotoxins:
→ deoxynivalenol
incidence: 3/21, conc.: 150-335 µg / kg, ∅ conc.: 272 µg / kg, country: Austria
incidence: 3 products analysed, ∅ conc.: 120 µg / kg, country: Canada
incidence: 1/1, conc.: 174 µg / kg, country: Germany
incidence: 2/2*, conc. range: 55-56 µg / kg, ∅ conc.: 55.5 µg / kg, country: Germany, *organic
incidence: 1/1, conc.: 33 µg / kg, country: Germany
ergocornine (→ ergot alkaloids)

incidence: 4/4, conc. range: 1.8-6 µg / kg, country: Canada
ergocristine
incidence: 4/4, conc. range: 9.6-31 µg / kg, country: Canada
ergometrine
incidence: 4/4, conc. range: 1.2-9.3 µg / kg, country: Canada
ergosine
incidence: 4/4, conc. range: 1.4-5.3 µg / kg, country: Canada
ergotamine
incidence: 4/4, conc. range: 5.5-23 µg / kg, country: Canada
α-ergokryptine
incidence: 4/4, conc. range: 2.4-7 µg / kg, country: Canada
→ nivalenol
incidence: 1/1, conc.: 3 µg / kg, country: Germany
→ ochratoxin A
incidence: 4/35, conc. range: 0.1-17.7 µg / kg, country: Germany
incidence: 2/15, conc. range: ≤ 1.8 µg / kg, country; ∅ conc.: 0.28 µg / kg, Germany
incidence: 11/11, conc. range: tr-20 µg / kg, country: Japan
incidence: 8/14, conc. range: ≤ 1.2 µg / kg, ∅ conc.: 0.3 µg / kg, country: Sweden
→ zearalenone
incidence: 1/21, conc.: 10 µg / kg, country: Austria
→ flour

Rye grits may contain the following
→ mycotoxins:
→ ochratoxin A
incidence: 2/15, conc. range: 1.7-1.8 µg / kg, ∅ conc.: 1.75 µg / kg, country: Germany
→ barley grits, → maize grits, → wheat grits

S

Sago (→ cassava starch)
may contain the following → mycotoxins:
→ aflatoxin B_1
incidence: 2*/65, ∅ conc. 150 µg / kg, country: Thailand, *total: ∅ conc.: 294 µg / kg AFB_1, AFB_2, AFG_1, AFG_2

Sago hemolysis This disease which has been reported in Papua New Guinea since 1974 affects both males and females. Several clinical signs such as severe → anemia, sudden onset of → jaundice, and dark red urine are suggestive of hemolysis. In addition, fever and vomiting occurred in some cases, mental confusion and loss of consciousness in the worst cases. A mortality rate of almost 20% has been observed although blood transfusions were given.
"Stale" sago was suggested as being the cause of the disease because the patients consumed it the day before the onset of symptoms. Microscopical examination revealed bacterial and fungal cells but no fungal hyphae which would indicate excessive fungal growth. Unsuspicious, non-pathogenic microorganisms such as → Paecilomyces *lilanicus* and *Tilletiopsis minor* (one colony of each), two yeast species and *Bacillus* spp. were isolated after plating the sago on agar. So far no toxin (→ mycotoxins) has been detected in the sago.

Saint Anthony's fire → Ergotism

Salami may contain the following → mycotoxins:
→ aflatoxin B_1
incidence: 1/1, conc.: 5 µg / kg, country: Germany
→ sausages

Sambutoxin is a mycotoxin (→ mycotoxins) (4-hydroxy-5-(-4hydroxyphenyl)-1-methyl-3-[(2R,5S,6S)-tetrahydro-5-methyl-6-[(1E,3R,5S)-1,3,5-trimethyl-1-heptenyl]-2H-pyran-2-yl]-2(1H)-pyridinone) which was first isolated in 1994 (see Figure Sambutoxin).

CHEMICAL DATA
Empirical formula: $C_{28}H_{40}NO_4$, molecular weight: 453

FUNGAL SOURCES
mainly → Fusarium sambucinum Fuckel and → Fusarium oxysporum Schlecht. emend. Snyd. & Hans.

NATURAL OCCURRENCE
→ potatoes
This mycotoxin was not only found in rotten Korean potatoes but also in potatoes from parts of Iran where humans showed a high incidence of esophageal cancer.

TOXICITY
hemorrhagic (→ hemorrhage) (stomach, intestines of rats), feed refusal, weight loss
LD_{50} : 29.6 µg / egg (chicken)

DETECTION
HPLC

Sarcoma is a tumor composed of connective-like tissue.

Sausages → Ochratoxin A seems to be the most important mycotoxin in sausages. The use of OTA-containing → meat / and / or organs is the main cause for the contamination of sausages.

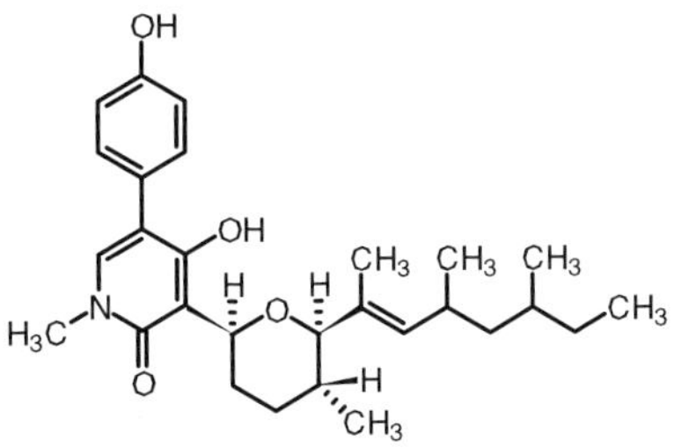

Sambutoxin

Aflatoxin contamination of sausages is primarily due to the use of mycotoxin contaminated → spices and / or the incorporation of aflatoxin producing fungi. Sausages may contain the following → mycotoxins:
→ aflatoxin B_1
incidence: 1/25, conc.: 7 µg / kg, country: Egypt
→ aflatoxin B_2
incidence: 1/25, conc.: 3 µg / kg, country: Egypt
incidence: 1*/5, conc.: 7 µg / kg, country: Germany, *German Rohwurst
→ aflatoxin G_2
incidence: 1/5, conc.: 30 µg / kg, country: Germany, *German Rohwurst
ochratoxin A
incidence: 20/125*, conc. range: 0.1-3.4 µg / kg, ∅ conc.: 0.9 µg / kg, country: Germany, *cooked, black pudding
incidence: 19/100*, conc. range: 0.1-1.7 µg / kg, ∅ conc.: 0.3 µg / kg, country: Germany, *liver-type
incidence: 19/100*, conc. range: 0.1-3.2 µg / kg, ∅ conc.: 0.8 µg / kg, country: Germany, *Bologna-type
incidence: 19/100*, ∅ conc.: 3.4 µg / kg, country: Germany, *scalding
incidence: 1/12*, conc.: 0.8 µg / kg, country: Switzerland, *scalding
incidence: 4/32*, conc.: ≤ 1.8 µg / kg, ∅ conc.: 0.6 µg / kg, country: UK, *black pudding
incidence: 25/206*, conc. range: 10-920 µg / kg, country: Yugoslavia, *total of smoked meat products
→ salami

Scabby grain intoxication → Red mold disease

Scented supar may contain the following → mycotoxins:
→ patulin
incidence: 1/147, conc.: nc, country: India

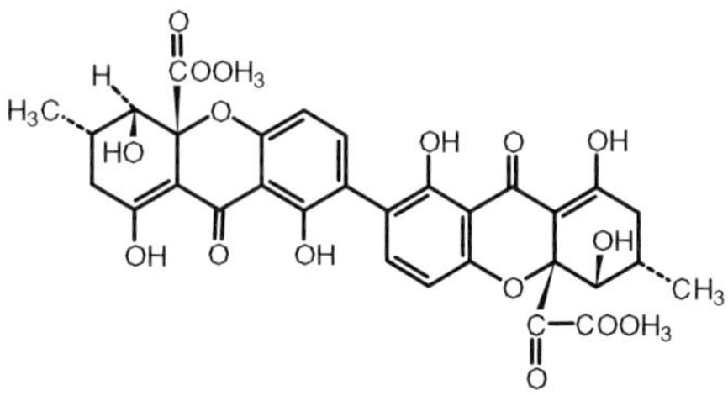

Secalonic acids. Secalonic acid D

Secalonic acids represent a group of six diastereoisomeric toxic fungal pigments (ergochromes, xanthone dimers) initially isolated in 1965 and 1966 from cultures of → Claviceps *purpurea*. The most important member is secalonic acid D (see Figure Secalonic acids).

CHEMICAL DATA
Empirical formula: $C_{32}H_{30}O_{14}$, molecular weight: 638 (all secalonic acids)

FUNGAL SOURCES
Secalonic acids are produced by the five fungal genera → Aspergillus, *Claviceps*, → Penicillium, → Phoma, and *Pyrenochaeta*. Secalonic acid D is the major toxic fungal metabolite of *P. oxalicum*.

NATURAL OCCURRENCE
→ maize dust 300-4500 µg secalonic acid D / kg, grain dust (secalonic acid D); Secalonic acids are produced on a variety of substrates (→ grains) suitable for human consumption such as → barley, maize, → rice, → sorghum, → soybeans, and → wheat.

TOXICITY
toxic to mice and rats, → teratogenic, possibly → mutagenic.
LD_{50} (po) : 24.6 mg / kg bw new borne rats

DETECTION
ELISA, HPLC, TLC

FURTHER COMMENTS
The six secalonic acids A-G (B = E) are known.
Secalonic acids may be involved in → dyspnea, grain fever and airway

obstruction of grain workers. Since secalonic acid D seem to be produced almost exclusively in stored grain (maize), proper grain storage should inhibit contamination. Even in fungal-contaminated maize little or no secalonic acid D could be detected prior to harvest.

Semi-hard cheese → cheese (semi-hard)

Semolina → maize grits

Septic angina → Alimentary toxic aleukia

Sesame oil may contain the following → mycotoxins:
→ ochratoxin A
incidence: 1/3, conc.: 0.4 μg / kg, country: UK
→ oil

Sesame seeds may contain the following → mycotoxins:
→ aflatoxin B_1
incidence: 4*/19, conc. range: 4-10 μg / kg, country: Germany, *moldy
incidence: 2*/75, Ø conc.: < 1 μg / kg, country: Thailand, *total Ø conc.: < 10 / kg AFB_1, AFB_2, AFG_1, AFG_2
→ nivalenol
incidence: 2/7, Ø conc.: 10 μg / kg, country: Yemen
→ ochratoxin A
incidence: 3/3, conc.: ≤ 0.4 μg / kg, country: UK

Sherbet may contain → aflatoxin M_1 if it is made from naturally AFM_1 contaminated → milk. The toxin remained stable during 8 months of frozen storage.

Sherry → wine

Shoshin-kakke → Acute cardiac beriberi

Shoyu may contain → aflatoxins if (i) → Aspergillus flavus Link or → Aspergillus parasiticus Speare are used to make koji for soy sauce (ii) the koji may be contaminated with an aflatoxin producer. The presence of *Lactobacillus delbrueckii* does not enable an aflatoxin free product to be made.
Shoyu may contain the following → mycotoxins:
→ aflatoxin G_2
incidence: 1/149, conc.: nc, country: Taiwan
aflatoxin
incidence: 1/nc, conc.: nc, country: Hong Kong
→ citrinin
incidence: nc, conc.: nc, country: China
→ Oriental fermentations

Shrimps (fried with pork, garlic, & chilli peppers)
may contain the following → mycotoxins:
→ aflatoxin B_1
incidence: 1/1, conc.: 207 μg / kg, country: Thailand, *total: 355 μg AFB_1, AFB_2, AFG_1, AFG_2 / kg
→ fish

Small grains Small grains (→ barley, → millet, → oats, → rice, → rye, → sorghum, → wheat) are only very rarely contaminated by → aflatoxins.
Small grains may contain the following → mycotoxins:
→ aflatoxins
incidence: 19/3489, Ø conc.: 5 μg / kg, country: USA

Snack foods may contain the following → mycotoxins:
→ deoxynivalenol
incidence: 25/44, conc. range: ≤ 450 μg / kg, country: USA
→ ochratoxin A
incidence: 4/11, conc. range: 0.1-0.49 μg / kg, country: Germany

→ zearalenone
incidence: 1/9, conc.: 2.9 µg / kg, country: USA

Soft drinks may contain the following → mycotoxins:
→ patulin
incidence: 2/24, conc. range: 2-10 µg / kg, country: Germany
→ apple juice, → breakfast drinks, → fruit juices, → grape juice

Solaniol → neosolaniol

Sordariales → Ascomycota

Sorghum (*Sorghum* spp.)
Good quality sorghum does not seem to be contaminated by → Alternaria mycotoxins. However, in weathered and discolored sorghum (U.S.) which was repeatedly wetted and then dried during rainy periods, the two → Alternaria metabolites → alternariol (AOH) and → alternariol methyl ether (AME) were detected. A correlation between the degree of grain discoloration and rainy days during plant growth (September and October) and the level of alternariols was established. Fluctuation of temperature, change in relative humidity and excessive rainfall seem to promote *Alternaria* infection and subsequent toxin production in the seeds. As the number of rain-free days increased the AOH/AME level decreased.
Contamination with → alternariols might be due to several separate invasions of the maturing and mature seeds. → Altenuene and → tenuazonic acid were not detected, while altertoxin I (→ altertoxin I-III) occurred only in trace amounts in weathered sorghum. It was suggested, that the time for mycelial growth of *Alternaria* spp. was not sufficient for the synthesis of these late-produced metabolites. Wet conditions during or shortly after ripening of the grain contribute to *Alternaria* mycotoxin contamination whereas the date of harvest was not decisive.
Sorghum may contain the following → mycotoxins:
→ aflatoxin B_1
incidence: 3/6, conc. range: 70-120 µg / kg, country: Thailand
incidence: 2/6, conc. range: 30-35 µg / kg, country: Tunisia
incidence: 10/788, Ø conc.: 12 µg / kg, country: USA
incidence: 6/533, conc. range: 3-19 µg / kg, country: USA
→ aflatoxin B_2
incidence: 2/6, conc. range: nc, country: Thailand
→ aflatoxin G_1
incidence: 1/6, conc.: nc, country: Thailand
incidence: 3/533, conc. range: 3-19 µg / kg, country: USA
→ aflatoxin (no specification)
incidence: 2/2, conc. range: 29 µg / kg, country: Philippines
→ aflatoxins
incidence: 2/8, conc. range: 2-16 µg / kg, Ø conc.: 9 µg / kg, country: Gambia
incidence: 26*/69, conc. range: 1-100 µg / kg (19 samples), 100-1000 µg / kg (5 sa), > 1000 µg / kg (1 sa), country: Uganda, * 16 samples contained AFB_1, 11 AFB_2, 13 AFG_1, 1 AFG_2
incidence: 4/786, conc. range: ≤ 50 µg / kg, Ø conc.: 17 µg / kg, country: USA
incidence: 6/533, conc. range: 3-19 µg / kg, country: USA
incidence: 103/200, conc. range: 1-100 µg / kg, country: USA
incidence: 2/66, conc. range: 13-50 µg / kg, Ø conc.: 61.5 µg / kg, country: USA
altenuene
incidence: 3/12, conc. range: 120-1500 µg / kg, Ø conc.: 670 µg / kg, country: USA
incidence: 5/20, conc. range: 20-700 µg / kg, Ø conc.: 264 µg / kg, country: India (*Sorghum bicolor* (L.) Moench)

alternariols* (alternariol and alternariol methyl ether)
incidence: 21/63, conc. range: tr-7900 µg / kg country: USA, *weathered, discolored sorghum
alternariol methyl ether
incidence: 7/20*, conc.: 600-1800 µg / kg, ∅ conc.: 1012 µg / kg, country: India, **Sorghum bicolor* (L.) Moench
altertoxin I
incidence: 3/12, conc. range: traces, country: USA
→ deoxynivalenol
incidence: 31/32, conc. range: 1540 µg / kg, ∅ conc.: 190 µg / kg, country: USA
→ nivalenol
incidence: 1/5, con.: 100 µg / kg, country: Yemen
tenuazonic acid
incidence: 5/20*, conc. range: 1300-5600 µg / kg, ∅ conc.: 3380 µg / kg, country: India, **Sorghum bicolor* (L.) Moench
→ zearalenone
incidence: 60/200, conc. range: 251-1500 µg / kg, country: USA
incidence: 57/197, conc. range: 400 µg / kg (4 samples), 400-900 µg / kg (16 sa), 1000-5000 µg / kg (35 sa), > 5000 µg / kg (2 sa), country: USA
→ cereals, → millet

Sorghum meal may contain the following → mycotoxins:
→ fumonisin B_1
incidence: 2/2, conc. range: 20 µg / kg, ∅ conc.: 20 µg / kg, country: Botswana
incidence: 1/1, conc.: 28,200 µg / kg, country: Burundi
→ maize meal

Soy sauce → shoyu

Soybean concentrate may contain the following → mycotoxins:
→ ochratoxin A
incidence: 1/2, conc. range: 50-500 µg / kg, country: UK

Soybean flour may contain the following → mycotoxins:
→ aflatoxins
incidence: 1/4, conc.: nc, country: UK
→ ochratoxin A
incidence: 1/4, conc. range: 50-500 µg / kg, country: UK
incidence: 4/21*, conc. range: < 50-500 µg / kg, country: UK, *defatted
→ flour

Soybean milk powder may contain the following → mycotoxins:
→ aflatoxin M_1
incidence: 1/8, conc.: 0.015-0.035 µg / kg, country: Italy

Soybeans Several factors such as high moisture content, number of splits and high total damage contribute to aflatoxin contamination of soybeans. However, generally soybeans are not a good substrate for aflatoxin production. The availability of zinc bound to phytic acid seems to be a decisive factor. With the breakdown of phytic acid due to heat or the addition of zinc increased aflatoxin production was observed in soybeans. Reddish and dark discolored soybean seeds indicate the potential presence of → Fusarium mycotoxins, especially → HT-2 toxin. Reddish seeds contained the highest mycotoxin concentrations with a maximum found in the seed coat. However, although in lower concentration HT-2 toxin was also present in the non-reddish seeds. The absence of reddish seeds therefore does not necessarily denote the absence of Fusarium mycotoxins.

Soybeans may contain the following → mycotoxins:
→ aflatoxin B_1
incidence: 2/866, conc. range: 7-10 µg / kg, ∅ conc.: 8.5 µg / kg, country: USA
incidence: 5/34, conc. range: < 5-20 µg / kg, country: USA
→ aflatoxin G_1
incidence: 1/866, conc.: 4 µg / kg, country: USA
→ aflatoxin (no specification)
incidence: 12/25, conc. range: ≤ 48 µg / kg, ∅ conc.: 18 µg / kg, country: Philippines
→ deoxynivalenol
incidence: 2?/30, conc. range: 490-1000 µg / kg, country: Canada
incidence: 1/2*, conc.: 36 µ / kg, country: Papua New Guinea, *imported
→ nivalenol
incidence: 1/2*, conc.: 50 µg / kg, country: Papua New Guinea, *imported
→ HT-2 toxin
incidence: 2?/30, conc. range: tr-330 µg / kg, country: Canada
→ ochratoxin A
incidence: 9/25, conc. range: < 50-500 µg / kg, country: UK
→ zearalenone
incidence: 2/17*, conc. range: > 200 µg / kg, country: Uruguay, *and by-products
→ beans, → cabbage, → cowpeas, → lentils, → pigeon peas, → peas, → vegetables

Spaghetti During cooking of spaghetti, average losses of → deoxynivalenol amounted to 43-53% of the amount present before cooking.
Spaghetti may contain the following → mycotoxins:
→ aflatoxin B_1
incidence: nc, conc. range: ≤ 12.5 µg / kg, country: Canada
deoxynivalenol
incidence: 7/7, conc. range: < 10-175 µg / kg, ∅ conc.: 89.3 µg / kg, country: Austria
incidence: 8/2*, conc. range: 2960-5020 µg / kg, country: Canada
* 2 wheat samples served for 8 different noodle preparation
→ cereals

Spelt may contain the following mycotoxins:
→ ochratoxin A
incidence: 1/22, conc.: 0.9 µg / kg, country: Germany

Spices (no specification)
During growth in tropical climates, spices are often exposed to extremely wide ranges of rainfall, temperature and humidity. Although these factors in addition to their botanical origin may contribute to a pre-harvest mycotoxin contamination of the spices in a tropical environment, the sun drying process represents the greatest potential for contamination. The typical ground drying in the open air with high temperatures and humidities favor mold growth, especially → Aspergillus spp., and subsequent mycotoxin (→ aflatoxins) production. During handling and storage mycotoxin contamination is also possible.
The largest amounts of spices are used in the → meat industry where they represent a potential hazard for mold and mycotoxin contamination of the endproducts. However, due to their essential oils which reduce mold growth as well as aflatoxin production spices are not an ideal substrate for aflatoxin formation. In addition, spices generally are consumed in small amounts and therefore contribute little to the total health hazard posed by → mycotoxins, especially aflatoxins.
Spices (mixed) may contain the following mycotoxins:

→ aflatoxin B_1
incidence: 5/37, conc. range: 0.2-0.8 µg / kg, country: Japan
→ aflatoxin B_2
incidence: 5/37, conc. range: 0.2 µg / kg, country: Japan
aflatoxins (AFB_1, AFB_2, AFG_1, AFG_2)
incidence: 1/4*, conc.: 4 µg / kg, country: UK
→ fumonisins (FB_1, FB_2)
incidence: nc/4*, conc. range: 13-17 µg / kg, country: UK
→ ochratoxin A
incidence: 5/108, conc. range: nc, country: India
incidence: 1/4*, conc.: 2.6 µg / kg, country: UK
→ zearalenone
incidence: nc/4*, conc. range: 3.2-5.2 µg / kg, country: UK
*five spice powder
According to Frisvad (1988) spices may be contaminated with the following mycotoxins: → aflatoxins, → citrinin, → cyclopiazonic acid, → luteoskyrin, → ochratoxin A, → penicillic acid, rubratoxin (→ rubratoxins), → rugulosin, → sterigmatocystin, → viomellein, → xanthomegnin.
The following spices showed a contamination with mycotoxins:
→ bay leaf, → cardamom, → cardamom, greater, → cayenne pepper, → chilli, → chilli pickles, → chilli powder, → chilli sauce, → coriander, → cumin, → curcuma, → curry, → curry paste, → fennel, → fenugreek, → nutmeg, → pepper, → tandoori, → turmeric

St. Nectaire cheese → cheese, St. Nectaire

Starch may contain the following → mycotoxins:
→ fumonisin B_1
incidence: 1/1, conc.: 283 µg / kg, country: USA
→ fumonisin B_2
incidence: 1/1, conc.: 70 µg / kg, country: USA

Sterigmatocystin as a furofuran (3*a*,12*c*-dihydro-8-hydroxy-6-methoxy-furo[3',2',4,5]furo[3,2-*c*]xanthene-7-one) is a precursor in the biosynthesis of → aflatoxin B_1 (see Figure Sterigmatocystin). It was originally isolated and named in 1954 (→ mycotoxins). In 1962 elucidation of its molecular structure followed.

Chemical Data
Empirical formula $C_{18}H_{12}O_6$, molecular weight: 324

Fungal Sources
mainly → Aspergillus versicolor (Vuill.) Tiraboshi and *Emericella nidulans*, further producers: e.g. → Aspergillus spp. (ca. 20 different species), → Emericella spp., → Eurotium spp. Sterigmatocystin is an intermediate in the biosynthesis of → aflatoxins by → Aspergillus flavus Link and → Aspergillus parasiticus Speare.

Natural Occurrence
→ barley, → breakfast cereals, → cheese, → cheese, Edam Cake, → cheese, Gouda, → cheese, Moravian Block, → coffee beans, → corn flakes, → fennel, → maize, → oil seed rape, → pecans, → pepper, → rice, → wheat
In general, isolation succeeded only from severely moldy substrates. Apart from that this mycotoxin is rarely found in nature. However, analytical methods for its detection are not as sensitive as for the → aflatoxins. Therefore, low concentrations in → food products may not always be detected. Residues in fresh meats are unlikely to be expected although in Canada sterigmatocystin has occasionally been detected in feeds. Here, a higher degree of sterigmatocystin contaminated → grains in storage has also been reported.
Sterigmatocystin has also been reported to be a contaminant of marihuana.

Toxicity
hepatotoxic, nephrotoxic, carcinogenic, → mutagenic, → teratogenic
The toxic effects are much the same as those of aflatoxin B_1 but it is less acutely toxic.
LD_{50} (po): 60-166 mg / kg bw rat
In rat the metabolized sterigmatocystin is primarily secreted via the gastrointestinal tract and to a minor degree via the urine and feces within 12-24 hours.

Detection
ELISA, GC-MS, HPLC, TLC

Possible Mycotoxicosis
Implication in the etiology of chronic liver disease in man in Africa is suggested.

Further Comments
Sterigmatocystin was the first known substance of natural origin which contained the dihydrofurobenzofuran system.
At least eight derivatives are known.
Production: A minimum → a_w of 0.80 is necessary for sterigmatocystin production on bread (*A. versicolor*), a_w 0.85 on agar media. The optimum a_w for production of this mycotoxin lays between 0.92-0.93.
Sterigmatocystin is only rarely found in → foods which are usually visibly moldy. Despite its considerable carcinogenicity, it does not seem to be a significant hazard to human health.
Reduction / elimination: In milled brown rice sterigmatocystin concentration decreased gradually with a decrease in milling yield.

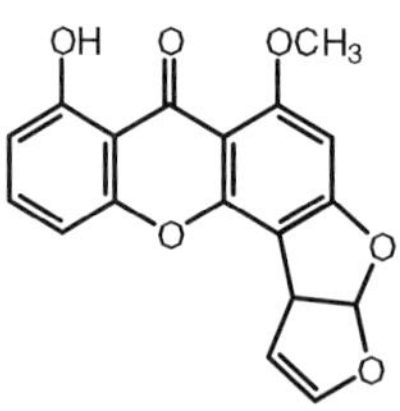

Sterigmatocystin

In contrast to the aflatoxins sterigmatocystin was stable in 2% ammonia solution.

Stilton cheese → cheese, Stilton

Storage fungi The original source of these fungi is the field. They represent those microorganisms which are most tolerant to low water availability and therefore primarily grow on stored cereal → grains. As xerophilic saprophytes they develop at relative humidities of 65-90% where free water is not available; e.g. a seed moistur of only ≈ 14% (cereals) is sufficient for initial growth of → Eurotium *halophilicum* and → Aspergillus restrictus G. Sm. *Eurotium* spp. represent the most important genus within the group of storage fungi, growing at → a_w values from 0.62 to 0.75. Members of the genera → Aspergillus and → Penicillium are responsible for mycotoxin (→ mycotoxins) contamination. Mycotoxin production set in if the moisture content of cereal seeds rises to 13-16%. However, water activities of less than a_w 0.70 which correspond to a moisture content of ≈ 15% (most → cereals) minimizes growth as well as mycotoxin production. Highest mycotoxin yields may occur at a water content of 20-25%.
→ field fungi

Sunflower seed oil may contain the following → mycotoxins:
→ aflatoxin B_1
incidence: 2/21, conc. range: 0.8-1 µg / kg, ∅ conc.: 0.9 µg / kg, country: Germany
→ aflatoxin G_1
incidence: 1/21, conc.: 0.3 µg / kg, country: Germany
→ coconut oil, → oil, → olive oil, → peanut oil

Sunflower seeds are a good substrate for aflatoxin production which may be due

to their high lipid content. However, the hard and thick seed coat impedes penetrability for aflatoxigenic fungi (→ Aspergillus flavus Link, → Aspergillus parasiticus Speare) and should be responsible for low toxin production. Broken seeds gave substantially higher mycotoxin yields than whole seeds almost comparable to other → oil seeds such as → peanuts and → soybeans.
To prevent aflatoxin contamination sunflower seeds should be stored with the seed coat. Dehulling should be carried out just prior to extraction of → oil. The phenomenon of a hard seed coat protecting seeds against fungal penetration is also known from e.g. certain varieties of peanuts and cotton seeds.
Sunflower seeds may contain the following → mycotoxins:
→ aflatoxin B_1
incidence: 1/4*, conc.: 10.5 µg / kg, country: Germany
→ aflatoxin B_2
incidence: 1/4*, conc.: 0.5 µg / kg, country: Germany
→ aflatoxin G_1
incidence: 1/4*, conc.: 0.4 µg / kg, country: Germany
→ aflatoxin G_2
incidence: 1/4*, conc.: 0.03 µg / kg, country: Germany
*moldy
→ aflatoxins (no specification)
incidence: 7*/136, conc. range: 5-19.9 µg / kg, country: Canada, *AFB_1, AFB_2, AFG_1, AFG_2
incidence: 9/135, conc. range: 25-230 µg / kg, country: Tunesia
→ alternariol
incidence: 37/50, conc. range: 35-792 µg / kg, ∅ conc.: 166 µg / kg, country: Argentina
incidence: 128/150, conc. range: 50-676 µg / kg, ∅ conc.: 189 µg / kg, country: Argentina
incidence: 2/2*, conc. range: 357-1840 µg / kg, ∅ conc.: 1090 µg / kg, country: Italy, *samples visibly affected by → Alternaria rot
→ alternariol methyl ether
incidence: 31/50, conc. range: 90-630 µg / kg, ∅ conc.: 114 µg / kg, country: Argentina
incidence: 70/150, conc. range: 30-836 µg / kg, ∅ conc.: 202 µg / kg, country: Argentina
incidence: 1/2*, conc.: 129 µg / kg, country: Italy, *samples visibly affected by → Alternaria rot
→ cyclopiazonic acid
incidence: 1/1*, conc.: 10,000 µg / kg, country: USA, *moldy
→ ochratoxin A
incidence: 4/25, conc. range: 0.2-0.49 µg / kg (2 samples), 1.5-9.99 µg / kg (2 sa), country: Germany
→ tenuazonic acid
incidence: 98/150, conc. range: 2500-15,796 µg / kg, ∅ conc.: 6459 µg / kg, country: Argentina

Sweet potatoes → tubers

Swine → pork

Swiss cheese → cheese, Swiss

T

T-2 toxin belongs to the group of naturally occurring → trichothecenes (3α-hydroxy-4,15-diacetoxy-8α-(3-methylbutyryloxy)-12,13-epoxytrichthec-9-ene) produced by different species of the genus → Fusarium (see Figure T-2 toxin). During the search for causatives of the → moldy corn toxicosis in 1966 besides → diacetoxyscirpenol this new trichothecene, named T-2 toxin, could be isolated. The molecular structure was established in 1968.

Chemical Data
Empirical formula: $C_{24}H_{34}O_9$, molecular weight: 466

Fungal Sources
F. acuminatum, → Fusarium avenaceum (Fr.) Sacc. (?), → Fusarium culmorum (W. G. Smith) Sacc. (?), *Fusarium equiseti* (Corda) Sacc. sensu Gordon, → Fusarium graminearum Schwabe, → Fusarium oxysporum Schlecht. emend. Snyd. & Hans. (?), → Fusarium poae (Peck) Wollenw., *F. semitectum*, → Fusarium sporotrichioides Sherb.

Natural Occurrence
→ barley, → beans, → beer, → curry, → ginger, → grains, → maize, → oats, → rye, → wheat
Although this mycotoxin is quite common in animal feed, T-2 toxin is a rare contaminant of → foods.

Toxicity
During metabolization of T-2 toxin into more hydrophilic compounds the trichothecane skeleton is not modified.
dermatotoxic (like → HT-2 toxin), emetic, → immunosuppressive, cancerogenic (?)
LD_{50} (po): 4 mg / kg bw rat
clinical symptoms: e.g. inflammation and hemorrhaging (→ hemorrhage) of the digestive tract, → edema, → leucopenia, degeneration of the bone marrow, and death (→ cattle, swine)
inhibition fo the initiation step of protein synthesis on polyribosomes

Detection
ELISA, GC, HPLC, MS, RIA, TLC

Possible Mycotoxicosis
→ alimentary toxic aleukie, → moldy corn toxicosis

Further Comments
In general, T-2 toxin is an uncommon fungal secondary metabolite because most cereal → grains are harvested appropriately.
Feeding studies reveal that T-2 toxin at levels typically encountered in contaminated feeds is extensively metabolized and rapidly eliminated from most of the host tissues (swine). The liver appears to be the primary site of residue accumulation. A higher transmission rate for T-2 toxin and T-2 metabolites into edible tissue occurred in the case of chick compared to pig. A hydroxy derivative and a deacetylated hydroxy derivative of the toxin are the major toxic metabolites found in tissue (*in vivo*). → Carry over into the → milk is much less than 1%.
A synergistic effect with → deoxynivalenol has been discussed.
Reduction / elimination: During the wet → milling of maize the major portion (almost 70%) of T-2 toxin initially present was found in the steep and process water, 4% (8%) occurred in the starch whereas the rest was detected in the germ, gluten, and fiber.
No residues could be detected in → oil prepared from the germ following the refining process.

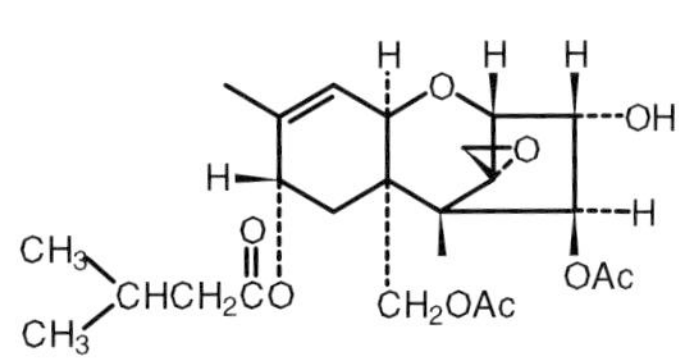

T-2 toxin

Maize syrup will contain only low T-2 toxin levels because of the acidic processing conditions.

Table wine → wine

Tachycardia Excessive increase in heart rate.

Taco → Tortilla

Tandoori may contain the following → mycotoxins:
→ aflatoxins (AFB_1, AFB_2, AFG_1, AFG_2)
incidence: nc/3, conc. range: 1.9-6.8 µg / kg, country: UK
→ fumonisins (FB_1, FB_2)
incidence: 1/3, conc.: 19 µg / kg, country: UK
→ nivalenol
incidence: nc/3, conc. range: 60-126 µg / kg, country: UK
→ ochratoxin A
incidence: nc/3, conc.: 2.2-23.9 µg / kg, country: UK
T2-triol
incidence: 1/3, conc.: 281 µg / kg, country: UK
→ spices

Tapioca and products containing tapioca are starchy foods made from → cassava. Tapioca may contain the following → mycotoxins:
→ ochratoxin A
incidence: 3/17, conc. range: < 5 µg / kg, country: UK
→ zearalenone
incidence: 6/17, conc. range: < 5 µg / kg, country: UK

Taro may contain the following → mycotoxins:
→ aflatoxin B_1
incidence: 1*/140, conc.: 30 µg / kg, country: Thailand, *total: 46 µg / kg AFB_1, AFB_2, AFG_1, AFG_2

Temperature Environmental factors are decisive for mycotoxin production and one of the most important besides the → a_w is temperature.
In general, → Penicillium spp. and → Fusarium spp. need lower temperatures for the synthesis of → mycotoxins, e.g. → patulin: 0-24 °C → Penicillium expansum Link, 4-31 °C *P. patulum*, → Fusarium mycotoxins: 1.5 to 4 °C → Fusarium sporotrichioides Sherb. than → Aspergillus spp. (no patulin production below 12 °C). A similar pattern was also observed for ochratoxin production by → Penicillium aurantiogriseum Dierckx?, → Pencillium viridicatum Westling? (= *Penicillium verrucosum* Dierckx), and *A. ochraceus*. *Penicillium* species are able to produce mycotoxins over a broader range of temperature than *Aspergillus* spp. Since *Penicillium* spp. prefer temperate climatic regions (Northern Europe, Canada) their mycotoxins predominate in → foods originating from these areas while *Aspergillus* species and their mycotoxins are more common in warmer climates (South-East Asia, Africa).

Tenuazonic acid is a 3-acetyl-5-*sec*-butyltetramic acid (3-acetyl-5-[(1S)-1-methylpropyl]-2H-pyrrolol-2-one (5S)-) produced by → Alternaria spp. and other fungi (see Figure Tenuazonic acid). It was first isolated in 1957 (→ mycotoxins) and probably possesses the highest toxicity of all → Alternaria mycotoxins.

CHEMICAL DATA
Empirical formula: $C_{10}H_{15}NO_3$, molecular weight: 197

FUNGAL SOURCES
Alternaria spp. (most important → Alternaria alternata (Fr.) Keissler), *A. citri*, *A. japonica*, *A. kikuchiana*, → Aspergillus spp. (→ Aspergillus nomius Kurtzman et al.), *Magnaporthe grisea* (anamorph: *Pyriculuria oryzae*), → Phoma *sorghina*.

Natural Occurrence
→ apples, → mandarin fruits, → olives, → pepper, → ragi, → sorghum, → sunflower seeds, → tomatoes, → tomato paste, → wheat

Toxicity
acutely very toxic, inhibition of protein synthesis, cardiovascular collapse, salivation, → anorexia, erythema, → convulsions, emesis, → tachycardia, massive gastrointestinal hemorrhages (→ hemorrhage) etc. and death;
antiviral, antibacterial, antifungal, phytotoxic and antitumor activity
LD_{50} (po): 81 / 168 mg / kg bw female / male mice

Detection
GC, HPLC, spectroscopy, TLC

Possible Mycotoxicosis
It is suggested that tenuazonic acid is involved in the etiology of a hematologic disorder named → onyalai.

Further Comments
Tenuazonic acid seems to be one of the most important mycotoxins among the *Alternaria* mycotoxins.
It has been reported that tenuazonic acid occurred as magnesium, calcium, sodium and potassium salts in unacidified cultures.

teratogenic is a substance, causing malformations.

Tercinin (Syn.: → patulin)

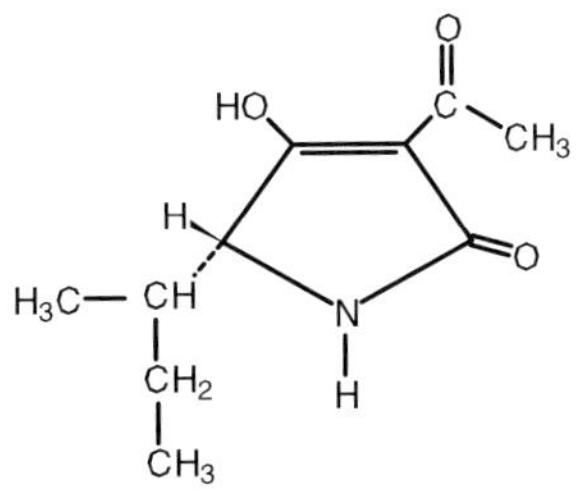

Tenuazonic acid

Tilsit cheese → cheese, Tilsit

Time In general, mycotoxin production starts at the same time as the formation of conidia with an increase up to the period of intense sporulation. Subsequently a decrease in mycotoxin (→ mycotoxins) synthesis occurs sometimes associated with a metabolization of these secondary fungal metabolites.

Toast → bread

Tomato ketchup may contain the following → mycotoxins:
→ aflatoxin B_1
incidence: 2/18, conc. range: ≈ 1 µg / kg, country: Germany

Tomato paste may contain the following → mycotoxins:
→ tenuazonic acid
incidence: 6/8, conc. range: 10-100 µg / kg, country: Canada
incidence: 8/nc (several brands), conc. range: 0.01-0.1 µg / kg, country: USA

Tomatoes Decay of the post-harvest tomato fruit (black rot lesion) is mainly due to → Alternaria alternata (Fr.) Keissler. This → black mold predominatly invades tomato tissue damaged by sun scald. Warm and rainy weather or dew formation on the fruit surface favors the disease. Tomatoes in the ripe stage are more susceptible than in the green stage. Substantial losses of tomatoes, especially those used for canning, have been reported. Fungal deterioration of the → fruits is often associated with the contamination of → Alternaria mycotoxins. In rotted tomatoes → alternariol, → aternariol methyl ether, and → tenuazonic acid are the most common → mycotoxins. Infections with → Aspergillus flavus Link, *A. niger* and *Rhizopus stolonifer* are of minor importance.

Tomatoes may contain the following → mycotoxins:
→ aflatoxin B_1
incidence: 1/8, conc.: 5 µg / kg, country: Germany
→ altenuene
incidence: 4/19, conc. range: < 100-1100 µg / kg, ∅ conc.: 100 µg / kg, country: USA
alternariol
incidence: 1*/nc, conc.: 1274 µg / kg, country: Italy
incidence: 6/19, conc. range: < 100-5.300 µg / kg, ∅ conc.: 300 µg / kg, country: USA
alternariol methyl ether
incidence: 2*/nc, conc. range: 37-268 µg / kg, country: Italy
incidence: 5/19, conc. range: < 100-800 µg / kg, ∅ conc.: 100 µg / kg, country: USA
tenuazonic acid
incidence: 2*/nc, conc. range: 24-7210 µg / kg, country: Italy
incidence: 73/142* (USA), conc. range: 400-1900 µg / kg (28 samples), 2000-70,000 µg / kg (45 sa), country: USA
incidence: 11/19, conc. range: < 100-139,000 µg / kg, ∅ conc.: 17,600 µg / kg, country: USA
*samples visibly affected by *Alternaria* rot

Tortilla chips Experimental studies show that aflatoxin losses (→ aflatoxins) during cooking are associated primarily with the alkaline conditions.
Tortilla chips may contain the following → mycotoxins:
→ fumonisin B_1
incidence: 8/12, conc. range: tr-216 µg / kg, country: Canada
incidence: 1/2, conc.: 60 µg / kg, country: Italy
incidence: 1/2, conc.: 30 µg / kg, country: USA
incidence: 2/2, conc. range: ca. 310-320 µg / kg, country: USA
→ fumonisin B_2
incidence: 1/2, conc.: 10 µg / kg, country: Italy
hydrolyzed fumonisin B_1
incidence: 2/2, conc.: present, country: USA
→ fumonisins (no specification)
incidence: 14/14*, conc. range: 200-1450 µg / kg, country: USA, *white
incidence: 1/1*, conc.: 400 µg / kg, country: USA, *yellow
incidence: 2/2*, conc. range: 400-1000 µg / kg, ∅ conc.: 700 µg / kg, country: USA, *blue
incidence: 2/2*, conc. range: 300-400 µg / kg, ∅ conc.: 350 µg / kg, country: USA, *organic blue
→ maize

Tortillas Tortillas as a staple food in Mexico and Central America are traditionally made from → maize. During cooking under alkaline conditions the high pH of the alkaline system seems to promote ionization of starch hydroxyl groups, producing Ca-starch crosslinks. This nixtamalization may hydrolyze → fumonisin B_1 to HFB_1.
The alkali processing during tortilla and tortilla-type food preparation causes an effective reduction in the amounts of → aflatoxins in contaminated maize. This might be due to the initial soaking of the maize in lime water and a chemical change by alkali.
Tortillas may contain the following → mycotoxins:
fumonisin B_1
incidence: 9/11*, conc. range: 24-612 µg / kg, ∅ conc.: 227 µg / kg, country: Canada, *dried
incidence: 7/7, conc. range: 210-1070 µg / kg, ∅ conc.: 601 µg / kg, country: Mexico

incidence: 1/2, conc.: 120 µg / kg, country: USA
incidence: 1/3, conc.: 60 µg / kg, country: USA
incidence: 50/52, conc. range: 12-672 µg / kg, ∅ conc.: 187 µg / kg, country: USA / Mexico
→ fumonisin B_2
incidence: 6/11*, conc. range: 26-218 µg / kg, ∅ conc.: 73.5 µg / kg, country: Canada, *dried
incidence: 6/7, conc. range: 50-180 µg / kg, ∅ conc.: 88.3 µg / kg, country: Mexico
incidence: 1/2, conc.: 30 µg / kg, country: USA
hydrolyzed fumonisin B_1
incidence: 5/7, conc. range: 10-50 µg / kg, ∅ conc.: 22 µg / kg, country: Mexico
incidence: 48/52, conc. range: 13-204 µg / kg, ∅ conc.: 82 µg / kg, country: USA/Mexico
→ fumonisins (FB_1, FB_2, FB_3)
incidence: 6/20*, conc. range: 10-31 µg / kg, ∅ conc.: 13 µg / kg, country: UK, *as well as taco and enchilada
fumonisins (no specification)
incidence: 4/5, conc. range: ≤ 800 µg / kg, country: USA

Tremorgenic mycotoxins There are only a few → mycotoxins that act on the level of the central nervous system in vertebrate animals. Members of the first class like → citreoviridin and steltoxin are responsible for respiratory arrest and → paralysis. The tremorgenic mycotoxins which all possess an indole moiety from tryptophan belong to the second class and induce trembling in vertebrate animals. Based on chemical similarity (nitrogen content) the tremorgens are classified into three groups: → penitrems A, B, and C (→ Penicillium spp.) as well as aflatrem (→ Aspergillus flavus Link) which was the first isolated fungal tremorgen (1964) contain only one nitrogen per molecule and belong to group A; fumitremorgins A & B (→ Aspergillus fumigatus Fres.) and verruculogens (→ Aspergillus spp., *Penicillium* spp.) contain three nitrogen atoms per molecule and belong to group B;

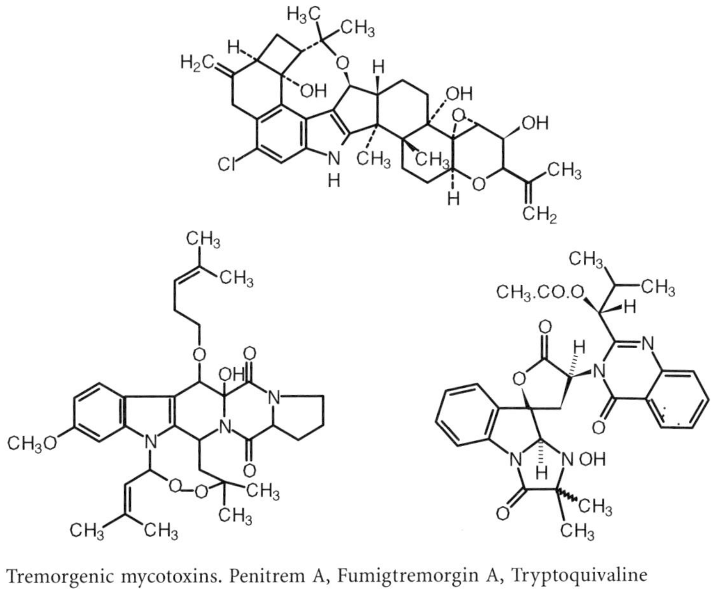

Tremorgenic mycotoxins. Penitrem A, Fumigtremorgin A, Tryptoquivaline

tryptoquivaline and tryptoquivalone (→ Aspergillus clavatus Desm.) contain four nitrogens per molecule and belong to group C (see Figure Tremorgenic mycotoxins). The members of the last group are comparatively less toxic than the remaining tremorgens.
Although tremorgenic compounds are apparently uncommon in nature different fungal genera (*Aspergillus*, → Claviceps, *Penicillium*) produce such mycotoxins. Informations e.g. about their natural occurrence are very limited. However, various naturally occurring neurological disorders, primarily of → cattle ("staggers" syndromes) (e.g. paspalum, ryegrass, and corn staggers)) closely resemble the disorders produced under experimental conditions with fungal tremorgens.

Tremortin A (Syn.: penitrem A, → penitrems)

Tremortin B (Syn.: penitrem B, → penitrems)

Trichocomaceae → Eurotiales

Trichothecenes represent a family of chemically related sesquiterpenoids which all possess a tetracyclic 12,13-epoxy-trichothec-9-ene ring system (→ mycotoxins). They can be substituted at positions C-3, C-4, C-7, C-8, and C-15 (see Figure Trichothecenes 1). In 1967 the formerly called scirpenes (spiroepoxy-containing sesquiterpenoid compounds) were named trichothecenes. This name derived from the fungus *Trichothecium*. Trichothecin was the first trichothecene isolated from *T. roseum* in 1949 but correct chemical structure was not elucidated until 1964. The more than 170 known trichothecenes may be divided into simple (non-macrocyclic) and macrocyclic compounds. The latter ones contain a macrocyclic ring linking at C-4 and C-5 with diesters or triesters, e.g. verrucarins, roridins and satratoxins. There is little evidence that these compounds naturally occur in human food. The non-macrocyclic trichothecenes are divided into three groups A, B and C (see Figure Trichothecenes 2). Members of the first and largest group like → T-2 toxin, → HT-2 toxin, → diacetoxyscirpenol, → monoacetoxyscirpenol and → neosolaniol do not contain a carbonyl group at C-8 (type A). Type B trichothecenes like → nivalenol, → deoxynivalenol, → fusarenon X and diacetylnivalenol are characterized by the presence of a ketone group at C-8. An epoxide at C-7-8 is characteristic for crotocin (type C).

CHEMICAL DATA
For detailed information see each single trichothecene

FUNGAL SOURCES
Macrocyclic trichothecenes are produced by genera such as *Stachybotrys*, *Myrothecium*, *Cylindrocarpon*, *Phomopsis*, *Verticimonosporium*. The fusaria mainly pro-

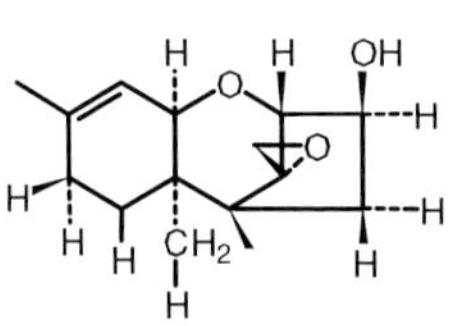

Trichothecenes 1. Trichothecene nucleus

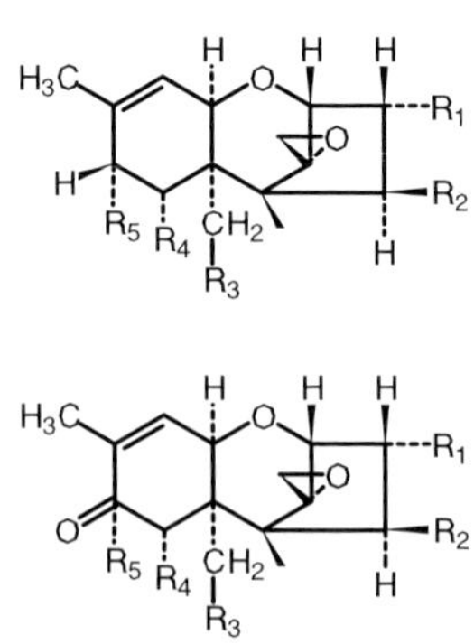

Trichothecenes 2. Basic molecular structure of type A and B trichothecenes

duce the food relevant non-macrocyclic trichothecenes of type A (e.g. *F. acuminatum*, → Fusarium equiseti (Corda) Sacc. sensu Gordon, → Fusarium poae (Peck) Wollenw., → Fusarium sambucinum Fuckel, → Fusarium sporotrichioides Sherb.) and B (e.g. *Fusarium cerealis*, → Fusarium culmorum (Wm. G. Smith) Sacc., → Fusarium graminearum Schwabe). Crotocin (type C) is produced by *Trichothecium roseum*.

Natural Occurrence

→ muesli , → muesli ingredients
For further information see each single trichothecene as well as the primarily contaminated → cereals such as → barley, → maize, → wheat and → cereal products

Toxicity

Although the *Fusarium* trichothecenes greatly vary in their toxicity they are acutely very toxic. T-2 toxin (type A) being probably the most toxic, and deoxynivalenol (type B) being among the least toxic. The A trichothecenes possess a 10 times higher toxicity than members of category B.
antibacterial, antiviral, antifungal, insecticidal (some), phytotoxic and cytostatic; primary mechanisms of toxicity are the inhibition of protein and DNA synthesis; clinical symptons: e.g. → hypothermia, reduced respiratory rate, diarrhea, skin irritation and necrosis, emesis, hemorrhaging (→ hemorrhage), hematological changes (cellular damage in the bone marrow, spleen, and thymus leading to reductions in leucocytes and platelets (→ aleukia) and reproductive problems
LD_{50}: see each single trichothecene
Swine and other monogastric animals (including humans) are most severely affected by these toxins. Affected animals are more susceptible to different fungal infections (e.g. *Cryptococcus*, *Candida*) and food-borne bacteria like *Listeria* and *Salmonella* which might be due to immuno suppression (→ immunossuppressive). A high tolerance to trichothecenes was established in the case of chicken and turkey whereas ruminants were almost insensitive. If the contaminated feed source is removed there is an excellent prognosis for recovery for all species suffering from chronic trichothecene-induced toxicoses.

Detection

ELISA, GC-MS (best method), HPLC-MS (after derivatization), LC-MS, RIA, TLC

Possible Mycotoxicosis

→ alimentary toxic aleukia, "Cobalt-beer" cardiomyopathy, → moldy corn toxicosis, → pellagra, → red mold toxicosis.

Further Comments

Deoxynivalenol is the most common trichothecene in food and feed. Nivalenol, T-2 toxin, and HT-2 toxin occur to lesser extents whereas diacetoxyscirpenol is rarely isolated. It seems that trichothecenes are not very stable in cereals and feeds for longer periods. Almost no toxin could be detected in Finnish cereals after 3-6 months storage at 4 °C. However, overwintered cereals in the USSR remained toxic for several years, although no Fusaria could be isolated from the stored grains.
The naturally occurring combinations diacetoxyscirpenol / fusarenon X, diacetoxyscirpenol / deoxynivalenol and T-2 toxin / diacetoxysirpenol should act synergistically in laboratory animals. T-2 toxin synergized the negative effects of deoxynivalenol in swine. The natural combination of T-2 toxin / HT-2 toxin is synergistic in some ratios and antagonistic in other ratios.
It was speculated that these stable and relatively small molecules were used as chemical warfare agents, termed "yellow rain", in south-east Asia. However, it became obvious that this yellow rain resulted from a mass defecation of the Asian giant honey bee, *Apis dorsata*.

Production: Temperatures below 10 °C favor the synthesis of trichothecenes whereas the greatest amounts are probably produced at low temperatures. Nevertheless, these mycotoxins have also been isolated from cereals grown in tropical (temperature 35 ± 5 °C) and subtropical areas. → Malt is contaminated with trichothecenes only in very low concentrations (traces).
Reduction / elimination: Trichothecenes are very hard to remove from contaminated grains under moderate conditions. A transmission (up to 50%) into the end-products like → wheat flour, → bread, crackers and → baby cereals is therefore possible. Since trichothecenes are heat stable at 120 °C they probably survive the baking processes.
Alkali is effective in the destruction of trichothecenes.

Trichothecin → trichothecenes

Triticale may contain the following → mycotoxins:
→ alternariol
incidence: 3/19, conc. range: 80-250 µg / kg, ∅ conc.: 155 µg / kg, country: Poland
→ alternariol methyl ether
incidence: 3/19, conc. range: 120-400 µg / kg, ∅ conc.: 229 µg / kg, country: Poland
→ 3-acetyl deoxynivalenol
incidence: 11*/50, conc. range: 1200-6000 µg / kg, ∅ conc.: 3600 µg / kg, country: Poland, *healthy and damaged kernels, winter triticale
→ citrinin
incidence: 2/4, conc. range: 0.3-0.7 µg / kg, ∅ conc.: 0.5 µg / kg, country: Switzerland
→ deoxynivalenol
incidence: 11/50, conc. range: 900-5900 µg / kg*, 2400-31,200 µg / kg**, ∅ conc.: 10,109 µg / kg, country: Poland, *healthy kernels, **damaged kernels
→ moniliformin
incidence: 3/3*, conc. range: 2600-15,700 µg / kg, ∅ conc.: 8700 µg / kg, country: Poland, *hand-selected, visible fungal damage
→ ochratoxin A
incidence: 9/10, conc. range: ≤ 5.6 µg / kg, ∅ conc. 2.7 µg / kg, country: Germany
→ cereals

Triticale flour may contain the following → mycotoxins:
ergometrine (→ ergot alkaloids)
incidence: 2/2, conc. range: 13-31 µg / kg, country: Canada
ergosine
incidence: 2/2, conc. range: 9.3-16 µg / kg, country: Canada
ergotamine
incidence: 2/2, conc. range: 28-111 µg / kg, country: Canada
ergocornine
incidence: 2/2, conc. range: 12-26 µg / kg, country: Canada
α-ergokryptine
incidence: 2/2, conc. range: 17-21 µg / kg, country: Canada
ergocristine
incidence: 2/2, conc. range: 50-190 µg / kg, country: Canada

Tryptophan The indole nucleus of this important amino acid is frequently found in → mycotoxins such as in the → cyclopiazonic acid, → ergot alkaloids, sporidesmins, and → tremorgenic mycotoxins.

Tubers (ubi, gabi, tugi, singkamas, sweet potatoes)
may be contaminated by → aflatoxins due to poor storage conditions.
Tubers may contain the following → mycotoxins:

aflatoxins (no specification)
incidence: 6/59, conc. range: > 20- ≤ 780 µg / kg, country: Philippines

Tugi → tubers

Turkey Experimental studies show that feed tissue ratios of → aflatoxin B_1 to AFB_1 and → aflatoxin M_1 are high for kidney and liver but low for muscle. Turkey possess a high tolerance against → trichothecenes.
Turkey may contain the following → mycotoxins:
→ ochratoxin A
incidence: 10/17, conc. range: ≤ 0.11 µg / kg, ∅ conc.: 0.02 µg / kg, country: Denmark
incidence: 3/17*, conc. range: ≤ 0.28 µg / kg, ∅ conc.: 0.04 µg / kg, country: Denmark, *liver
→ meat

Turkey "X" disease In 1960 a severe outbreak of the Turkey "X" disease occurred at 500 locations in Great Britain (mainly East Anglia and southern England) killing about 100,000 → turkey poults. In addition, thousands of ducklings (→ duck) and young → pheasants also died. Brazilian groundnut meal ("Rosetti meal") was the toxic factor which served as a protein source in the feed. The toxic factor was produced by → Aspergillus flavus Link and → Aspergillus parasiticus Speare which resulted in the name aflatoxin.
Using thin layer chromatography, the toxic factor could be separated into four distinct spots. These spots were named after their fluorescent color (**b**lue, **g**reen) whereas the subscripts described their relative chromatographic mobility (→ aflatoxin B_1, → aflatoxin B_2, → aflatoxin G_1 and → aflatoxin G_2). Although the → aflatoxins were responsible for at least the → hepatic lesions and the high mortality they do not reproduce all signs of this disease, e.g. the strange attitudes of the head and neck. Therefore, it was suggested and proved that other → mycotoxins like → cyclopiazonic acid an other metabolite of *A. flavus* was also involved in Turkey "X" disease.
The Turkey "X" disease represents a turning point in mycotoxin research which greatly enhanced the scientific interest in the study of mycotoxins.

Turmeric is a dried rhizome of tropical origin. Way of mycotoxin contamination is not yet clear.
Turmeric may contain the following → mycotoxins:
→ aflatoxin B_1
incidence: 2/15, ∅ conc.: 12 µg / kg, country: Egypt
incidence: 5/9, conc. range: 21-165 µg / kg, country: India
incidence: 6/7, conc. range: tr-3.8 µg / kg, country: Canada
→ aflatoxin B_2
incidence: 5/9, conc. range: 12-150 µg / kg, country: India
→ aflatoxin G_1
incidence: 2/15, ∅ conc.: 8 µg / kg, country: Egypt
incidence: 5/9, conc. range: 20-125 µg / kg, country: India
→ aflatoxin G_2
incidence: 5/9, conc. range: 14-125 µg / kg, country: India
→ citrinin
incidence: 2/9, conc. range: 48-52 µg / kg, ∅ conc.: 50 µg / kg, country: India
→ rubratoxin
incidence: 1/9, conc.: 375 µg / kg, country: India
→ spices

U

Ubi → tubers

Urov disease → Kashin-Beck disease

V

Vegetables (no specification)
may contain the following → mycotoxins:
→ aflatoxin B_1
incidence: 4/51, conc. range: < 5 µg / kg, country: Germany
→ aflatoxin
incidence: 3/100*, conc. range: 2-20 µg / kg (2 samples), > 20 µg / kg (1 sa), country: Uruguay, *dried
→ ochratoxin A
incidence: 6/7*, conc. range: 245-7444 µg / kg, country: Tunesia, *chickpea, bean, lentil (dried)
→ beans, → cabbage, → cowpeas, → lentils, → peas, → pigeon peas, → soybeans, → tomatoes

Vermouth → wine

Viomellein is structurally similar to → xanthomegnin but is asymmetric due to the hydroxyl group at the 1´ position and the lack of a ketone group at the 4´ position (3,3´,4,4´-tetrahydro-9´,10,10´-trihydroxy-7,7´-dimethoxy-3,3´-dimethyl-[8,8´-Bi-1H-naphtho[2,3-c]pyran]-1,1´,6,9-tetrone). It is the second most naturally occurring fungal xanthoquinone (→ mycotoxins) (see Figure Viomellein).

CHEMICAL DATA
Empirical formula: $C_{30}H_{24}O_{11}$, molecular weight: 560

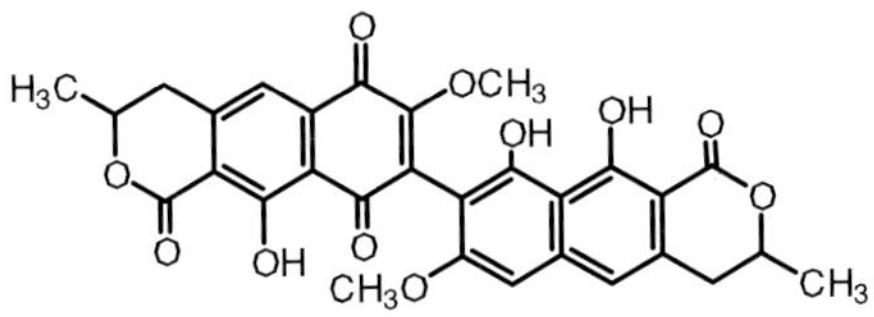

Viomellein

FUNGAL SOURCES
→ Penicillium aurantiogriseum Dierckx, → Penicillium crustosum Thom, *P. simplicissimum*, → Penicillium viridicatum Westling, *Eupenicillium javanicum*, → Aspergillus ochraceus group

NATURAL OCCURRENCE
→ barley, → oil seed rape, → wheat
In → cereals, it often co-occurs with → xanthomegnin and it may be associated with → ochratoxin A and → citrinin.

TOXICITY
toxicity similar to that of xanthomegnin, hepatotoxic, nephrotoxic (lesions)

DETECTION
HPLC, TLC

Vomitoxin (Syn.: → deoxynivalenol)

Vulvo-vaginitis → F-2 toxicosis

W

Walnuts → Aflatoxins seem to be the most important → mycotoxins in walnuts whereas infection with aflatoxin-producing fungi is due to specific types of insects. It was estimated that the average probability of aflatoxin contamination in walnuts is one walnut in 28,250 → nuts. The removal of visibly damaged nuts immediately after harvest and subsequent cool and dry storage conditions effectively prevent aflatoxin contamination.
Walnuts may contain the following mycotoxins:
→ aflatoxin B_1
incidence: 4/97, conc. range: < 5 µg / kg (3 samples), conc.: 7 µg / kg (1 sa), country: Germany
incidence 3*/12, conc. range: 5-500,000 µg / kg, country: Germany, *moldy
incidence: 1/14* conc.: 8 µg / kg, country: Norway, *imported
→ aflatoxin B_2
incidence: 1/14* conc.: traces , country: Norway, *imported
→ aflatoxin G_1
incidence: 1/14* conc.: 4 µg / kg, country: Norway, *imported
→ aflatoxin G_2
incidence: 1/14* conc.: traces , country: Norway, *imported
aflatoxins
incidence: 10*/156, conc. range: 5-24.9 µg / kg (9 samples), > 25 µg / kg (1 sa), country: Canada, *AFB_1, AFB_2, AFG_1, AFG_2
incidence: 15/20, conc. range: 15-25 µg / kg, country: Egypt
incidence: 4/97, conc. range: < 5 µg / kg (3 samples), conc.: 18 µg / kg (1 sa), country: Germany
incidence: 8/330, conc. range: 2-70 µg / kg, ∅ conc.: 27 µg / kg, country: USA
incidence: 2/27, conc. range: 29-41 µg / kg, ∅ conc.: 35 µg / kg, country: USA
incidence: 2/4, conc. range: ≤ 8 µg / kg, ∅ conc.: 4 µg / kg, country: USA
penitrem A (→ penitrems)
incidence: 1/1*, conc.: nc, country: USA, *visibly moldy
→ zearalenone
incidence: 1/20, conc.: 125 µg / kg, country: Egypt
incidence: 3/60, conc. range: 50-450 µg / kg, country: France
→ nuts

Water activity → a_w

Wheat is one of the most important of the cereal crops grown for human consumption. During moist weather periods the maturing seeds may be heavily invaded by → Fusarium spp., especially → Fusarium graminearum Schwabe the causal factor for *Fusarium* head blight. → Fusarium culmorum (W. G. Smith) Sacc. and → Fusarium avenaceum (Fr.) Sacc. are also very common on wheat. A reddish discoloration of the kernels may be associated with trichothecene contamination (→ trichothecenes).
Compared to the testa the wheat embryo is an excellent substrate for aflatoxin production of → Aspergillus flavus Link. However, → aflatoxins do not play an important role in mycotoxin contamination of wheat.
Wheat may contain the following → mycotoxins:
3-acetoxynivalenol
incidence: 3/27, conc. range: < 200 µg / kg, country: Finland
→ 3-acetyldeoxynivalenol
incidence: 5/10, conc. range: 15-731 µg / kg, ∅ conc.: 363 µg / kg, country: China
incidence: 9/40, conc. range: 12-67 µg / kg, ∅ conc.: 31 µg / kg, country: Finland

incidence: 50/84, conc. range: 3-18 µg / kg, ∅ conc.: 7 µg / kg, country: Germany
incidence: nc/9, conc. range: 100-30,000 µg / kg, country: Poland
incidence: 2/3*, conc. range: 100 µg / kg, ∅ conc.: 100 µg / kg, country: Poland, *healthy and damaged kernels
incidence: 13/13*, conc. range: 100-3000 (5600) µg / kg, ∅ conc.: 790 µg / kg, country: Poland, *healthy and damaged kernels
→ 15-acetyldeoxynivalenol
incidence: 3/3*, conc. range: 100-2000 µg / kg, ∅ conc.: 675 µg / kg, country: Poland, *healthy and damaged kernels
→ aflatoxin B_1
incidence: 40/545, ∅ conc.: 16.3 µg / kg, country: Croatia
incidence: 3/7*, conc. range: 10-15 µg / kg, country: Germany, *moldy
incidence: 1/11*, conc.: traces, country: UK, *moldy
incidence: 23/31*, conc. range: 0.8-17 µg / kg, ∅ conc.: 3.37 µg / kg, country: USA, *scabby
incidence: 3/1.528, conc.: 11 µg / kg, country: USA
incidence: 2/531, conc. range: 7 µg / kg, ∅ conc.: 7 µg / kg, country: USA
→ aflatoxin G_1
incidence: 2/531, conc. range: 2 µg / kg, ∅ conc.: 2 µg / kg, country: USA
aflatoxin (no specification)
incidence: 29/123*, conc. range: 2-20 µg / kg (28 samples), > 20 µg / kg (1 sa), country: Uruguay, *and by-products
aflatoxins (no specification)
incidence: 10/30, conc. range: 15-263 µg AFB_1 / kg, 10-107 µg AFB_2 / kg, 12-95 µg AFG_1 / kg, 22-90 µg AFG_2 / kg, country: India
→ alternariol
incidence: 27/33*, conc. range: ≤ 1050 µg / kg, ∅ conc.: 152 µg / kg, country: Australia, *weather-damaged
incidence: 2/105, conc. range: 6-12 µg / kg, ∅ conc.: 9 µg / kg, country: Germany
incidence: 1/5, conc.: 590 µg / kg, country: Poland
incidence: 9/49, conc. range: 20-600 µg / kg, ∅ conc.: 131 µg / kg, country: Poland
→ alternariol methyl ether
inicdence: 24/33*, conc. range: ≤ 46 µg / kg, ∅ conc.: 14.4 µg / kg, country: Australia, *weather damaged
incidence: 12/199, conc. range: 4-200 µg / kg, ∅ conc.: 37.3 µg / kg, country: Germany
incidence: 7/49, conc. range: 20-1600 µg / kg, ∅ conc.: 305 µg / kg, country: Poland
→ citrinin
incidence: 10/15, conc. range: 70-80,000 µg / kg, country: Canada
incidence: 1/66, conc.: 2000 µg / kg, country: Poland
incidence: 11/11*, conc. range: tr-4800 µg / kg, country: UK, *moldy
→ deoxynivalenol
incidence: 3/20, ∅ conc.: 15 µg / kg, country: Argentina
incidence: 56/60, conc. range: 100-9250 µg / kg, ∅ conc.: 1798 µg / kg, country: Argentina
incidence: 32/40, conc. range: 300-4500 µg / kg, ∅ conc.: 1060 µg / kg, country: Argentina
incidence: 11/12, conc. range: ≤ 6700 µg / kg, ∅ conc.: 1800 µg / kg, country: Australia
incidence: 3/4, ∅ conc.: 360 µg / kg, country: Austria
incidence: 11/32, conc. range: 80-2110 µg / kg, ∅ conc.: 580 µg / kg, country: Austria
incidence: 4/16, conc. range: 27-1280 µg / kg, ∅ conc.: 449 µg / kg, country: Austria
incidence: 3/3*, conc. range: 465-4450 µg / kg, ∅ conc.: 3062 µg / kg, country: Austria, *durum

incidence: 1/2, conc.: 211 µg / kg, country: Bulgaria
incidence: 55/199, conc. range: 20-1320 µg / kg, country: Canada
incidence: 9/10, conc. range: 25-3475 µg / kg, ∅ conc.: 1257 µg / kg, country: Canada
incidence: 11/208, conc. range: 20-3200 µg / kg, country: Canada
incidence: 40/53*, conc. range: 50-3650 µg / kg, ∅ conc.: 434 µg / kg, country: Canada, *suspected
incidence: 412/560, conc. range: 10-5670 µg / kg, ∅ conc.: 460 µg / kg, country: Canada
incidence: 86/258, conc. range: 10-1510 µg / kg, ∅ conc.: 210 µg / kg, country: Canada
incidence: 270/1493*, conc. range: 10-10,500 µg / kg, ∅ conc.: 430 µg / kg, country: Canada, *hard
incidence: 5/5*, conc. range: 20-100 µg / kg (1 sample), 101-500 µg / kg (3 sa), > 500 µg / kg (1 sa), country: Canada, *soft
incidence: 1/5, conc.: 1710 µg / kg, country: China
incidence: 4/4, ∅ conc.: 4284 µg / kg, country: China
incidence: 5/10, conc. range: 73-1051 µg / kg, ∅ conc.: 349 µg / kg, country: China
incidence: 25/27, conc. range: 1-6300 µg / kg, country: Finland
incidence: 37/40, conc. range: 8-356 µg / kg, ∅ conc.: 81 µg / kg, country: Finland
incidence: 8/10*, conc. range: 10-68 µg / kg, ∅ conc.: 35 µg / kg, country: Finland, *imported from Canada, Saudi-Arabia, USA
incidence: 1/2, conc.: 86 µg / kg, country: France
incidence: 1/1, conc.: 5000 µg / kg, country: France
incidence: 45*/51, conc. range: ≤ 1200 µg / kg, ∅ conc.: 420 µg / kg, country: Germany, *conventional
incidence: 38*/50, conc. range: ≤ 1000 µg / kg, ∅ conc.: 486 µg / kg, country: Germany, *ecological
incidence: 2/6, ∅ conc: 712 µg / kg, country: Germany
incidence: 14/44, conc. range: 10-5600 µg / kg, ∅ conc.: 810 µg / kg, country: Germany
incidence: 92/106*, conc. range: 70-43,800 µg / kg, ∅ conc.: 3960 µg / kg, country: Germany, *moldy
incidence: 43/45, conc. range: 40-750 µg / kg, ∅ conc.: 190 µg / kg, country: Germany
incidence: 140/154, conc. range: 40-3240 µg / kg, ∅ conc.: 170 µg / kg, country: Germany
incidence: 24/29*, conc. range: 10-2000 µg / kg, country: Germany, *food grade wheat and wheat products
incidence: 2/8 ∅ conc.: 700 µg / kg, country: Germany
incidence: 5/123, conc. range: 10-1300 µg / kg, country: Germany
incidence: 7/10*, conc. range: 36-340 µg / kg, ∅ conc.: 176 µg / kg, country: Germany, *organic produce
incidence: 2/2, conc. range: 36-370 µg / kg, ∅ conc.: 203 µg / kg, country: Germany
incidence: 81/84, conc. range: 4-20,538 µg / kg, ∅ conc.: 1632 µg / kg, country: Germany
incidence: 1/1, conc.: 9 µg / kg, country: Greece
incidence: 2/2, ∅ conc.: 671 µg / kg, country: Hungary
incidence: 1/12, conc.: 120 µg / kg, country: Italy
incidence: 2/17, conc. range: 90-280 µg / kg, country: Japan
inicdence: 4/6, ∅ conc.: 23 µg / kg, country: Japan
incidence: 95/101, conc. range: 10-12,400 µg / kg, ∅ conc.: 1178 µg / kg, country: Japan
incidence: 1/1, conc.: 440 µg / kg, country: Japan

incidence: 11/18, conc. range: ND-1800 µg / kg, ∅ conc.: 800 µg / kg, country: Japan
incidence: 8/11, conc. range: 100-9180 µg / kg, ∅ conc.: 1290 µg / kg, country: Japan
incidence: 2/3, conc. range: 160-370 µg / kg, ∅ conc.: 260 µg / kg, country: Japan
incidence: 18/18*, conc. range: 740-6920 µg / kg, ∅ conc.: 3812 µg / kg, Japan, *scabby wheat
incidence: 5/9, conc. range: ≤ 170 µg / kg, ∅ conc.: 42 µg / kg, country: Korea
incidence: 1/10, conc.: 61 µg / kg, country: Nepal
incidence: 78/90, conc. range: ≤ 11,950 µg / kg, country: New Zealand
incidence: 13/42, conc. range: ≤ 310 µg / kg, ∅ conc.: 95 µg / kg, country: Poland
incidence: nc/9, conc. range: 200-30,400 µg / kg, country: Poland
incidence: 3/3*, conc. range: 2000-38,000 µg / kg, ∅ conc.: 16,216 µg / kg, country: Poland, *healthy and damaged kernels
incidence: 11/13*, conc. range: 400-39,600 µg / kg, ∅ conc.: 14,540 µg / kg, country: Poland, *healthy and damaged kernels
incidence: 1/2, conc.: 30 µg / kg, country: Portugal
incidence: 1/2, conc.: 26 µg / kg, country: Scotland
incidence: 31/43, conc. range: ≤ 1180 µg / kg, ∅ conc.: 240 µg / kg, country: Sweden
incidence: 8/14, conc. range: 110-1180 µg / kg, ∅ conc.: 400 µg / kg, country: Sweden
incidence: 23/29, conc. range: 60-360 µg / kg, ∅ conc.: 190 µg / kg, country: Sweden
incidence: 12/22, conc. range: ≤ 2500 µg / kg, ∅ conc.: 480 µg / kg, country: Taiwan
incidence: 9/12, conc. range: 45-2450 µg / kg, ∅ conc.: 562 µg / kg, country: Taiwan
incidence: 3/10, conc. range: 26-505 µg / kg, ∅ conc.: 245 µg / kg, country: Taiwan
incidence: 13/13, conc. range: 20-231 µg / kg, ∅ conc.: 115 µg / kg, country: The Netherlands
incidence: 1/35, conc.: 90 µg / kg, country: UK
incidence: 20/31, conc. range: 4-312 µg / kg, ∅ conc.: 31 µg / kg, country: UK
incidence: 32/199, conc. range: 20-400 µg / kg, country: UK
incidence: 23/33*, conc. range: 20-1320 µg / kg, country: UK, *imported
incidence: 34/205, conc. range: 20-500 µg / kg, country: UK
incidence: 1/35, conc.: 90 µg / kg, country: UK
incidence: 6/55, conc. range: 80-750 µg / kg, ∅ conc.: 340 µg / kg, country: UK
incidence: 75/123, conc. range: tr-500 µg / kg (38 samples), conc. range: 500-1000 µg / kg (32 sa), conc. range: 1000-2000 µg / kg (4 sa), conc. range: > 2000 µg / kg (1 sa), country: USA
incidence: 31/33*, conc. range: 120-5500 µg / kg ∅ conc.: 1782 µg / kg, country: USA, *scabby
incidence: 132/247, conc. range: ≤ 2650 µg / kg, ∅ conc.: 570 µg / kg, country: USA
incidence: 14/27, conc. range: 600-3800 µg /kg, ∅ conc.: 2800 µg / kg, country: USA
incidence: 23/116, ∅ conc.: 100 µg / kg, country: USA
incidence: 12/14, conc. range: 20-100 µg / kg (7 samples), 101-500 µg / kg (4 sa), > 500 µg / kg (1 sa), country: USA
incidence: 156/157, conc. range: 200-43,000 µg / kg, country: USA
incidence: 201/207, conc. range: 400-4000 µg / kg, country: USA
incidence: 120/206, conc. range: 900-7600 µg / kg, country: USA

incidence: 333/483, conc. range: 500-18,000 µg / kg, ∅ conc.: 2000 µg / kg, country: USA
incidence: 1/7, conc.: 5 µg / kg, country: Yemen
incidence: nc/5, conc. range: 3400-8000 µg / kg, country: Yugoslavia
4,7-dideoxynivalenol
incidence: 3/3*, conc. range: 100-150 µg / kg, ∅ conc.: 113 µg / kg, country: Poland, *healthy and damaged kernels
→ diacetoxyscirpenol
incidence: 20/53*, conc. ranges: ≤ 80 µg / kg, country: Canada, *suspected
incidence: 1/87, conc.: 50 µg / kg, country: Germany
incidence: 5/59, conc. range: 300-2000 µg / kg, country: Germany
incidence: 3/100, conc. range: nc, country: France
incidence: 1/nc, conc.: 50 µg / kg, country: USSR
ergocristine (→ ergot alkaloids)
incidence: 1/1*, conc.: 0.2-0.3 µg / kg, country: Canada, *uncooked
→ fusarenon X
incidence: 3/27, conc. range: < 40 µg / kg, country: Finland
incidence: 3/55, conc. range: 140-570 µg / kg, ∅ conc.: 350 µg / kg, country: UK
→ HT-2 toxin
incidence: 24/208, conc. range: 60-590 µg / kg, country: Canada
incidence: 10/53*, conc range: < 50 µg / kg, country: Canada, *suspected
incidence: 2/27, conc. range: 8-40 µg / kg, ∅ conc.: 24 µg / kg, country: Finland
incidence: 2/87, conc. range: 50-60 µg / kg, ∅ conc.: 55 µg / kg, country: Germany
incidence: 1/80, conc.: 150 µg / kg, country: Germany
incidence: 6/84, conc. range: 3-20 µg / kg, ∅ conc.: 10 µg / kg, country: Germany
incidence: 1/2, conc.: 200 µg / kg, country: Hungary
→ moniliformin
incidence: 6*/6, conc. range: 500-17,100 µg / kg, ∅ conc.: 8660 µg / kg, country: Poland, *hand-selected, visible fungal growth
→ neosolaniol
incidence: 1/nc, conc.: 200 µg / kg, country: USSR
→ nivalenol
incidence: 3/4, ∅ conc.: 25 µg / kg, country: Austria
incidence: 1/2, conc.: 32 µg / kg, country: Bulgaria
incidence: 1/208, conc.: 60 µg / kg, country: Canada
incidence: 4/10, conc. range: 4-40 µg / kg, ∅ conc.: 23 µg / kg, country: Canada
incidence: 1/5, conc.: 6644 µg / kg, country: China
incidence: 3/4, ∅ conc.: 162 µg / kg, country: China
incidence: 8/10, conc. range: 8-373 µg / kg, ∅ conc.: 118 µg / kg, country: China
incidence: 3/27, conc. range: < 1000 µg / kg, country: Finland
incidence: 2/2, ∅ conc.: 42 µg / kg, country: France
incidence: 2/2, ∅ conc.: 274 µg / kg, country: Germany
incidence: 3/44, conc. range: 10-50 µg / kg, ∅ conc.: 30 µg / kg, country: Germany
incidence: 2/8, ∅ conc.: 270 µg / kg, country: Germany
incidence: 16/29, conc. range: 10-120 µg / kg, country: Germany
incidence: 22/84, conc. range: 3-32 µg / kg, ∅ conc: 9 µg / kg, country: Germany
incidence: 1/1, conc.: 2 µg / kg, country: Greece
incidence: 1/2, conc.: 4 µg / kg, country: Hungary
incidence: 95/101, conc. range: 3-7300 µg / kg, ∅ conc.: 942 µg / kg, country: Japan

incidence: 4/17, conc. range: 20-580 μg / kg, country: Japan
incidence: 6/6, ∅ conc.: 391 μg / kg, country: Japan
incidence: 1/1, conc.: 160 μg / kg, country: Japan
incidence: 11/18, conc. range: ND-1000 μg / kg, ∅ conc.: 400 μg / kg, country: Japan
incidence: 8/11, conc. range: ND-3580 μg / kg, ∅ conc.: 450 μg / kg, country: Japan
incidence: 2/3, conc. range: ND-20 μg / kg, ∅ conc.: 10 μg / kg, country: Japan
incidence: 7/18*, conc range: 47-435 μg / kg, ∅ conc.: 205 μg / kg, country: Japan, *scabby wheat
incidence: 9/9, conc. range: ≤ 3200 μg / kg, ∅ conc.: 534 μg / kg, country: Korea
incidence: 9/10, ∅ conc.: 135 μg / kg, country: Korea
incidence: 5/10, ∅ conc.: 70 μg / kg, country: Nepal
incidence: 81/90, conc. range: ≤ 1270 μg / kg, country: New Zealand
incidence: 37/48, conc. range: ≤ 350 μg / kg, ∅ conc.: 48 μg / kg, country: Poland
incidence: 1/3*, conc.: 10 μg / kg, country: Poland, *healthy and damaged kernels
incidence: 6/12, conc. range: 26-169 μg / kg, ∅ conc.: 74 μg / kg, country: Taiwan
incidence: 10/22, conc. range: ≤ 170 μg / kg, ∅ conc.: 54 μg / kg, country: Taiwan
incidence: 4/10, conc. range: 5-52 μg / kg, ∅ conc.: 22 μg / kg, country: Taiwan
incidence: 12/13, conc. range: 7-203 μg / kg, ∅ conc.: 38 μg / kg, country: The Netherlands
incidence: 17/31, conc. range: 4-670 μg / kg, ∅ conc.: 101 μg / kg, country: UK
→ ochratoxin A
incidence: 1/61, conc.: 160 μg / kg, country: Austria
incidence: 4/41, conc. range: 5-100 μg / kg, country: Austria
incidence: 14/18*, conc. range: 30-27,000 μg / kg, country: Canada, *heated
incidence: 4/4*, conc. range: 20-100 μg / kg, country: Canada, *visible moldy
incidence: 119/402*, conc. range: 0.05-4.9 μg / kg (110 samples), 5-25 μg / kg (6 sa), > 25- ≤ 51 μg / kg (3 sa)*, ∅ conc.: 0.7 μg / kg, country: Denmark, *conventional
incidence: 29/73*, conc. range: 0.05-4.9 μg / kg (25 samples), 5-25 μg / kg (3 sa), > 25- ≤ 36 μg / kg (1 sa), ∅ conc.: 1.2 μg / kg, country: Denmark, *ecological
incidence: 17/45*, conc. range: 0.05-4.9 μg / kg (16 samples), 5-25 μg / kg (1 sa), ∅ conc.: 0.9 μg / kg, country: Denmark, *conventional, imported
incidence: 72/194, conc. range: 0.8-37 μg / kg, country: Denmark
incidence: 17/36*, conc. range: 1.2-21 μg / kg, country: Denmark, *ecological
incidence: 1/3, conc.: 10 μg / kg, country: Egypt
incidence: 3/97, conc. range: ≤ 4.9-24.9 μg / kg, country: Germany
incidence: 1/64, conc.: 0.4 μg / kg, country: Germany
incidence: 8/64, conc. range: 0.1-137.3 μg / kg, ∅ conc.: 17.9 μg / kg, country: Germany
incidence: 94/719, conc. range: 0.1-12.5 μg / kg, country: Germany
incidence: 3/97, conc. range: 0.4-15.4 μg / kg, country: Germany
incidence: 1/30, conc.: nc, country: India
incidence: 10/10*, conc.: ≤ 2.6 μg / kg, ∅ conc.: 1.4 μg / kg, country: Italy, *soft wheat
incidence: 2/34, conc. range: 188-430 μg / kg, ∅ conc.: 309 μg / kg, country: Italy
incidence: 2/66, conc. range: 160-1000 μg / kg, ∅ conc.: 580 μg / kg, country: Poland
incidence: 28/239, conc. range: 5-2400 μg / kg, country: Poland

incidence: 1/209, conc.: 1.8 μg / kg, country: Saudi Arabia
incidence: 5/5, conc. range: ≤ 0.8 μg / kg, country: Spain
incidence: 2/24, conc. range: ≤ 0.6 μg / kg, country: Spain
incidence: 6/35, conc. range: ≤ 4.9-8.6 μg / kg, country: Sweden
incidence: 7/27, conc. range: ≤ 4.1 μg / kg, country: The Netherlands
incidence: 6/38*, conc. range: 0.1-4.2 μg / kg, country: The Netherlands, *imported
incidence: 8/28, conc. range: 34-360 μg / kg, country: Tunesia
incidence: 43/44*, conc. range: 0.1-11,064 μg / kg, country: Tunesia, *and derived food
incidence: 2/8, conc. range: ≤ 2 μg / kg, country: UK
incidence: 2/129, conc. range: ≤ 15 μg / kg, country: UK
incidence: 22/250, conc. range: ≤ 4.9-31.6 μg / kg, country: UK
incidence: 10/18, conc. range: ≤ 4.9-12 μg / kg, country: UK
incidence: 10/30, conc. range: ≤ 1.2 μg / kg, country: UK
incidence: 8/25, conc. range: ≤ 4.9-13.9 μg / kg, country: UK
incidence: 2/9, conc. range: ≤ 0.2 μg / kg, country: UK
incidence: 15/101, conc. range: < 25-2700 μg / kg, country: UK
incidence: 9/11*, conc. range: < 50-3500 μg / kg, country: UK, *moldy
incidence: 11/577*, conc. range: 5-115 μg / kg, country: USA, *hard red winter
incidence: 9/848, conc. range: 20-114 μg / kg, country: USA
incidence: 56/383, conc. range: 0.03-15,410 μg / kg, ∅ conc.: 2.04 μg / kg, country: USA
incidence: 11/130*, conc. range: 14-135 μg / kg, country: Yugoslavia, *area with endemic nephropathy
incidence: 3/40, conc. range: 12-55 μg / kg, ∅ conc.: 34.3 μg / kg, country: Yugoslavia

→ rubratoxin
incidence: 1/30, conc.: 245 μg / kg, country: India

→ sterigmatocystin
incidence: 1/18, conc.: ca. 300 μg / kg, country: Canada
incidence: 2/30, conc. range: 110-145 μg / kg, ∅ conc.: 128 μg / kg, country: India
incidence: 7/11*, conc. range: tr-400 μg / kg, country: UK, *moldy

→ tenuazonic acid
incidence: 33/33*, conc. range: ≤ 220 μg / kg, ∅ conc.: 50.1 μg / kg, country: Australia, *weather-damaged

→ T-2 toxin
incidence: 3/208, conc. range: ≤ 18 μg / kg, country: Canada
incidence: 11/53*, conc. range: ≤ 200 μg / kg, country: Canada, *suspected
incidence: 2/24, conc. range: 3-8 μg / kg, country: Finland
incidence: 1/100, conc.: nc, country: France
incidence: 1/87, conc.: 100 μg / kg, country: Germany
incidence: 4/21, conc. range: 23-45 μg / kg, ∅ conc.: 25 μg / kg, country: Germany
incidence: 4/81, conc. range: 200-500 μg / kg, country: Germany
incidence: 22/84, conc. range: 3-249 μg / kg, ∅ conc.: 82 μg / kg, country: Germany
incidence: 2/2, conc. range: 200-1900 μg / kg, country: Hungary
incidence: 3/7*, conc. range: 2000-4000 μg / kg, country: India, *moldy
incidence: 8/57, conc. range: 13-63 μg / kg, country: Poland
incidence: 1/nc, conc.: 500 μg / kg, country: USSR

→ viomellein
incidence: 8/11*, conc. range: 300-1800 µg / kg, ∅ conc.: 962 µg / kg, country: UK, *moldy
vioxanthin
incidence: 8/11*, conc. range: 200-1200 µg / kg, ∅ conc.: 443 µg / kg, country: UK, *moldy
→ xanthomegnin
incidence: 8/11*, conc. range: 120-1100 µg / kg, ∅ conc.: 390 µg / kg, country: UK, *moldy
→ zearalenone
incidence: 20/20, ∅ conc.: 10 µg / kg, country: Argentina
incidence: 9/10, conc. range: 2-21 µg / kg, ∅ conc.: 9 µg / kg, country: Canada
incidence: 4/4, ∅ conc.: 78 µg / kg, country: China
incidence: 5/10, conc. range: 5-25 µg / kg, ∅ conc.: 15 µg / kg, country: China
incidence: 2/40, conc. range: 12-32 µg / kg, ∅ conc.: 22 µg / kg, country: Finland
incidence: 8*/51, conc. range: ≤ 18 µg / kg, ∅ conc.: 6 µg / kg, country: Germany, *conventional
incidence: 18*/50, conc. range: ≤ 105 µg / kg, ∅ conc.: 24 µg / kg, country: Germany, *ecological
incidence: 1/6, conc.: 5 µg / kg, country: Germany
incidence: 2/2, ∅ conc.: 2 µg / kg, country: Germany
incidence: 15/92, conc. range: 0.5-290 µg / kg, ∅ conc.: 30 µg / kg, country: Germany
incidence: 58/106, conc. range: ≤ 1560 µg / kg, ∅ conc.: 80 µg / kg, country: Germany
incidence: 12/48, conc. range: 5-20 µg / kg, ∅ conc.: 10 µg / kg, country: Germany
incidence: 3/8, ∅ conc.: 4 µg / kg, country: Germany
incidence: 19/159, conc. range: 10-2000 µg / kg, country: Germany
incidence: 1/2, conc.: 10 µg / kg, country: Germany
incidence: 67/84, conc. range: 1-8036 µg / kg, ∅ conc.: 178 µg / kg, country: Germany
incidence: 1/12, conc.: 4 µg / kg, country: Italy
incidence: 1/6, conc.: 1 µg / kg, country: Japan
incidence: 18/18, conc. range: 8-706 µg / kg, ∅ conc.: 189 µg / kg, country: Japan
incidence: 2/10*, conc. range: 8-40 µg / kg, ∅ conc.: 5 µg / kg, country: Korea, *polished
incidence: 5/9, ∅ conc.: 141 µg / kg, country: Korea
incidence: 2/10, ∅ conc.: 4 µg / kg, country: Nepal
incidence: 48/151, conc. range: ≤ 460 µg / kg, country: New Zealand
incidence: 1/48, conc.: 76 µg / kg, country: Poland
incidence: 2/3*, conc. range: 10-2000 µg / kg, ∅ conc.: 1005 µg / kg, country: Poland, *healthy and damaged kernels
incidence: 5/13*, conc. range: 25-600 µg / kg, ∅ conc.: 425 µg / kg, country: Poland, *healthy and damaged kernels
incidence: 2/4, ∅ conc.: 16 µg / kg, country: Portugal
incidence: 2/2, conc. range: 3-10 µg / kg, ∅ conc.: 6.5 µg / kg, country: Scotland
incidence: 9/12, conc. range: 4-32 µg / kg, ∅ conc.: 16 µg / kg, country: Taiwan
incidence: 7/13, conc. range: 2-174 µg / kg, ∅ conc.: 45 µg / kg, country: The Netherlands
incidence: 4/31, conc. range: 1-3 µg / kg, ∅ conc.: 1 µg / kg, country: UK
incidence: 5/106*, conc. range: 100-200 µg / kg (2 samples), > 200 µg / kg (3 sa), country: Uruguay, *and by-products
incidence: 14/27, conc. range: 400-3700 µg / kg, ∅ conc.: 950 µg / kg, country: USA
incidence: 1/116, conc.: 5000 µg / kg, country: USA

incidence: 3/33*, conc. range: 35-115 µg / kg, Ø conc.: 80 µg / kg, country: USA, *scabby
incidence: 18/112, conc. range: 400 µg / kg (1 sample), 400-900 µg / kg (2 sa), 1000-5000 µg / kg (13 sa), > 5000 µg / kg (2 sa), country: USA
incidence: 4/7, conc. range: 2 µg / kg, country: Yemen
α-zearalenol (→ zearalenol)
incidence: 4/84, conc. range: 8-71 µg / kg, Ø conc.: 23 µg / kg, country: Germany
β-zearalenol
incidence: 1/1, conc.: 12 µg / kg, country: Germany
→ cereals

Wheat (coarse ground) may contain the following → mycotoxins:
→ deoxynivalenol
incidence: 1/1*, conc.: 1820 µg / kg, country: Papua, New Guinea, *imported
→ zearalenone
incidence: 1/1*, conc.: 1040 µg / kg, country: Papua New Guinea, *imported

Wheat (intermediate products): → aflatoxin B_1
incidence: 35/475, Ø conc.: 11.1 µg / kg, country: Croatia

Wheat bran may contain the following → mycotoxins:
→ deoxynivalenol
incidence: 2/3, conc. range: 170-450 µg / kg, Ø conc.: 310 µg / kg, country: Austria
incidence: 1/2*, conc.: 45 µg / kg, country: Papua New Guinea, *imported
incidence: 14/27, Ø conc.: 3400 µg / kg, country: USA
→ nivalenol
incidence: 1/2*, conc.: 19 µg / kg, country: Papua New Guinea, *imported
→ ochratoxin A
incidence: 1/1, conc.: 3 µg / kg, country: China
incidence: 6/57, conc. range: 5-20 µg / kg, country: Denmark
incidence: 39/57, conc. range: 0.5-12 µg / kg, country: Denmark
incidence: 10/15, conc. range: 0.1-26 µg / kg, country: Denmark
incidence: 74/120*, conc. range: 0.05- 4.9 µg / kg (72 samples), 5-12 µg / kg (2 sa), country: Denmark, *conventional
incidence: 15/22*, conc. range: 0.05-2.6 µg / kg, Ø conc.: 0.6 µg / kg, country: Denmark, *ecological
incidence: 1/41, conc.: 0.1 µg / kg, country: Germany
incidence: 3/5, conc. range: 0.2-0.8 µg / kg, Ø conc: 0.4 µg / kg, country: Switzerland
incidence: 3/7, conc. range: ≤ 2.5 µg / kg, country: The Netherlands
→ zearalenone
incidence: 14/27, Ø conc.: 2050 µg / kg, country: USA
→ bran

Wheat flour During → milling → deoxynivalenol accumulated in the → wheat bran whereas lower levels (≈ 5%) were found in the break → flour.
Wheat flour may contain the following → mycotoxins:
→ acetyldeoxynivalenol
incidence: 4/12*, conc. range: 600-2400 µg / kg, country: India, *moldy, refined
→ aflatoxin B_1
incidence: 21/238, Ø conc.: 4.13 µg / kg, country: Croatia
incidence: 1/83, conc.: 25.6 µg / kg, country: Malaysia
→ aflatoxin B_2
incidence: 4/83, conc. range: 11.3-253 µg / kg, Ø conc.: 75.2 µg / kg, country: Malaysia
→ aflatoxin G_1
incidence: 3/83, conc. range: 25-289 µg / kg, Ø conc.: 135 µg / kg, country: Malaysia

→ aflatoxin G_2
incidence: 11/83, conc. range: 16.3-436 µg / kg, ∅ conc.: 153 µg / kg, country: Malaysia
→ deoxynivalenol
incidence: 61/61, conc. range: 250-9000 µg / kg, ∅ conc.: 1309 µg / kg, country: Argentina
incidence: 54/54, ∅ conc.: 1210 µg / kg, country: Argentina
incidence: 6/6, conc. range: 400-800 µg / kg, ∅ conc.: 467 µg / kg, country: Argentina
incidence: 11/47, conc. range: 27-830 µg / kg, ∅ conc.: 229 µg / kg, country: Austria
incidence: 43 products analysed, ∅ conc. 400 µg / kg, country: Canada
incidence: 7/7, ∅ conc.: 129 µg / kg, country: China
incidence: 5/5, conc. range: 11-690 µg / kg, ∅ conc.: 180 µg / kg, country: China
incidence: 42/44, conc. range: $\leq$ 580 µg / kg, ∅ conc.: 130 µg / kg, country: Germany
incidence: 4/4*, conc. range: 41-180 µg / kg, ∅ conc.: 102 µg / kg, country: Germany, *organic produce
incidence: 3/3, conc. range: 60-90 µg / kg, ∅ conc.: 75 µg / kg, country: Germany
incidence: 9/12*, conc. range: 430-4850 µg / kg, country: India, *moldy, refined
incidence: 2/5*, conc. range: 346-8380 µg / kg, country: India, *moldy, refined
incidence: 26/36, conc. range: 2-239 µg / kg, country: Japan
incidence: 1/1*, conc.: 1720 µg / kg, country: Papua New Guinea, *imported, fine-ground biscuit flour
incidence: 1/1*, conc.: 2270 µg / kg, country: Papua New Guinea, *imported, raw flour
incidence: 44/50, conc. range: ND-460 µg / kg, country: USA
incidence: 2/27, conc. range: ND-2000, ∅ conc.: 1500 µg / kg, country: USA
ergometrine (→ ergot alkaloids)
incidence: 4/4, conc. range: 0.3-0.7 µg / kg, country: Canada
ergosine
incidence: 4/4, conc. range: 0.4-0.7 µg / kg, country: Canada
ergotamine
incidence: 4/4, conc. range: 0.3-2.3 µg / kg, country: Canada
ergocornine
incidence: 4/4, conc. range: 0.7-1.3 µg / kg, country: Canada
α-ergokryptine
incidence: 4/4, conc. range: 0-1.1 µg / kg, country: Canada
ergocristine
incidence: 4/4, conc. range: 0.4-4 µg / kg, country: Canada
→ nivalenol
incidence: 2/12*, conc. range: 30-100 µg / kg, country: India, *moldy, refined
incidence: 12/36, conc. range: 4-84 µg / kg, country: Japan
incidence: 1/1*, conc.: 310 µg / kg, country: Papua New Guinea, *imported, finge-ground biscuit flour
→ ochratoxin A
incidence: 3/23*, conc. range: 0.2-0.5 µg / kg, country: Germany, *whole meal
incidence: 12/13, conc. range: 0.1-1.9 µg / kg, ∅ conc.: 0.49 µg / kg, country: Switzerland
→ T-2 toxin
incidence: 2/12*, conc. range: 550-800 µg / kg, country: India, *moldy, refined
→ zearalenone
incidence: 5/7, ∅ conc.: 4 µg / kg, country: China
incidence: 2/5, conc. range: 2-3 µg / kg, ∅ conc.: 2.5 µg / kg, country: China
incidence: 3/4*, conc. range: 5.1-10 µg / kg, ∅ conc.: 6.9 µg / kg, country: Germany, *organic produce
incidence: 2/3, conc. range: 11-12 µg / kg, ∅ conc.: 11.5 µg / kg, country: Germany
incidence: 3/27, conc. range: 1-6 µg / kg, country: Japan

incidence: 1/1*, conc.: 250 µg / kg, country: Papua New Guinea, *imported, raw flour
incidence: 2/27, Ø conc.: 100 µg / kg, country: USA
→ flour, → milling

Wheat grits may contain the following → mycotoxins:
→ citrinin
incidence: 2/4*, conc. range: 0.3-0.7 µg / kg, Ø conc.: 0.5 µg / kg, country: Switzerland, *durum wheat
→ deoxynivalenol
incidence: 1/1, conc.: 160 µg / kg, country: Germany
→ ochratoxin A
incidence: 4/4*, conc. range: 0.8-2.7 µg / kg, Ø conc.: 1.65 µg / kg, country: Switzerland, *durum wheat
→ barley grits, → maize grits, → rye grits

Wheat products may contain the following → mycotoxins:
→ deoxynivalenol
incidence: 545/1257*, conc. range: 9-4060 µg / kg, Ø conc.: 260 µg / kg, country: Canada, *→ flour, → bran, → bread, cookies, crackers, cakes, pasta, etc.
→ ochratoxin A
incidence: 10/10*, conc. range: 0.2-3.5 µg / kg, Ø conc.: 1.37 µg / kg, country: Switzerland, *durum
→ figazzas, → libritos

Whey powder A storage period of 40 days did not change the → aflatoxin M_1 of lyophylized whey powder to any significant degree.
Whey powder may contain the following → mycotoxins:
aflatoxin M_1
incidence: 28/74, conc. range: 0.5-6.5 µg / kg, country: France
incidence: 88/88, conc. range: < 0.1-0.6 µg / kg, country: UK

White cheese → cheese (white)

Wine → Ochratoxin A seems to be the most important mycotoxin in wine. Red wine and red → grape juice originating from the more southern and warmer regions of Europe and northern Africa are primarily affected. This may be due to the enhanced growth of OTA-producing → Aspergillus species over → Penicillium spp. and / or different practices in grape cultivation (e.g. pesticides, cultivars) and wine making (e.g. period and storage condition of the harvested grapes, maceration type, kind of fermentation). In addition, growth of OTA-producing molds in barrels and / or tanks or any other equipment as well as the failure to remove moldy → fruits before processing might be responsible for the higher incidence and concentration of OTA in these wines. Since the climatic conditions are warm but less humid than in central Europe, it was suggested that OTA contamination of the grapes mainly occurs after harvest. Lower levels ($\leq$ 0.005 µg OTA / l) have been found in red wines originating from the more central parts of Europe (Switzerland, Burgundy, Germany).
It is assumed that OTA is probably formed prior to alcoholic fermentation since ethanol and the generally anaerobic conditions inhibit mold growth. In addition, red grape juices as well as the red wines contained similar concentrations. No significant degradation occurs during wine making and storage.
White wines contained less OTA than rosé and these less than red wines.
Besides OTA a contamination of red wines with → ochratoxin C (ethyl ester of OTA) has been reported. This ochratoxin might be of fungal origin or an artifact.
Wine may contain the following → mycotoxins:
→ aflatoxins (no specification)

incidence: 2/33, conc. range: < 1 µg / l, country: Germany
→ ochratoxin A
incidence: 14/41*, conc. range: ≤ 1.2 µg / l, country: Germany, *white, partly imported
incidence: 6/14*, conc. range: ≤ 2.4 µg / l, country: Germany, *rosé, partly imported
incidence: 40/89*, conc. range: ≤ 7 µg / l, country: Germany, *red, partly imported
incidence: 22/24*, conc. range: < 0.005-0.178 µg / l, Ø conc.: 0.011 µg / l, country: Switzerland, *white table wine, partly imported
incidence: 77/79*, conc. range: < 0.005-0.388 µg / l, Ø conc.: 0.039 µg / l, country: Switzerland, *red table wine, partly imported
incidence: 13/15*, conc. range: < 0.005-0.123 µg / l, Ø conc.: 0.011 µg / l, country: Switzerland, *rosé table wine, imported
incidence: 2/3*, conc. range: < 0.049-0.451 µg / l, Ø conc.: 0.290 µg / l, country: Switzerland, *Malaga, imported
incidence: 2/2*, conc. range: < 0.044-0.337 µg / l, Ø conc.: 0.191 µg / l, country: Switzerland, *Marsala, imported
incidence: nc/6*, conc. range: ≤ 0.17 µg / l, Ø conc.: 0.011 µg / l, country: Switzerland, *Port wine, imported
incidence: 2/2*, conc. range: < 0.029-0.054 µg / l, Ø conc.: 0.041 µg / l, country: Switzerland, *Sherry, imported
incidence: 2/2*, Ø conc.: 0.003 µg / l, country: Switzerland, *Vermouth, imported

Wort In an experimental study an 8-day fermentation (*Saccharomyces cerevisiae*) of wort containing → ochratoxin A, → fumonisin B_1 and → fumonisin B_2 at 25 °C caused mycotoxin losses in the range of 2-13%, 3-28% and 9-17%, respectively. While some OTA was taken up by the yeast (≤ 21%) almost no uptake occurred in the case of the → fumonisins (FB_1 < 1%, FB_2 < 2%). No decrease in mycotoxin (→ mycotoxins) concentration was observed if a yeast-free wort was used.
In a further study it could be shown that OTA does not survive the malting process. If OTA was added at the start of the mashing process, simulating the use of OTA contaminated adjuncts, the finished → beer contained OTA in the range of 2-28%.
→ beer

X

Xanthomegnin is a lactone (3,3´,4,4´-tetrahydro-10,10´-dihydroxy-7,7´-dimethoxy-3,3´-dimethyl-[8,8´-bi-1H-naphtho[2,3-c]pyran]-1,1´,6,6´,9,9-´hexone) mycotoxin (→ mycotoxins) which was first isolated from *Trichophyton megninii* in 1963 (see Figure Xanthomegnin).

Chemical Data
Empirical formula: $C_{30}H_{22}O_{12}$; molecular weight: 574

Fungal Sources
The penicillia are the main sources of xanthoquinones, e.g. → Penicillium aurantiogriseum Dierckx, → Penicillium crustosum Thom, *P. simplicissimum*, → Penicillium verrucosum Dierckx, → Penicillium viridicatum Westling, *Eupenicillium javanicum*, → Aspergillus ochraceus group, *Trichophyton* spp.

Natural Occurrence
→ barley, → oil seed rape, → wheat
Xanthomegnin may be found in ca. 50% of ochratoxin A suspected → cereals and feed samples.

Toxicity
hepatotoxic, nephrotoxic.
strong uncoupler of oxidative phosphorylation (increased rate of respiration)

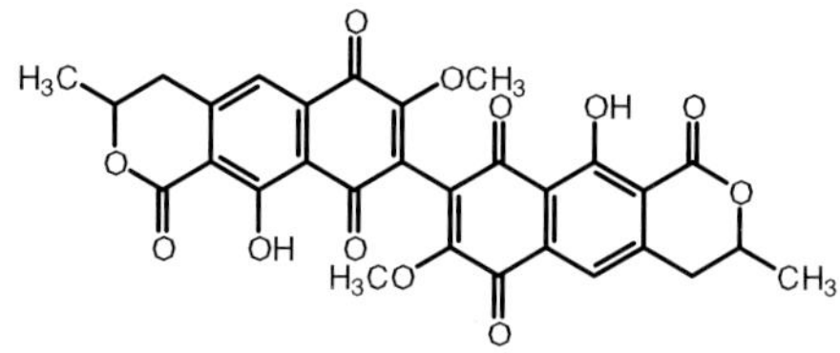

Xanthomegnin

Detection
HPLC, TLC

Possible Mycotoxicosis
Xanthomegnin in combination with viomellein, ochratoxin A and citrinin may also be involved in kidney diseases of human and animals. Fungi producing these nephrotoxins are often co-occurring.

Further Comments
Xanthomegnin is often associated with → viomellein. A simultaneous occurrence with → ochratoxin A and → citrinin is possible.

Y

Yeasts (fermentative)
Speed of fermentation (*Saccharomyces cerevisiae*) is depressed by → T-2 toxin, → diacetoxyscirpenol, aflatoxin (→ aflatoxins) (in decreasing order). A similar effect has been observed with → patulin. T-2 toxin also inhibits yeast growth. However, a substantial reduction in patulin concentration (< 1% of the original levels) occurred during fermentation of → apple juice.

Yellow rice disease is an intoxication which mainly occurred in the 19th and early 20th centuries as well as shortly after World War II in Japan. Many human deaths have been reported due to the consumption of moldy (yellow) → rice imported form south-east Asia which had been declared unfit for human consumption. The syndrome involved → acute cardiac beriberi. Moldy rice toxins should mainly be responsible for emesis, ascending → paralysis, → convulsions and respiratory arrest. Death may occur. More than 15 kinds of molds have been incriminated in the yellow rice syndrome but → Penicillium islandicum Sopp (→ islanditoxin, → luteoskyrin), → Penicillium citrinum Thom (→ citrinin), → Penicillium citreonigrum Dierckx (synonym *P. citreo-viride*) (→ citreoviridin), and *P. rugulosum* (→ rugulosin) are the most important. Their → mycotoxins primarily act on the liver but other organs such as the kidneys may also be affected.

Yogurt Although the contamination rate of yogurt with → aflatoxin M_1 due to the → carry over of → aflatoxin B_1 from the feed into the → milk (AFM_1) seems to be low, this aflatoxin is the most important mycotoxin in this kind of foodstuff.
Different reports concerning the behavior and influence of AFM_1 and aflatoxin B_1 in / on yogurt do exist. The results are as follows: (i) no influence of yogurt manufacture and refrigerated storage on AFM_1 content, (ii) variable increases of AFM_1 content in yogurt, (iii) a high reduction of AFM_1 in yogurt, (iiii) complete transformation of AFB_1 in its hydroxy derivative AFB_{2a}. AFB_1 caused a delay in curdling.
In addition, AFM_1 caused thickening of the cell walls of *Lactobacillus bulgaricus* and *Streptococcus thermophilus*. A change in cell shape from coccoid to oval (*S. thremophilus*) and shortening of cell chain length (*L. bulgaris*) was also observed.
Yogurt may contain the following → mycotoxins:
aflatoxin M_1
incidence: 44/54, conc. range: 0.05-0.47 µg / kg, Ø conc.: 0.2 µg / kg, country: Germany
incidence: 91/114, conc. range: < 0.001-0.496 µg / kg, Ø conc.: 0.018 µg / kg, country: Italy
incidence: 1/1*, conc.: 0.19 µg / kg, country: Syria, *Koshk (sundried mixture of parboiled → wheat and yogurt)
milk

Z

Zearalenol (Syn.: α-zearalenol) is a hydroxylated derivative of → zearalenone due to zearalenone reductases present in animal tissues. Formation by *F. semitectum* has been reported. It is used as a growth promoter in livestock due to its anabolic potential. Apparently no residues accumulate in animal tissues and it does not exert potent uterotropic effects. α-Zearalenol possess a ten-times higher estrogenic activity than zearalenone whereas the β-isomer is considerably less active (similar or slightly less than that of zearalenone). Zearalenol may be of concern to food hygienists if it is transmitted into → milk and other edible tissues.

Zearalenone (Syn.: F-2 toxin) is a 6-(10-hydroxy-6-oxo-*trans*-1-undedenyl)-β-resorcyclic acid lactone which is produced by → Fusarium spp., primarily → Fusarium graminearum Schwabe and → Fusarium culmorum (W. G. Smith) Sacc. (see Figure Zearalenone). Originally (1962) this mycotoxin which was recovered from cultures of *Giberella zea* (sexual stage of *Fusarium roseum*) was called F-2 toxin. Determination of molecular structure followed in 1966.

Chemical Data

Empirical formula: $C_{18}H_{22}O_5$, molecular weight: 318

Fungal Sources

Fusarium spp.: e.g. → Fusarium avenaceum (Fr.) Sac. (?), *F. culmorum,* → Fusarium equiseti (Corda) Sacc. sensu Gordon, *F. graminearum,* → Fusarium moniliforme Sheldon, → Fusarium oxysporum Schlecht. emend. Snyd. & Hans., → Fusarium sambucinum Fuckel, *F. semitectum,* → Fusarium sporotrichioides Sherb.

Natural Occurrence

→ bananas, → barley, → barley malt, → beans, → beer, → beer, joala, → beer, opaque maize, → beer, pito, → bread, → breakfast cereals, cereals, → chilli powder, → chilli sauce, → coriander, → corn flakes, → curry, → curry paste, → fennel, → fermented products, → foods, → garlic pickle, → grains, → job's-tears, → maize, → maize flour, → maize malt, → maize meal, → maize, brewers, → millet, → millet meal, → muffin mix, → oats, → oil, → oil seeds, → pepper, → pop corn, → rice, → rye, → rye bran, → rye flour, → snack food, → sorghum, → spices, → tapioca, → walnuts, → wheat

Zearalenone is commonly found in food, mainly in → cereals and → cereal products of the temperate regions.

Zearalenone is of worldwide importance since it occurs in maize in international trade.

High levels of zearalenone in cereals usually accumulate during storage of mature, *Fusarium* infected grains that have not sufficiently dried because of wet weather at harvest or in grains that were stored wet (e.g. maize: moisture content > 22%). Beside this, zearalenone production has been reported on grains in the field, during harvest, commercial grain processing, and / or subsequently during storage of any food- or feedstuff containing the grain.

According to the mean zearalenone levels naturally found in feed transmission of this mycotoxin into tissues and → milk of ruminants generally does not pose a significant human health risk. A normal daily intake (cow) of 50-165 mg zearalenone from protein rations did not result in any detectable residues. Although experimental → carry over of zearalenone residues into → poultry products was shown, rates of carry over due to naturally contaminated feed may be neglected. Residues of zearalenone in → meat, milk and eggs seem to be negligible.

Toxicity

not acutely toxic (20,000 mg / kg oral application did not cause deaths in mice and rats), hyperestrogenic in swine; → cattle are less, → poultry are minimal affected; weakly → teratogenic (pigs), → mutagenic (?), possibly carcinogenic (class 2B carcinogen, IARC)
LD_{50} (po): > 4000 - > 10,000 mg / kg bw rat / LD_{50} sodium chloride (po): 3750 mg / kg bw rat)
hyperestrogenic syndromes: e.g. uterine enlargement, swelling of the vulva (vulvovaginitis), mammary glands and nipples, prolapse of the vagina or rectum, prolonged or interrupted estrus, pseudopregnancy, infertility especially prepubertal gilts but other species like rats, mice or monkeys are also affected
Transmission of zearalenone via sow's milk to piglets cause estrogenism in the young pig.
Since the very high LD_{50} of zearalenone it might better be called a non-steroidal fungal hormon (estrogen), rather than a direct mycotoxin. Besides estrogenic zearalenone also possesses anabolic activity.

Detection

ELISA, GC-MS, HPLC, LC-MS, TLC

Possible Mycotoxicosis

Although an estrogenic syndrome in humans could not be correlated with the consumption of foods containing zearalenone, this mycotoxin has been implicated in several incidents of precocious pubertal changes in children (→ premature thelarche).

Further Comments

This mycotoxin seems to be a suitable indicator for the presence of other → Fusarium mycotoxins in cereals such as → trichothecenes (e.g. deoxynivalenol, → nivalenol).
Temperatures between 12 and 14 °C are required for significant zearalenone formation but production also occurs at temperatures below 10 °C and even below freezing.
Zearalenone often co-occurs with deoxynivalenol in grain worldwide. At low concentrations the effect of pure zearalenone is antagonized by the presence of pure deoxynivalenol whereas zearalenone slightly enhanced the effects of deoxynivalenol over a range of concentrations.
One strain of *F. semitectum* not only produced zearalenone but also → zearalenol and 8´-hydroxyzearalenone.
Acremonium species of New Zealand produce zearalenols. These fungi may be important in maize grown in subtropical countries.
Reduction / elimination: Cleaning removed only 3-10% of zearalenone. Wet → milling of maize led to an accumulation in the gluten (49-56%) > solubles > fiber > germ whereas zearalenone was not present in the starch fraction. The steeping procedure did not destroy zearalenone. During dry-milling, high levels were found in the maize germ, degermer fines, bran meal, hull, and high fat fractions. Low zearalenone levels (10-22%) occurred in the prime products (grits, low-fat meal, and flour).
Sieving of coarsely ground barley, wheat and maize caused substantial reductions in zearalenone (and → deoxynivalenol) concentrations.
Zearalenone possess a relatively high heat **stability** - most survived a temperature of 180 °C for 30 min - and it is insensitive to hydrolytic cleavage.
Making → bread caused losses in the range of 34-40% of the zearalenone originally present in → wheat flour; instant → noodles 48-62%, and → biscuits 16-27%.
Zearalenone can survive the process of **brewing** corn, corn malt and other substrates whereas only little destruction of the mycotoxin occurred. The recovered solids contained about twice the levels of zearalenone originally present in maize.

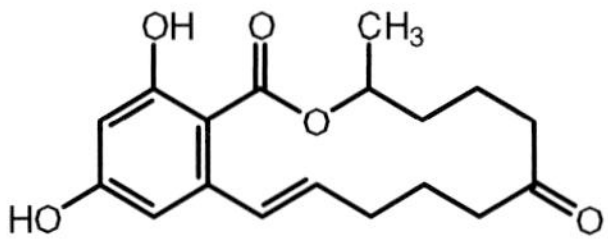

Zearalenone

The stability of zearalenone during **fermentation** is further documented by its natural occurrence in maize → beer. No zearalenone was found in ethanol resulting from the distillation of fermented maize naturally contaminated with zearalenone. *Saccharomyces cerevisiae* converted zearalenone largely to β-zearalenol and, to a minor degree, to α-zearalenol.

Zwieback may contain the following
→ mycotoxins:
→ ochratoxin A
incidence: 6/9, conc. range: 0.1-0.49 μg / kg (5 samples), 0.50-1.49 μg / kg (1 sa), country: Germany
→ bread

Mycotoxin legislation. Maximum tolerated levels of mycotoxins in foodstuffs, dairy products and animal feedstuffs (FAO 1997, modified)

Country / food & feedstuff	Commodity	(Sum of) Mycotoxin(s)	Level µg / kg	Responsible authority	Remarks
Antiuga & Barbuda: no regulations					
Argentina: situation 1991, see also Mercosur					
food	baby food	AFB_1	0		
	groundnut, maize and by-products	AFB_1 AFB_1, B_2, G_1, G_2	5 20		
dairy	liquid milk, powdered milk milk products	AFM_1 AFM_1	0.05 0.5		
feed	soya meal	AFB_1	30		
Australia, adopted by all states and territories					
food	all foods	AFB_1, B_2, G_1, G_2 Phomopsin	5 5	Natl Food Auth	
	peanut butter, nuts and the nut proportion of products	AFB_1, B_2, G_1, G_2	15	"	
Austria (see European Union):					
Food	all foods	AFB_1 AFB_2, G_1, G_2	1 5	Min Pub Health	
	milling and shelled products and derived products	AFB_1 AFB_2, G_1, G_2	2 5	"	
	children's foods (in prepared foods)	AFB_1, B_2, G_1, G_2, M_1	0.02	"	
	wheat, rye	OTA DON ZEA	5 500 60		Guideline level
	durum wheat	OTA DON ZEA	5 750 60		
	fruit juice	Patulin	50	Min Pub Health	
dairy	milk(products)	AFM_1	0.05	"	
	whey powder, whey paste	AFM_1	0.4	"	Calcd on dry matter
	whey, liquid whey products	AFM_1	0.025	"	
	cheese	AFM_1	0.25	"	
	butter	AFM_1	0.02	"	
	pasteurized fresh milk for infants / children; children's food	AFM_1	0.01	"	Calcd on reconstituted product
	powdered milk(products), condensed milk, milk concentrates	AFM_1	0.4	"	Calcd on dry matter
feed	see European Union				

(continued)

Country / food & feedstuff	Commodity	(Sum of) Mycotoxin(s)	Level µg / kg	Responsible authority	Remarks
Bahamas: situation 1991; no national regulations; FDA regulations are used					
food	all foods, all grains	AFB_1, B_2, G_1, G_2	20		
Bahrain: no regulations					
Barbados: situation 1991					
Foods	all foods	AFB_1, B_2, G_1, G_2	20		
Dairy	milk	AFM_1	0.05		
Feed	all feedstuffs	AFB_1, B_2, G_1, G_2	50		
Belgium (see European Union):					
Food	peanuts	AFB_1	5	Min Pub Health	
Dairy	milk	AFM_1	0.05	"	
Feed	see European Union				
Belize:					
Food	maize, groundnut	AFB_1, B_2, G_1, G_2	20		Situation 1991
Bolivia: situation 1991; no regulations					
Bosnia and Herzegowina: situation 1981					
Food	wheat, maize, rice, cereals	AFB_1, G_1	1	Fed Comm Labour Health Soc Welf	
	beans	AFB_1, G_1	5	"	
feed	feedstuffs	?	?		
Brazil: situation 1987; proposals; see also Mercosur					
food	all foodstuffs	AFB_1 AFB_1, B_2, G_1, G_2	15 30		
	imported foodstuffs	AFB_1 AFB_1, B_2, G_1, G_2	5 10		
	industrially prepared foodstuffs for children from 0-2 years and for school meals	AFB_1, B_2, G_1, G_2	3		
	rice, barley, beans, maize	OTA	50		
	maize	ZEA	200		
	maize, groundnut	AFB_1, G_1	30		Situation 1991
dairy	milk(products)	AFM_1	0.5		Situation 1987, proposal
	imported milk(products)	AFM_1	0.1		Situation 1987, proposal
feed	peanut meal (export)	AFB_1, B_2, G_1, G_2	50		Situation 1977

(continued)

Country / food & feedstuff	Commodity	(Sum of) Mycotoxin(s)	Level µg / kg	Responsible authority	Remarks
Bulgaria: situation 1992					
food	peanut(product)s, kernel(product)s, cocoa beans, cocoa butter, cocoa powder	AFB_1, B_2, G_1, G_2	5		
	grain(products), cereal(products)	AFB_1, B_2, G_1, G_2	2.5		
dairy	liquid milk products powdered milk powdered milk for dietetics and infant feeding cheese and similar products	AFM_1 AFM_1 AFM_1 AFM_1	0.5 0.1 0 0.5		
Canada:					
food	nut(product)s	AFB_1, B_2, G_1, G_2	15	Health Can	Calcd on the nut meat portion
	uncleaned soft wheat	DON	2000	"	
feed	animal feeding stuffs	all aflatoxins	20	Agric Food Canada	
	diets for cattle/poultry	DON HT-2 toxin	5000 100		Recommendation
	diets for swine/young calves/lactating dairy animals	DON HT-2 toxin	1000 25		Recommendation
	feedstuffs for reproducing animals	all mycotoxins	0		Recommendation
Chile: situation 1991					
feed	feedstuffs feedstuffs	AFB_1 AFB_1, B_2, G_1, G_2 AFB_1 AFB_1, B_2, G_1, G_2	20 50 5 20		
China (People's Republic of China):					
food	rice, edible oils	AFB_1	10	Min Health	
	maize, peanut(product)s, maize, peanut oil	AFB_1	20	"	
	wheat, barley, oats, beans, sorghum, other grains, fermented foodstuffs	AFB_1	5	"	
dairy	cow milk, milk products (calcd. on the basis of milk)	AFB_1	0.5	"	
feed	compound feed for chickens	AFB_1	10	St Tech Sup Bur	

(continued)

Country / food & feedstuff	Commodity	(Sum of) Mycotoxin(s)	Level µg / kg	Responsible authority	Remarks
	compound feed for laying hens, compound feed and mixed feed for fattening pigs	AFB_1	20	St Tech Sup Bur	
	maize, peanut cake, peanut residues	AFB_1	50	St Tech Sup Bur	
Colombia: situation 1991					
food	foods cereals (sorghum/millet)	AFB_1, B_2, G_1, G_2 AFB_1, B_2, G_1, G_2	20 30		
	oil seeds	AFB_1, B_2, G_1, G_2	10		
feed	cattle feed	AFB_1, B_2, G_1, G_2	50		
	sesame seeds	AFB_1, B_2, G_1, G_2	20		
	poultry feedstuffs	AFB_1, B_2, G_1, G_2	20		
Costa Rica: situation 1991					
Food	maize	AFB_1, B_2, G_1, G_2	35		
Feed	maize	AFB_1, B_2, G_1, G_2	50		
Côte d'Ivoire: situation 1987; proposals, types of aflatoxins not precisely stated					
feed	straight feedstuffs	AFB_1, B_2, G_1, G_2	100	Min Pub Health Min Animal Prod Min Commerce	
	complete feedstuffs	AFB_1, B_2, G_1, G_2	10	"	
	complete feedstuffs for pigs / poultry (except young animals / ducks)	AFB_1, B_2, G_1, G_2	38	"	
	complete feedstuffs for cattle / sheep, goats	AFB_1, B_2, G_1, G_2	75		
	complete feedstuffs for dairy cattle	AFB_1, B_2, G_1, G_2	50		
Cuba: situation 1991					
food	foods	AFB_1, B_2, G_1, G_2	5		
	cereals, groundnuts	AFB_1, B_2, G_1, G_2	5		
feed	feedstuffs, raw materials for feedstuffs	AFB_1, B_2, G_1, G_2	5		
	feedstuffs, raw materials for feedstuffs	AFB_1, B_2, G_1, G_2	5		

(continued)

Country / food & feedstuff	Commodity	(Sum of) Mycotoxin(s)	Level µg / kg	Responsible authority	Remarks
Cyprus: situation 1992					
food	cereals, pulses, dried fruit, sesame and foods produced exclusively from these, caraway seed, poppy seed, seeds used in bakery products and confectionery	AFB_1, B_2, G_1, G_2	5		
dairy	milk, dairy products	all mycotoxins	0.5		
Czech Republic:					
food	all foods	AFB_1 AFB_2, G_1, G_2	5 10	Min Health	
	infant's /children's foods	AFB_1 AFB_2, G_1, G_2	1 2	"	
	all foods	Patulin OTA	50 20	"	
	children's foods	Patulin OTA	30 5	"	
	infant's foods	Patulin OTA	20 1	"	
dairy	all foods	AFM_1	5		
	milk	AFM_1	0.5	"	
	infant's / children's foods	AFM_1	1	"	
	infant's foods on milk basis	AFM_1 AFB_1 AFB_2, G_1, G_2	0.1 0.1 0.2		Calcd on reconstituted product
Denmark (see European Union):					
food	peanut(product)s	AFB_1 AFB_1, B_2, G_1, G_2	2 4	"	
	brazil nuts	AFB_1 AFB_1, B_2, G_1, G_2	2 4	"	
	dried figs	AFB_1 AFB_1, B_2, G_1, G_2	2 4	"	
	pig kidney	OTA	25	Dan Vet Serv	whole carcass condemned; visibly damaged kidneys are analyzed chemically
	pig kidney	OTA	10		viscera condemned; visibly damaged kidneys are analyzed chemically
	cereal(product)s	OTA	5		
feed	see European Union				

(continued)

Country / food & feedstuff	Commodity	(Sum of) Mycotoxin(s)	Level µg / kg	Responsible authority	Remarks
Dominican Republic: situation 1991					
food	maize(product)s, peanut, soya, tomato(products)	AFB_1, G_1	0		
	imported maize	AFB_1, B_2, G_1, G_2	20		
Ecuador: situation 1991; no regulations					
Egypt:					
food	peanut(product)s, oil seed(product)s, cereal(product)s	AFB_1, B_2, G_1, G_2 AFB_1	10 5		
	maize	AFB_1 AFB_1, B_2, G_1, G_2	10 20		
	starch and its derivatives	AFB_1 AFB_1, B_2, G_1, G_2	0 0		
dairy	milk, dairy products	AFG_1, G_2, M_1, M_2 AFM_1	0 0		
feed	animal and poultry fodders	AFB_1 AFB_1, B_2, G_1, G_2	10 20		
European Union: All European Union tolerances refer to a commodity content of 12%; United Kingdom has extra regulation for feedstuff ingredients. 1st January 1999: 2 µg / kg AFB_1 and 4 µg / kg sum of AFB_1, B_2, G_1 and G_2 for cereals, peanuts, nuts, dried fruits and their products intended for direct human consumption or use as an ingredient in foodstuff. 8 µg / kg AFB_1 and 15 µg / kg sum of AFB_1, B_2, G_1 and G_2 for peanuts and 5 µg / kg AFB_1 and 10 µg / kg sum of AFB_1, B_2, G_1 and G_2 for nuts and dried fruits to be subjected to sorting, or other physical treatment, before human consumption or use as an ingredient in foodstuffs. 0.05 µg / kg AFM_1 in milk(products).					
feed	straight feedstuffs	AFB_1	50	various	
	straight feedstuffs: peanut(products), copra(products), cotton seed(products), palmnut(products), babassu(products), maize(products)	AFB_1	20	"	
	complete feedstuffs for pigs and poultry (except young animals)	AFB_1	20	"	
	complete feedstuffs for cattle / sheep / goats (except dairy cattle / calves / lambs)	AFB_1	50	"	
	complete feedstuffs for dairy cattle	AFB_1	5	"	
	complete feedstuffs for calves and lambs	AFB_1	10	"	

(continued)

Country / food & feedstuff	Commodity	(Sum of) Mycotoxin(s)	Level µg / kg	Responsible authority	Remarks
	other complete feedstuffs	AFB_1	10	"	
	complementary feed-stuffs for pigs and poultry (except young animals)	AFB_1	30	"	
	complementary feed-stuffs for catle / sheep / goats (except dairy animals / calves / lambs)	AFB_1	50	"	
	other complementary feedstuffs	AFB_1	5	"	
	raw materials: groundnut(product)s, copra(products), palmnut(products), cotton seed(products), babassu(products), maize(products)	AFB_1	200	"	
Finland (see European Union):					
food	all foods	AFB_1, B_2, G_1, G_2	5	Min Trade Ind Natl Brd Trade Cons Int	
	all foods	Patulin	50	Min Trade Ind Natl Brd Trade Cons Int	
feed	see European Union			Min Agr For	
France (see European Union):					
food	all foods	AFB_1	10		
	peanuts, pistachio nuts, almonds, oil seeds, children foods	AFB_1	1	Min Consump	Not intended for the production of vegetable oils
	wheat meal	AFB_1	3	"	
	wheat bran	AFB_1	10	"	
	vegetable oils, cereals, wheat meal (complete)	AFB_1	5	"	
	apple juice (products)	Patulin	50	"	
	cereals, vegetable oils	ZEA	200	"	
	cereals	OTA	5	"	
dairy	milk, milk powder(calcd on reconstituted product)	AFM_1	0.05	"	
	milk, milk powder(calcd on reconstituted product) for infants under 3 years	AFM_1	0.03	"	
feed	see European Union				

(continued)

Country / food & feedstuff	Commodity	(Sum of) Mycotoxin(s)	Level µg / kg	Responsible authority	Remarks
Germany (see European Union):					
food	all foods	AFB_1 AFB_1, B_2, G_1, G_2	2 4	Bundes Ges	
	enzyme(preparation)s intended for the production of foodstuffs	AFB_1, B_2, G_1, G_2	0.05	"	
	foods for infants and young children	AFB_1, B_2, G_1, G_2	0.05	"	
dairy	milk	AFM_1	0.05	"	
	foods for infants and young children	AFM_1	0.01	"	
feed	see European Union			Min Agr For	
Greece (see European Union):					
food	peanuts, hazelnuts, walnuts, cashewnuts, pistachio nuts, almonds, pumpkin seeds, sunflower seeds, pine, seeds, apricot seeds	AFB_1 AFB_1, B_2, G_1, G_2	5 10	Min Agr	
	maize, dried figs, dried apricots, dried prunes, dates, raisins	AFB_1 AFB_1, B_2, G_1, G_2	5 10	"	
	raw coffee beans apple juice, apple products	OTA Patulin	20 50	"	
feed	see European Union				
Guatemala: situation 1991					
Food	maize, kidney beans, rice, sorghum	AFB_1, B_2, G_1, G_2	20		
	groundnuts, groundnut butter	AFB_1, B_2, G_1, G_2	20		Guide value until regulation is approved
Feed	concentrate	AFB_1, B_2, G_1, G_2	20		Guide value until regulation is approved
Honduras: situation 1991					
Food	all foods	AFB_1, B_2, G_1, G_2	1		
	maize (grounded or whole grain)	AFB_1	1		
	baby food	AFB_1, B_2, G_1, G_2 AFM_1	0.01 0.02		
Dairy	milk(products)	AFM_1	0.05		
	cheeses	AFM_1	0.25		

(continued)

Country / food & feedstuff	Commodity	(Sum of) Mycotoxin(s)	Level µg / kg	Responsible authority	Remarks
Hong Kong:					
Food	foods	AFB_1, B_2, G_1, G_2, M_1, M_2, P_1, aflatoxicol	15	Dep Health	
	peanut(products)	AFB_1, B_2, G_1, G_2, M_1, M_2, P_1, aflatoxicol	20	"	
Hungary:					
Food	all foods	AFB_1	5	Min Health	Situation 1987
	groundnut kernels	AFB_1	30		Situation 1987
	preserved foods	all mycotoxins	0	Min Health	Situation 1992
	groundnuts	AFB_1, B_2, G_1, G_2	5		Situation 1992
India: situation 1987					
food	all foods	AFB_1	30	Min Health Fam Welf Dept Health	
feed	peanut meal (export)	AFB_1	120	Min Fd Cvl Supp Dept Civil Supp	
Indonesia:					
food	peanuts, maize, herbs, seeds			Min Health	Proposal in preparation
feed	copra in cow / pig / duck / sheep feed	AFB_1, B_2, G_1, G_2	1000	Dir Anim Husb	Proposal ultimo 1994; includes max %-ages of raw material in various feed-stuffs for all cow / pig / duck / sheep feedstuffs
	groundnut / sesame seed / rape seed meal	AFB_1, B_2, G_1, G_2	200	"	Proposal ultimo 1994
	cassava in chicken feed	AFB_1, B_2, G_1, G_2	120	"	Proposal ultimo 1994
	capok seed / coconut meal in chicken feed, co-conut meal in cow / pig / duck / sheep feed	AFB_1, B_2, G_1, G_2	100	"	Proposal ultimo 1994
	sunflower seed meal in chicken feed	AFB_1, B_2, G_1, G_2	90	"	Proposal ultimo 1994
	soya bean / capok seed / fish / meat / bone meal / rice / maize bran, leucae-na (?), maize / wheat pol-lar (?), and sorghum in cow / pig / duck / sheep feed, maize / meat / bone / cotton seed meal in chicken feed	AFB_1, B_2, G_1, G_2	50	"	Proposal ultimo 1994

(continued)

Country / food & feedstuff	Commodity	(Sum of) Mycotoxin(s)	Level µg / kg	Responsible authority	Remarks
	soya bean / leucaena (?) / fish / meat / bone meal, rice / maize bran, wheat pollar (?), sorghum, copra in chicken feed	AFB_1, B_2, G_1, G_2	20	"	Proposal ultimo 1994
Iran: no regulations					
Iraq: no regulations					
Ireland (see European Union):					
food	all foods	AFB_1 AFB_1, B_2, G_1, G_2	5 30		Situation 1987
feed	see European Union				
Israel:					
food	nut(product)s, peanut(product)s, maize flour (products), fig(products)	AFB_1 AFB_1, B_2, G_1, G_2	5 15		Proposal
	apple juice	Patulin	50		
	cereal(product)s, pulse(product)s	OTA	50		Proposal
dairy	milk, milk powder (calcd on the basis of milk)	AFM_1	0,05		Proposal
feed	according to European Union				Situation 1987
	grain for feed	AFB_1 OTA T-2 toxin DAS	20 300 100 1000		Situation 1991
Italy (see European Union):					
food	all foods	AFB_1 AFB_1, B_2, G_1, G_2	5 10	ISS	
	dried figs	AFB_1 AFB_1, B_2, G_1, G_2	5 10	Min Health	
	spices	AFB_1 AFB_1, B_2, G_1, G_2	20 40	ISS	
feed	see European Union				
Jamaica: situation 1991					
Food	food, grains	AFB_1, B_2, G_1, G_2	20		
Japan:					
Food	all foods	AFB_1	10	Min Health Welf	

(continued)

Country / food & feedstuff	Commodity	(Sum of) Mycotoxin(s)	Level µg / kg	Responsible authority	Remarks
Feed	peanut meal (import)	AFB_1	1000	Min Agr Forest-Fish	Not more than 2% in feed for dairy cattle; not more than 4% in feed for chicken (over 4 weeks of age), swine (over 30 kg) and cattle (over 3 months of age, except dairy cattle); not for use in feed for other livestock
Jordan: situation 1981					
Food	almonds, cereals, maize, peanuts, pistachio nuts, pine nuts, rice	AFB_1 AFB_1, B_2, G_1, G_2	15 30	Min Health	
Feed	all feedstuffs	AFB_1 AFB_1, B_2, G_1, G_2	15 30	"	
Kenya: situation 1981					
Food	peanut(product)s, vegetable oils	AFB_1, B_2, G_1, G_2	20	Min Health	
Kuwait: no regulations					
Luxembourg (see European Union):					
Food	peanut(product)s	AFB_1	5	Min Pub Health	Situation 1981
Feed	see European Union				
Macedonia: situation 1981					
Food	wheat, maize, rice, cereals	AFB_1, G_1	1	Fed Comm Labour Health Soc Welf	
	beans	AFB_1, G_1	5	"	
Feed	feedstuffs				
Malawi: situation 1987					
food	peanuts (export)	AFB_1	5		
Malaysia: situation 1987					
food	all foods	AFB_1, B_2, G_1, G_2	35		
Mauritius: situation 1987					
Food	all foods	AFB_1 AFB_1, B_2, G_1, G_2, M_1, M_2	5 10		
	groundnuts	AFB_1 AFB_1, B_2, G_1, G_2	5 15		

(continued)

Country / food & feedstuff	Commodity	(Sum of) Mycotoxin(s)	Level µg / kg	Responsible authority	Remarks
Mercosur: (Argentina, Uruguay, Brazil, and Paraguay); proposals for common regulations, probably effective in a few years, will overrule national regulations					
Food	maize kernels (whole, pieces, ground, peeled), maize flour / meal, peanuts (in shell, raw, roasted) peanut cream, peanut butter	AFB_1, B_2, G_1, G_2	20		
Dairy	liquid milk	AFM_1	0.5		
	milk powder	AFM_1	5		
Mexico:					
food	flours	all aflatoxins	20		
feed	cereals for bovine and porcine fattening feedstuffs	AFB_1, B_2, G_1, G_2	200		Situation 1991; less than 10% of cereals in feedstuffs
	feedstuffs for dairy / cattle / poultry	AFB_1, B_2, G_1, G_2	0		Situation 1991
Morocco: currently no regulations; Codex Alimentarius is followed					
Netherland, The (see European Union)					
food	all foods and food ingredients except groundnuts used for the preparation of peanut oil	AFB_1	5	Min VWS	
	cereal(product)s, pulse(product)s, legume(product)s	all mycotoxins	0	Min VWS C Board	
dairy	milk(products), milk powder (calcd on reconstituted product)	AFM_1	0.05	Min VWS	
	cheese	AFM_1	0.2	"	
	butter	AFM_1	0.02	"	
	infant foods on milk basis	AFM_1	0.05	"	As a proportion of the milk basis in infant food
feed	see European Union				
New Zealand: situation 1987					
Food	all foods	AFB_1, B_2, G_1, G_2	5		
	peanut butter, shelled nuts, nut portion of products containing nuts	AFB_1, B_2, G_1, G_2	15		
Nicaragua: situation 1991: no regulations					
Nigeria: situation 1987					
Food	all foods	AFB_1	20	FDA	
	infant foods	AFB_1	0	"	

(continued)

Country / food & feedstuff	Commodity	(Sum of) Mycotoxin(s)	Level µg / kg	Responsible authority	Remarks
Dairy	fluid milk	AFM_1	1	"	
Feed	feedstuffs	AFB_1	50	"	
Norway: situation 1987					
Food	all foodstuffs	AFB_1, B_2, G_1, G_2	5		
	brazil nuts, buckwheat	AFB_1, B_2, G_1, G_2	5		
	apple juice (concentrated)	Patulin	50		Calcd on reconstituted product
Feed	mixed feedstuffs depending on type of animal	AFB_1	10–50	Min Agr	Groundnut meal and cottonseed meal are not allowed entry
Oman: situation 1987					
Feed	complete feedstuffs	AFB_1	10	Min Comm Ind	Maximum content referred to a moisture content of 12%
	complete feedstuffs for poultry	AFB_1	20	"	Maximum content refered to a moisture content of 12%
Panama: situation 1991: no regulations					
Peru: situation 1991; no national regulations, Codex Alimentarius proposals used					
Food	all foodstuffs	AFB_1, B_2, G_1, G_2	10		
Feed	all feedstuffs	AFB_1, B_2, G_1, G_2	10		
	complementary products for milk, animal products, feedstuffs	AFB_1	10		
	cereals for porcine growing feedstuffs	AFB_1, B_2, G_1, G_2	100		Situation 1991
Philippines:					
Food	nut(products)	AFB_1, B_2, G_1, G_2	20		
Feed	poultry feedstuffs	AFB_1, B_2, G_1, G_2	20	Bur Anim Husb	
	livestock feedstuffs	AFB_1, B_2, G_1, G_2	50	"	
Poland:					
Food	all foods	AFB_1	0	Min Publ Health	
feed	feedstuffs, feedstuff ingredients, complete feedstuffs for cattle / sheep / goats	AFB_1	50		
	complete feedstuffs for pigs / poultry / dairy cows	AFB_1	20		

(continued)

Country / food & feedstuff	Commodity	(Sum of) Mycotoxin(s)	Level µg / kg	Responsible authority	Remarks
Portugal: situation 1987 (see European Union):					
Food	all foodstuffs	AFB_1	20	Min Pub Health Min Agr Min Commerce	Situation 1987
	peanuts	AFB_1	25	"	Situation 1987
	infant foods	AFB_1	5	"	Situation 1987
Feed	see European Union				
Qatar: no regulations					
Romania: situation 1987					
Food	all foods	AFB_1	0	Min Pub Health Min Agr	
	all foods	Patulin OTA ZEA	30 5 30	"	
Dairy	milk, dairy products	AFM_1	0	"	
Feed	all feedstuffs	AFB_1, B_2, G_1, G_2	50	"	
	all feedstuffs	Patulin OTA DON Stachyobotrio-toxin Chaetomin	30 5 5 0 0	"	
Russia:					
Food	animal fats	AFB_1 AFM_1	0 0.5	Min Health	
	bottled / canned / potted fruits and berries	Patulin	50	"	
	bottled / canned / potted vegetables	AFB_1 Patulin	5 50	"	
	casein	AFB_1 AFM_1	0 5	"	
	cereals (wheat of hard and strong types), flour, wheat bran	AFB_1 ZEA T-2 toxin DON	5 1000 100 1000	"	
	fruits, berries and vegetables (bottled / canned / potted juices and puree), cacao, cacao powder, chocolate, coffee, eggs, dehydrated egg, meat and poultry (fresh / chilled / frozen / tinned / potted / bottled), sausage and culinary products from meat and poultry, subproducts of farming animals and poultry, sweets	AFB_1	5	"	

(continued)

Country / food & feedstuff	Commodity	(Sum of) Mycotoxin(s)	Level µg / kg	Responsible authority	Remarks
	leguminous, protein isolators and concentrators, vegetable oil	AFB_1 ZEA	5 1000	"	
	nut(kernel)s	AFB_1 ZEA	? 1000	"	
Dairy	milk, sour dairy products, concentrated milk, cheese and cottage cheese products, cow butter	AFB_1 AFM_1	0 0.5	"	
Salvador, El: situation 1991					
Food	foods	AFB_1, B_2, G_1, G_2	20		
Feed	all feedstuffs	AFB_1	10		
	supplementary feeds for porcine / poultry / dairy cattle; single composite feedstuffs; bovine / caprine / ovine feedstuffs	AFB_1	20		
Saudi Arabia: no regulations					
Senegal: situation 1987					
Feed	peanut products (straight feedstuffs)	AFB_1	50	Min Commerce Min Pub Health	
	peanut products (feedstuff ingredients)	AFB_1	300	"	
Serbia: situation 1981					
Food	wheat, maize, rice, cereals	AFB_1, G_1	1	Fed Comm Labour Health Soc Welf	
	beans	AFB_1, G_1	5	"	
feed	feedstuffs	?	?		
Singapore: situation 1987					
Food	all foods	AFB_1 AFB_1, B_2, G_1, G_2	0 0	Min Env	
South Africa:					
Food	all foods	AFB_1 AFB_1, B_2, G_1, G_2	5 10	Dept Health	
Spain (see European Union):					
Food	all foods	AFB_1 AFB_1, B_2, G_1, G_2	5 10	Min Pub Health Cons	
Feed	see European Union				
Sri Lanka:					
Food	foods	all aflatoxins	30		

(continued)

Country / food & feedstuff	Commodity	(Sum of) Mycotoxin(s)	Level µg / kg	Responsible authority	Remarks
	foods intended for children up to 3 years of age	all aflatoxins	1		
Dairy	milk(products)	all aflatoxins	1		
Suriname: situation 1991					
Food	maize	AFB_1, B_2, G_1, G_2	30		
	groundnut(products), legumes	AFB_1	5		
Feed	feedstuffs	AFB_1, B_2, G_1, G_2	30		
Sweden: replacement of Swedish feestuff regulations with EU regulations to be reconsidered near 31-12-97 (see European Union)					
Food	all foods	AFB_1, B_2, G_1, G_2	5	Natl Food Adm	
	berries, fruits, juices	Patulin	50	"	Situation 1987
Dairy	liquid milk products	AFM_1	0.05	"	
Feed	feedstuff ingredients	AFB_1	50	"	
	feedstuff ingredients for dairy cattle	AFB_1	10	"	
	cereal grains and forages as feedstuff ingredients for dairy cattle	AFB_1	1	"	
	mixed feedstuffs (excluding forages) for dairy cattle	AFB_1	3	"	
	complete feedstuffs	AFB_1	10	"	
	complete feedstuffs for cattle / sheep / goats except dairy cattle / lambs / kids	AFB_1	50	"	
	complete feedstuffs for pigs and poultry except young animals	AFB_1	20	"	
	complete feedstuffs (including forages) for dairy cattle	AFB_1	1.5	"	
	complete feedstuffs for poultry	OTA	200	"	
	complete feedstuffs for pigs	OTA	100	"	
Switzerland:					
Food	all foods (except maize / cereals / herbs)	AFB_1 AFB_2, G_1, G_2	1 5	Lab Cantons	
	maize cereals (granular or ground)	AFB_1 AFB_2, G_1, G_2	2 5	"	
	herbs	AFB_1 AFB_2, G_1, G_2	5 5	"	

(continued)

Country / food & feedstuff	Commodity	(Sum of) Mycotoxin(s)	Level µg / kg	Responsible authority	Remarks
	babie's / infant's food	AFB_1, B_2, G_1, G_2	0,01	"	Calcd on reconstituted product
	cereal(product)s	OTA	2	Bund Amt Ges	Provisional
	maize(products)	Fumonisin B_1+B_2	1000	"	Provisional
	fruit juice	Patulin	50	Lab Cantons	
Dairy	milk(products)	AFM_1	0.05	"	
	whey(products)	AFM_1	0.025	"	
	cheese	AFM_1	0.25	"	
	butter, baby / infant food	AFM_1	0.02	"	
feed	prohibit feeding cattle with peanut bruise	?	?	For Viehw	
Taiwan, Province of China: situation 1991					
food	cereals	AFB_1, B_2, G_1, G_2	50	Dept Health Council Agr	
feed	feed, oilseed meals for feed under 4 % of mixed feed	AFB_1, B_2, G_1, G_2	1000		
Thailand: situation 1987					
food	all foods	AFB_1, B_2, G_1, G_2	20	Min Pub Health	
Trinidad & Tobago: situation; no national regulations, Codex Alimentarius proposals used					
Food	foods	AFB_1, B_2, G_1, G_2	10		
Feed	feedstuffs	AFB_1, B_2, G_1, G_2	10		
	complementary products	AFB_1	10		
	ice cream	all mycotoxins	0		Situation 1992
UAE (United Arab Emirates): no regulations					
UK (United Kingdom) (see European Union):					
food	nut(product)s, dried fig (product)s	AFB_1, B_2, G_1, G_2	4	Min Agr Fish Fd	
feed	see European Union				
	groundnut, copra, palmkernel, cottonseed, babassu, maize and derived products (raw materials)	AFB_1	20	"	Levels refer to a moisture content of 12%
Uruguay: see also Mercosur					
Food	foods and spices	AFB_1, B_2, G_1, G_2	20	Min Pub Health	
	texturized soy protein products: flour, starch, concentrate, isolate	AFB_1, B_2, G_1, G_2	30	"	
	peanuts, dried fruit(product)s	AFB_1, B_2, G_1, G_2	30	"	

(continued)

Country / food & feedstuff	Commodity	(Sum of) Mycotoxin(s)	Level µg / kg	Responsible authority	Remarks
	cocoa beans	AFB_1, B_2, G_1, G_2	10	"	
	infant foods, produced industrially	AFB_1, B_2, G_1, G_2	3	"	
	rice, barley, beans, coffee, maize	OTA	50	"	
	maize, barley	ZEA	200	"	
	fruit juice	Patulin	50	"	
dairy	milk(products)	AFM_1	0.5	"	
USA: United States of America					
food	all foods	AFB_1, B_2, G_1, G_2	20	FDA	
	finished wheat products	DON	1000	"	
dairy	whole milk, low fat milk, skim milk	AFM_1	0.5	"	
feed	feedstuff(ingredient)s	AFB_1, B_2, G_1, G_2	20	"	
	cottonseed meal intended for beef cattle / swine / poultry feedstuffs (regardless of age or breeding status)	AFB_1, B_2, G_1, G_2	300	"	
	maize and peanut prodcuts intended for breeding beef cattle / swine or mature poultry	AFB_1, B_2, G_1, G_2	100	"	
	maize and peanut products intended for finishing swine of 100 pounds or greater	AFB_1, B_2, G_1, G_2	200	"	
	maize and peanut products intended for finishing beefcattle	AFB_1, B_2, G_1, G_2	300	"	
	grains and grain by-products destined for ruminating beef and feedlot cattle older than 4 months and for chickens (not exceeding 50% of the cattle or chicken total diet)	DON	10,000	"	
	grains and grain by-products (not exceeding 40% of the diet)	DON	5000	"	
	grains and grain by-products destined for swine (not exceeding 20% of their diet)	DON	5000	"	

(continued)

Country / food & feedstuff	Commodity	(Sum of) Mycotoxin(s)	Level µg / kg	Responsible authority	Remarks
Venezuela: situation 1991					
Food	rice flour	AFB_1, B_2, G_1, G_2	5		
Feed	feedstuffs	AFB_1, B_2, G_1, G_2	20		
Zimbabwe:					
Food	foods	AFB_1 AFG_1	5 4	Min Agr	
	groundnuts, maize, sorghum	AFB_1 AFG_1	5 4	"	
dairy	feedstuffs	AFB_1, B_2, G_1, G_2		"	Levels vary with type of animal
	poultry feed	AFB_1, G_1	10 10	"	

AFB_1 = aflatoxin B_1, DAS = diacetoxyscirpenol, DON = deoxynivalenol, OTA = ochratoxin A, ZEA = Zearalenone

Literature

Anonymous (1978) Interactions of Mycotoxins in Animal Production. Proceedings of a Symposium July 13, 1978 Michigan State University. National Academy of Science, Washington DC

Anonymous (1989) Mycotoxins: Economic and Health Risks. Council for Agricultural Science and Technology, Ames, Iowa

Applebaum RS, Brackett RE, Wiseman DW, Marth EH (1982) Aflatoxin: Toxicity to dairy cattle and occurrence in milk and milk products - a review. J Food Prot 45:752–777

Arora DK, Mukerji KG, Marth EH (1991) (Eds) Handbook of Applied Mycology, Vol. 3, Foods and Feeds. Marcel Dekker, New York etc

Battaglia R, Hatzold T, Kroes R (Eds) (1996) Occurrence and Significance of Ochratoxin A in Food. 10–12 January 1996 Aix-en-Provence, France, ILSI, Europe. Food Add & Contam 13, Supplement:1–57

Betina V (Ed) (1984) Mycotoxins - Production, Isolation, Separation and Purification. Elsevier, Amsterdam etc

Betina V (Ed) (1989) Mycotoxins - Chemical, Biological and Environmental Aspects. Elsevier, Amsterdam etc

Beuchat LR (Ed) (1987) Food and Beverage Mycology. 2nd Ed, AVI, New York

Bhatnagar D, Lillehoj EB, Arora DK (eds) (1992) Handbook of Applied Mycology. Marcel Dekker, mc, New York

Bullerman LB (1979) Significance and mycotoxins to food safety and human health. J Food Protect 42:65–86

Champ BR, Highley E, Hocking AD, Pitt JI (1991) Fungi and Mycotoxins in Stored Products. Proceedings of an International Conference held at Bangkok, Thailand, 23–26 April 1991. ACIAR Proceedings No 36

Chelkowski J (Ed) (1989) *Fusarium* - Mycotoxins, Taxonomy and Pathogenicity. Elsevier, Amsterdam

Chelkowski J (Ed) (1991) Cereal Grain. Mycotoxins, Fungi and Quality in Drying and Storage. Elsevier, Amsterdam

Cole RJ, Cox RH (Eds) (1981) Handbook of Toxic Fungal Metabolites. Academic Press, New York etc

Creppy EE, Castegnaro M, Dirheimer G (Eds) (1993) Human Ochratoxicosis and its Pathologies. John Libbey Eurotext, Montrouge, France

Douglas King A Jr, Schade JE (1984) *Alternaria* toxins and their importance in food. J Food Protect 47:886–901

Doyle ME, Steinhart CE, Cochrane BA (1994) Food Safety 1994. Marcel Dekker, New York etc

Eaton DL, Groopman JD (Eds) (1994) The Toxicology of Aflatoxins: Human Health, Veterinary, and Agricultural Significance. Academic Press Inc, San Diego

Egmond HP van (Ed) (1989) Mycotoxins in Dairy Products. Elsevier Applied Science, London, New York

Egmond HP van, Speijers GJA (1994) Survey of data of the incidence and levels of ochratoxin A in food and animal feed worldwide. Natural Toxins 3:125–144

Ellis WO, Smith JP, Simpson BK, Oldham JH (1991) Aflatoxins in food: occurrence, biosynthesis, effects of organisms, detection, and methods of control. Crit Rev Food Sci Nutr 30:403–439

FAO (1997) Food and Nutrition Paper 64. Worldwide Regulations for Mycotoxins. A compendium. Rome

Flannigan B (1991) Mycotoxins. In: D´Mello JP, Duffus CM, Duffus JH (Eds) Toxic Substances in Crop Plants, pp 226–257. Royal Society of Chemistry, Cambridge

Forgacs J, Carll WT (1962) Mycotoxicoses. Adv Vet Sci 7:273–382

Frank HK (1974) Aflatoxine. Bildungsbedingungen, Eigenschaften und Bedeutung für die Lebensmittelwirtschaft. BEHR´S Verlag, Hamburg

Frank HK (1992) Citrinin. Z Ernährungswiss 31:164–177

Frisvad J (1988) Fungal species and their specific production of mycotoxins. In: Samson RA, Reenen-Hoeckstra E.S (Eds) Introduction of Food-borne Fungi, pp 239-249. Centraalbureau voor Schimmelcultures, Baarn, Netherlands

Frisvad J (1989) The connection between the penicillia and aspergilli and mycotoxins with special emphasis on misidentified isolates. Arch Environ Cont Toxicol 18:452–467

Galvano F, Galofaro V, Galvano G (1996) Occurrence and stability of aflatoxin M_1 in milk and milk products: a worldwide review. J Food Protect 59:1079–1090

Gedek B (1989) Mykotoxine. In: Gemeinhardt, H (Ed) Endomykosen. Gustav Fischer Verlag, Jena

Gelda CS, Luyt LJ (1977) Survey of total aflatoxin content in peanuts, peanut butter, and other foodstuffs. Ann Nutr Alim 31:477–483

Goldblatt LA (Ed) (1969) Aflatoxin. Scientific Background, Control, and Implications. Academic Press, New York London

Hawksworth DL, Kirk PM, Sutton BC, Pegler DN (Eds) (1995) Ainsworth & Bisby's Dictionary of the Fungi. 8th Ed, CAB International, Wallingford

Hesseltine CW (1974) Natural occurrence of mycotoxins in cereals. Mycopath Mycol Appl 53:141–153

Höhler D (1998) Ochratoxin A in food and feed: occurrence, legislation and made at action. Z Ernährungswiss 37:2–12

IARC (1993) IARC monographs on the evaluation of carcinogenic risk of chemicals to humans. Ochratoxin A. Nr. 56:489–521

IARC (1993) IARC monographs on the evaluation of carcinogenic risk of chemicals to humans. Toxins derived from *Fusarium moniliforme*: fumonisins B_1 and B_2 and Fusarin C. Nr. 56:445–466

Jackson LS, de Vries JW, Bullerman LB (Eds) (1996) Fumonisins in Food. Plenum Press, New York

Jarvis B (1976) Mycotoxins in food. In: Skinner, FA, Carr, JG (Eds) Symposium: Microbiology in Agriculture, Fisheries & Food, pp 251–267. Academic Press, London

Jelinek CF, Pohland AE, Wood G (1989) Worldwide occurrence of mycotoxins in foods and feeds – an update. J Assoc Off Anal Chem 72:223–230

Kiermeier F (1973) Mykotoxine in Milch und Milchprodukten. Z Lebensm Unters Forsch 151:237–240

Krogh P (1987) Mycotoxins in Food. Academic Press, London etc

Kubacki SJ (1986) The analysis and occurrence of patulin in apple juice. In: Steyn PS, Vlegaar R (Eds) Mycotoxins and Phycotoxins. A collection of invited papers presented at the sixth International IUPAC Symposium of Mycotoxins and Phycotoxins, Pretoria. Bioactive Molecules, pp 293–304. Elsevier, Amsterdam etc

Kurata H, Ueno Y (1984) Toxigenic Fungi – Their Toxins and Health Hazard. Developments in Food Science 7. Elsevier, Amsterdam etc

Kühl H (1910) Über eine eigenartige Veränderung der Paranuss. Pharm Zentralhalle 106

Lacey J (Ed) (1985) Trichothecenes and other Mycotoxins. John Wiley and Sons, New York

Marasas WFO, Nelson PE (Eds) (1987) Mycotoxicology: Introduction to the Mycology, Plant Pathology, Chemistry, Toxicology, and Pathology of Naturally Occurring Mycotoxicoses in Animals and Man. The Pennsylvania State University Press, University Park, PA

Marasas WFO, Nelson PE, Tousson TA (1984) Toxigenic *Fusarium* Species, Identity and Mycotoxicology. The Pennsylvania State University Press, University Park, PA

Marasas WFO (1995) Fumonisins: their implications for human and animal health. Natural Toxins 3:193–198

McKee LH (1995) Microbial contamination of spices and herbs: a review. Lebensm-Wiss Technol 28:1–11

Miller JD (1995) Fungi and mycotoxin in grain: implications for stored product research. J Stored Prod Res 31:1–16

Miller JD, Trenholm HL (1994) Mycotoxins in Grain. Compounds Other Than Aflatoxin. Eagan Press, St. Paul, Minnesota

Ministry of Agriculture, Fisheries and Food (1980) Survey of Mycotoxins in the United Kingdom. The Fourth Report of the Steering Group on Food Surveillance. The Working Party on Mycotoxins. Food Surveillance Paper No. 4. London, HMSO

Ministry of Agriculture, Fisheries and Food (1987). Survey of Mycotoxins in the United Kingdom. The Eighteenth Report of the Steering Group on Food Surveillance. The Working Party on Naturally Occurring Toxicants in Food: Sub-Group on Mycotoxins. Food Surveillance Paper No. 18. London, HMSO

Moreau C, Moss M (Eds) (1979) Moulds, Toxins and Food. John Wiley and Sons, New York

Mücke W, Lemmen Ch (1999) Schimmelpilze. Vorkommen, Gesundheitsgefahren, Schutzmaßnahmen. Ecomed, Landsberg/Lech

Natori S, Hashimoto K, Ueno Y (Eds) (1989) Mycotoxins and Phycotoxins. Elsevier, Amsterdam etc

Pestka JJ (1986) Fungi and Mycotoxins in Meats. In: Pearson AM, Dutson TR (Eds) Advances in Meat Research, pp 277–309. Mac Millan Publishers, Michigan

Pitt JI (1979) The Genus *Penicillium* and its Teleomorphic States *Eupenicillium* and *Talaromyces*. Academic Press, London

Pohland AE, Nesheim S, Friedman L (1992) Ochratoxin A: a review (Technical report). Pure & Appl Chem 64:1029–1046

Pohland AE (1993) Mycotoxins in review. Food Add Cont 10:17–28

Purchase IFH (1971) Mycotoxins in Human Health. Macmillan, London

Purchase IFH (1974) (Ed) Mycotoxins. Elsevier, Amsterdam

Purchase KA (1998) Mycotoxins. In: Watson D (Ed) Natural Toxins in Food, pp 147–181. Academic Press, London

Reiss J (Ed) (1981) Mykotoxine in Lebensmitteln. Gustav Fischer Verlag, Stuttgart, New York

Reiss J (1998) Schimmelpilze. Lebensweise, Nutzen, Schaden, Bekämpfung. 2. Aufl. Springer, Berlin Heidelberg New York

Rodricks JV (Ed) (1976) Mycotoxins and Other Fungal Related Food Problems. Advances in Chemistry Series 149. American Chemical Society, Washington, DC

Rodricks JV, Hesseltine CW, Mehlmann MA (Eds) (1977) Mycotoxins in Human and Animal Health. Pathotox Publishers Inc, Park Forest South, IL

Roth L, Frank H, Kromann K (1990) Giftpilze – Pilzgifte. Ecomed, Landsberg/Lech

Samson RA, Hoekstra ES, Frisvad JC, Filtenborg O (1998) Introduction to Food-borne Fungi. Centraalbureau voor Schimmelcultures, Baarn, Netherlands

Sauer DB (Ed) (1992) Storage of Cereal Grains and Their Products: 4th Ed. American Association of Cereal Chemists, St. Paul, Minnesota

Scott, PM (1990) Trichothecenes in grains. Cereal Foods World 35:661–666

Scott, PM (1997) Multi-year monitoring of Canadian grains and grain-based foods for trichothecenes and zearalenone. Food Add Cont 14:333–339

Sharma RP, Salunkhe DK (1991) Mycotoxins and Phytoalexins. CRC Press, Boca Raton, Florida

Shepard GS, Thiel PG, Stockenström S, Sydenham EW (1996). Worldwide survey of fumonisin contamination of corn and corn-based products. JAOAC 79:671–687

Smith JE, Moss MO (1985) Mycotoxins. Formation, Analysis and Significance. John Wiley and Sons, New York

Stoloff L. (1976): Incidence, distribution, and disposition of products containing aflatoxins. Proc Am Phytopathol Soc 3:156–172

Tanaka T, Hasegawa A, Yamamoto S, Lee U-S, Sugiura Y, Ueno Y (1988) Worldwide contamination of cereals by the *Fusarium* mycotoxins nivalenol, deoxynivalenol, and zearalenone. 1. Survey of 19 counries. J Agric Food Chem 36:979–983

Ueno Y (Ed) (1983) Trichothecenes – Chemical, Biolgical and Toxicological Aspects. Kodansha/Elsevier, Tokyo

Ueno Y (1985) The toxicology of mycotoxins. Crit Rev Toxicol 14:99–132

Watson DH (1985) Toxic Fungal Metabolites in Food. CRC Crit Rev Food Sci Nutr 22:177–198

Weidenbörner M (1998) Lebensmittel-Mykologie. BEHR´S Verlag, Hamburg

Weidenbörner M (1999) Lexikon der Lebensmittelmykologie. Springer-Verlag, Berlin etc

WHO (1979) Environmental Health Criteria 11. Mycotoxins. World Health Organization, Geneva

Wilson DM, Abramson D (1992) Mycotoxins. In: Sauer DB (Ed) Storage of Cereal Grains. 4th Ed. pp 341–391. Amercian Association of Cereal Chemists, St. Paul, Minnesota

Wyllie TD, Morehouse LG (1978) (Eds) Mycotoxic Fungi, Mycotoxins, Mycotoxicoses. An Encyclopedic Handbook. Vol. 3 Marcel Dekker Inc, New York

Recommended Journals

Applied and Environmental Microbiology
Applied Microbiology
Biochemistry
Experientia
Food Additives and Contaminants
Food and Cosmetics Toxicology
Food Microbiology
Industries, Alimentaires et Agricoles
International Journal of Food Microbiology
Journal of Agricultural and Food Chemistry
Journal of the American Chemical Society
Journal of Food Microbiology
Journal of Food Protection
Journal of Food Safety
Journal of Food Science
Journal of Natural Toxins
Journal of Stored Product Research
Journal of the American Oil Chemists Society
Journal of the Association of Officials of Analytical Oil Chemists International
Microbiological Research
Microbiology
Mycologia
Mycological Research
Mycotoxin Research
Mycopathologia
Mycopathologia et Mycologia Applicata
Nature
Naturwissenschaften
Phytochemistry
Science
Tetrahedron Letters
Zeitschrift für Ernährungswissenschaft
Zeitschrift für Lebensmittel-Untersuchung und -Forschung

Printing and Binding: Stürtz AG, Würzburg